# Genetic Improvement of Rice for Salt Tolerance

Rakesh Kumar Singh • M. Prakash •
Raj Kumar Gautam • S. L. Krishnamurthy •
S. Thirumeni
Editors

# Genetic Improvement of Rice for Salt Tolerance

Springer

*Editors*
Rakesh Kumar Singh
International Centre for Biosaline Agriculture
Dubai, United Arab Emirates

M. Prakash
Department of Genetics & Plant Breeding
Annamalai University
Annamalai Nagar, Tamil Nadu, India

Raj Kumar Gautam
National Bureau of Plant Genetic Resource
Delhi, India

S. L. Krishnamurthy
ICAR
Central Soil Salinity Research Institute
Karnal, India

S. Thirumeni
Department of Plant Breeding and Genetics
Pandit Jawaharlal Nehru College of Agriculture and Research Institute
Karaikal, Puducherry (U.T.), India

ISBN 978-981-99-3829-2 ISBN 978-981-99-3830-8 (eBook)
https://doi.org/10.1007/978-981-99-3830-8



This Springer imprint is published by the registered company Springer Nature Singapore Pte Ltd.
The registered company address is: 152 Beach Road, #21-01/04 Gateway East, Singapore 189721, Singapore

Dedicated to
***Dr. D. Senadhira***
***Prof. T.J. Flowers***
***Dr. B. Mishra***

*Someone has aptly quoted* "Salt is the flavor of the ocean; when it becomes the flavor of the land, it spells trouble for the tiller, miller and the ruler." *The occurrence of excessive salts poses an inhospitable environment to the growth, development, and economic production of crop plants. Rice crop not only meets the calorific and nutritional requirements of the major population of the globe but is also the most preferable crop to begin with reclamation and sustain the agricultural activities and production in salt-affected soils.*

*Being research workers on salinity breeding in rice, it is our sacred duty to dedicate this book to three great scientists: the late Dr. Dharamawansa Senadhira, Prof. Timothy J. Flowers, and Dr. Banwari Mishra who made legendary contributions in this domain.*

Late Dr. D. Senadhira was born in Sri Lanka and served as Rice Breeder, Coordinator of Rice Breeding Program, and Deputy Director

Research for Department of Agriculture, Sri Lanka, before joining International Rice Research Institute (IRRI) in 1984 as Rice Breeder. He led the Flood Prone and Tidal-wetland Rice Research Program of IRRI from 1996 and developed numerous lines of improved germplasm shared with partners. He died due to an unfortunate accident on 7th July 1998 in Bangladesh and left a rich scientific legacy.

Professor TJ Flowers, Emeritus Professor in Plant Physiology, School of Life Sciences, University of Sussex, UK, is a globally renowned researcher in the field of effect of salt stress on plants, especially rice. His work concentrated on the mechanisms of salt tolerance in plants from a physiological viewpoint.

Dr. B. Mishra started his research career as rice breeder at ICAR-Central Soil Salinity Research Institute, Karnal, India. He rose to the ranks of Head, Division of Crop Improvement, CSSRI, Karnal; Director, ICAR-Indian Institute of Rice Research, Hyderabad; Director, Indian Institute of Wheat and Barley Research, Karnal; and Vice Chancellor, Shere Kashmir University of Agricultural Science and Technology, Jammu, India.

*They made remarkable beginning, inspired and facilitated an eventful journey to understand and harness the physiological, molecular, and genetic mechanisms of tolerance for the development of salt-tolerant rice varieties through various research networks. The cultivation and adoption of these salt-tolerant varieties have made prominent impact by elevating the socioeconomic conditions of farmers inhabiting the disadvantaged salt affected areas, especially in south and southeast Asian countries.*

# Foreword 1

The world today is preparing, or should be preparing, for a near future where many of the conditions that have led to relative prosperity and food security are threatened by the impact of increased greenhouse gas emissions that end up in our atmosphere and seas. Continued population growth also presents another significant challenge to the prosperity of present and future generations. Dr. R. K. Singh and his colleagues have brought together an impressive collection of research progress summaries and forward vision on one of the most important stresses confronting global food production under a changing climate: soil salinity.

Soil salinity has plagued farmers for centuries in South and Southeast Asia. Irrigation schemes are particularly vulnerable to progressive salt increases in the surface where evaporation can lead to increased salt concentration in the crop rooting zone. Many of the world's most important rice production regions are found in coastal highly fertile river deltas. Unfortunately, these lands are threatened by sea water incursion as sea level rises and the resulting salinity can be extremely toxic to crops. In both inland irrigation schemes and coastal production systems, very large areas have already been abandoned because of increased soil salinity. There is no doubt that without aggressive action addressing salinity, global food security could be significantly at risk.

This volume is a timely milestone for the considerable progress that the scientific community is making in addressing management of saline soils. In this impressive volume there is comprehensive treatment of what a number of leading agricultural institutions are doing in the areas of adaptation to or mitigation of soil salinity.

Several chapters go into detail on the agronomic and other practices that may help farmers prevent, manage, or reduce the salt concentrations in their soils. However, these treatments can be expensive and time consuming and may not be practical for many small, resource-poor farmers.

The real strength of this volume is the extensive treatment of genetic improvement for salinity tolerance. Several authors document the significant advances that have already been made in developing salt-tolerant varieties. But, most currently available varieties were developed when the genetic tools to do so were rudimentary by today's standards. So, with these advances as a foundation much of the book treats the use of new genetic tools to identify new donors, chromosome regions, and genes that can serve as new sources of tolerance to salinity. Most important challenge now is to combine salt tolerance with other sources of resistance to abiotic and biotic stresses. Eventually these may well be combined with genes conferring better nutritional value such as iron, zinc, and beta carotene.

Students and researchers interested in embarking on research in this area will find this volume to be a very useful starting point. Others wishing to understand how different major research thrusts evolve and may be targeted to link with other programs would do well to look into the many case studies that are presented here. Finally, those engaged in policy making around food security should use this work as a lesson in the difficulties in addressing serious problems and the continuity of support that is required to make significant progress.

Director General (Emeritus) IRRI
Los Baños, Philippines

Robert S. Zeigler

# Foreword 2

High-Yielding Varieties (HYVs) of rice first developed in the 1960s and 1970s have contributed greatly to food security during a time of high population growth and rice shortage. To achieve the benefits from these varieties, it was assumed they would need an optimum growing environment, including good water control and high soil fertility. Farmers in the unfavorable growing areas could not take full advantage of these varieties and could also be at a disadvantage considering their lower yields and declining price of rice. Breeding programs began giving more attention to developing higher-yielding varieties for these more unfavorable environments. Nevertheless, there was limited adoption of varieties from these efforts until recently.

The outlook for areas affected by stresses such as drought, flooding, and salinity improved greatly with the technology of QTL (Quantitative Trait Locus) mapping and the identification of good donors for stress tolerance alleles. The widespread cultivation of mega varieties like Swarna and IR64 in South Asia provided recipients for the newly identified QTLs. Furthermore, breeders began achieving success with the creation and adoption of varieties tolerant of drought, submergence, and salinity.

In 2007 we formed the STRASA (Stress Tolerant Rice for Africa and South Asia) project with significant funding from the Bill & Melinda Gates Foundation. In India, Bangladesh, and Nepal, national governments, NGOs, and private partners contributed greatly to the networks for developing varieties tolerant to drought, submergence, and salinity.

The many accomplishments of the STRASA project are documented on its web site (https://strasa.irri.org). In view of its formal ending in 2019, it is timely to take stock of the current state of knowledge and the status of the current efforts in developing new salt-tolerant varieties and associated technologies. The present volume documents decades of work on this important problem and will be useful for those working in the field as well as others encountering the subject for the first time. The authors are commended for their efforts to bring this volume to fruition.

Department of Plant Sciences, University of California, Davis David Mackill
Davis, CA, USA

# Acknowledgments

We also appreciate and acknowledge the generous funding, coordination, and capacity building support provided by IRRI, Philippines, through various research projects like BMGF-STRASA, CPWF-CP7, and BMZ Germany. The enabling support provided by Indian Council of Agriculture Research, State Agricultural Universities, and other research organizations is also sincerely acknowledged.

Editors

# Contents

# About the Editors

**Rakesh Kumar Singh** is Program Leader of Crop Diversification and Genetics section at the International Centre for Biosaline Agriculture (ICBA) based at Dubai, UAE. He is leading research solutions for marginal environments in sub-Saharan Africa, Central Asia, and Middle East and North Africa (MENA) region. Before joining ICBA, Dr. Singh worked as Senior rice breeder at IRRI from 2005 to lead the IRRI program on salinity breeding and problem soils. He served as regional plant breeding coordinator for eastern and southern Africa from 2009 to 2012 where he led the rice-based agriculture research in eastern and southern Africa. From 2012 to 2018, he led the breeding work on rainfed lowlands for SE Asia and rice breeding for problem soils from IRRI headquarters at Los Banos in the Philippines. He has guided 14 PDF, 7 PhD, 8 MS, and numerous OJT and interns, and trained numerous researchers across the globe. He is an Adjunct Professor of plant breeding and genetics at the University of the Philippines at Los Banos.

Dr. Singh has been conferred with many national and international honors and awards. Few of them are Fellow of the most prestigious Agricultural Academy of India, NAAS (National Academy of Agricultural Sciences); Fellow of most prestigious Mohd. Bin Rashid Academy of Sciences (MBRAS) of UAE; Fellow of Indian Society of Coastal Agricultural Research; Fellow of Indian Society of Genetics and Plant Breeding; and National recognition and honor for outstanding service to the nation by Indian Council of Agricultural Research (ICAR), DARE, Government of India, for the development of Basmati CSR30 rice variety. Dr. Singh has been instrumental for the development and commercial release of more than 35 rice varieties globally in south and south east Asia (India, Bangladesh, Myanmar, Vietnam, Indonesia, and Philippines) and eastern Africa (Tanzania, Uganda, and Kenya). He has published more than 115 research articles in refereed journals and also published more than 20 books and book chapters. He has extensive international experience and visited more than 60 countries.

**M. Prakash**, born in 1966, obtained his B.Sc. (Ag.) and M.Sc. (Ag.) from Tamil Nadu Agricultural University, Coimbatore, India, in 1988 and 1990, respectively. He did his Ph.D. from Annamalai University. He has served as the Head of the Department of Genetics and Plant Breeding from 2006 to 2012 and presently serves as the Controller of Examinations. He has put in 30 years of total teaching, research, and extension service and 16 years of service in the cadre of Professor.

He has guided 8 Ph.D. scholars and 25 PG scholars. His credentials include publication of more than 150 refereed research papers, ten books, two seminar proceedings, and 15 book chapters including one book chapter published in SESAME by globally reputed CRC Press. He organized 3 international and 15 national seminars/conferences/workshops, 1 refresher course, and completed five research projects. He has earned the status of Departmental Research Support with Special Assistance Programme of University Grants Commission (UGC-SAP-DRS) to the Department with financial assistance of Rs. 50 lakhs at phase I (2009–2014) and Rs. 1.025 crore at phase II (2015–2020) level as Program Coordinator.

He was awarded with National Merit Scholarship, Government of India, during 1981–1988, ICAR Junior Fellowship in Plant Physiology during 1988–1990, Crop Research Award in 2004, Best Researcher Award in Annamalai University in 2010, J J Chinoy Gold Medal Award of Indian Society of Plant Physiology in 2017, Dr. B. P. Pal Memorial NABS Best Scientist Award of National Academy of Biological Sciences, Chennai, in 2017, and Best Researcher (Publication) Prize of Annamalai University in 2016–2017, 2018–2019, and 2021. His academic excellence was well recognized by the award of Fellowships like FELLOW of Indian Society of Plant Physiology (FISPP), New Delhi, in 2015, FELLOW of National Academy of Biological Sciences (FNABS), Chennai, 2015, and FELLOW of National Environmental Science Academy, New Delhi, in 2016.

**Raj Kumar Gautam** presently working as Principal Scientist and Head, Division of Germplasm Evaluation at ICAR-NBPGR, New Delhi, has earlier served as Head, Division of Crop Improvement & Protection at ICAR-CIARI, Port Blair, A&N Islands. He has contributed to the development of 14 high-yielding, bacterial blight and salt-tolerant rice varieties for different states in India. Based on the amount of seed distribution, these varieties are estimated to have occupied more than 16 lakh ha area, thus causing significant socio-economic impacts in different states. A total of 12 unique germplasm materials were registered with ICAR-NBPGR and PPVFRA, New Delhi. Besides his peer-reviewed research publications in more than 70 journals, significant contributions were made for the development, documentation, and popularization of agro-technologies and innovations, including farmers' participatory variety selection and seed production (more than 650 quintals of quality seed produced and distributed).

As team leader, he facilitated PPVFRA Plant Genome Saviour Community Award of Rs. 10.00 lakh to Karen community of remote Andaman & Nicobar Islands for conservation of traditional land races of rice. He also got ICAR Fakhruddin Ali Ahmed Award for Outstanding Research in Tribal Farming System Research-2018. He is Fellow of National Academy of Biological Sciences, Indian Society of Genetics & Plant Breeding, and Indian Society of Coastal Agricultural Research.

**S. L. Krishnamurthy** is senior scientist in plant breeding. He started his career as a scientist at the ICAR Central Soil Salinity Research Institute (the Indian Council of Agricultural Research) in Karnal, India. He had risen through the ranks to become a senior scientist. He leads research in salt tolerance in rice at ICAR CSSRI Karnal.

In his past roles, Dr. S. L. Krishnamurthy led the teams, which developed and released ten salt-tolerant rice varieties in India. He developed five rice genetic stocks for salt tolerance. He holds a Ph.D. in Genetics and Plant Breeding from the University of Agricultural Sciences Bengaluru (UAS B), India. He has published more than 80 research papers in international peer-reviewed journals and over 12 book chapters.

**S. Thirumeni**, after graduating from Kerala Agricultural University, Thrissur, completed his postgraduation and doctoral programs in Plant Breeding and Genetics from Tamil Nadu Agricultural University, Coimbatore. Apart from teaching courses in Genetics and Plant Breeding to the students of Agriculture for over 30 years, he took a keen interest in understanding the genetics of salt and submergence tolerance of rice and consequently breeding tolerant rice varieties. He has identified the gene action and interaction of salt tolerance, employing classical genetic tools in rice, cotton, finger millet, and black gram. In rice, using molecular tools and genetic resources (germplasm and mapping populations), he mapped the chromosomal regions harboring salt tolerance genes through Genome Wide Association Studies and QTL analysis.

As a plant breeder, he has contributed to the development of several promising salt and submergence-tolerant rice strains and release of two crop varieties. His most significant contribution is the development of Near Isogenic Lines (NILs) through Marker-Assisted Backcross Breeding (MABB) in the background of popular rice varieties for improving salt tolerance (ADT 45) and submergence tolerance (ADT 39 & ADT 46) while retaining all the other agronomically desirable traits housed in the mentioned rice varieties. He has successfully completed seven externally funded research projects focusing on improving salt and submergence tolerance in rice and trained over 22 PG scholars in their thesis research. He has published more than 50 research papers in journals of repute and fellow of National Academy of Biological Sciences.

# Genetic Improvement of Rice for Salt Tolerance

# 1

Prakash Muthu Arjuna Samy, Rakesh Kumar Singh, Glenn B. Gregorio, Raj Kumar Gautam, S. L. Krishnamurthy, and S. Thirumeni

**Abstract**

Soil salinity and sodicity caused by occurrence of excessive salts pose daunting challenge for crop cultivation. Among various crops, rice (*Oryza sativa*) is the befitting crop for cultivation under salt affected conditions due to its inherent adaptive advantages. Moreover, rice has huge social and economic importance due its ability to feed a sizeable part of global population especially in Asia. Among various management options to augment crop productivity under salt stressed soils, harnessing the intra-specific genetic variation in the form of salt tolerant varieties is the most simple, economic and eco-friendly approach. In this context, considerable progress has been made for development and deployment of successful and salt tolerant varieties especially through conventional breeding tools. These scientific achievements and leads came up through investments in

P. M. A. Samy (✉)
Department of Genetics and Plant Breeding, Faculty of Agriculture, Annamalai University, Chidambaram, Tamil Nadu, India

R. K. Singh
International Center for Biosaline Agriculture, Dubai, United Arab Emirates

G. B. Gregorio
Southeast Asian Regional Center for Graduate Study and Research in Agriculture, Los Banos, Philippines

R. K. Gautam
ICAR-National Bureau of Plant Genetic Resources, New Delhi, India

S. L. Krishnamurthy
ICAR-Central Soil Salinity Research Institute, Karnal, India

S. Thirumeni
Department of Plant Breeding and Genetics, Pandit Jawaharlal Nehru College of Agriculture and Research Institute, Karaikal, Puducherry (U.T.), India



R. K. Singh et al. (eds.), *Genetic Improvement of Rice for Salt Tolerance*,
https://doi.org/10.1007/978-981-99-3830-8_1

multi-disciplinary research and institutional networking which was enabled by various national and international funding agencies. Rice is genomically the most explored cultivated crop. Though salt tolerance is a polygenically controlled trait, scientific leads have been attained in mapping of genes/QTLs conferring tolerance to salinity especially at seedling stage. Nevertheless, mapping the genomic regions conferring agronomic advantages at adult stages are the most sought after and potential targets for genetic improvement of rice for saline and sodic environments in future. The prevalence of biotic stresses and other mineral deficiencies/toxicities are also the associated problems in salt affected areas. Therefore, harnessing the synergy of modern genomic, gene editing and phenomics technologies can lead to the evolution of robust and multiple stress tolerant rice varieties. In this regard, research investments through multi-disciplinary and institutional partnerships can bring about better livelihood, economic and environmental security from the salt affected regions.

**Keywords**

Rice · Salinity · Genetic improvement · Abiotic stress

## 1.1 Introduction

Climate change has proven to be one of the major obstacles to the growth and sustainability of different life forms on Earth. Agriculture, including deforestation and land use, contributes to 21% of greenhouse gas (GHG) emissions and is one of the prominent sectors threatened by climate change. However, the agricultural sector has a high mitigating potential to diminish its carbon footprint, unlike energy and transportation, which have little flexibility. Rice cultivation is considered as one of the biggest factors responsible for GHGs, but being a staple cereal for more than half of the global population, its productivity cannot be compromised. Salt stress is the second major abiotic stress, after drought, limiting rice production worldwide, including that in India. Although various geographically diverse agroecosystems are likely to be affected directly or indirectly, coastal areas and islands are the ecosystems most vulnerable to the ill effects of climate change. Further, sea-level rise due to the melting of glaciers is causing intrusion of seawater in the cultivable transient coastal areas, thus adversely affecting the sustainability of agricultural activities, production, and livelihood in these vulnerable areas, especially in South and Southeast Asian countries. In addition, the increasing canal networks in the inland irrigated areas are known to cause secondary salinization that threatens agriculture. Similarly, sodicity/alkalinity is another form of salt stress that affects crop production mostly in arid inland areas.

The presence of excess salts in soil beyond the critical concentration is detrimental to crop growth. The source of this excess salt could be soil or irrigation water. The content and composition of salts (ions), in tandem with types of soil and weather conditions prevailing during the cropping period, determine the severity of salt stress. Although about 1 billion ha of land are globally affected by salinity and sodicity, about 6.7 million ha are affected by salts to varying degrees in India. Similarly, other Asian countries, such as Bangladesh, Vietnam, Indonesia, Sri Lanka, Iran, etc., are affected by salts where rice is grown as a mainstay crop. Salt stress is the second most occurring abiotic stress globally. As per global estimates, 1.128 billion ha of land are affected by salinity, with 683 million ha (60%) categorized as saline, 288 million ha (26%) as sodic, and 157 million ha (14%) as saline and sodic. Excluding forests, wetlands, unsuitable lands, and high-biodiversity areas, 971 million ha of land that can be used for agriculture are affected by salt (Wicke et al. 2011).

As the majority of crop plants globally are glycophytes, salinity invariably affects them negatively, which is manifested as decreased growth and economic yield. Several predominant mechanisms adversely affect normal plant growth. Plants resist the harm from excess salts through a combination of mechanisms such as osmotic balance, salt exclusion, organ-level partition, tissue tolerance, intracellular compartmentation, etc. These mechanisms are extensively reviewed by Singh and Flowers (2010).

The problem of salty soil can be addressed either by adopting suitable soil management practices or by breeding salt-tolerant rice varieties. The first option requires good-quality water besides soil amendments. Various management options, including the use of gypsum for the reclamation of sodic soils and drainage interventions to flush out excess salts, have been worked out and recommended to render such lands fit for agricultural cultivation. However, these options invariably involve economic constraints, environmental concerns, and community participation (which may be lacking). The second approach, developing salt-tolerant varieties, is a long-term approach and takes a couple of years to reach optimum productivity. Alternatively, a third approach, a hybrid one, which includes the cultivation of salt-tolerant varieties along with soil amendments and other management options to minimize the effect of excessive salt in the soil, is economically viable, environmentally sustainable, and widely acceptable. The dissemination of salt-tolerant seeds for cultivation in salt-affected areas is easy, and its adoption is faster than other options during crop cultivation.

Plant responses to salinity stress are highly complex phenomena that can mold a plant to acclimatize to salinity stress by modifying the various developmental, biochemical, and molecular-physiological processes. Under salinity stress conditions, plant responses have been separated into two main categories. The initial phase is an ion-independent response that occurs in minutes to days to cause a negative effect on stomatal closure and inhibit the expansion of shoot cells (Isayenkov and Maathuis 2019). The second phase is an ion-dependent response that occurs in days and/or weeks to build up ion cytotoxicity, which leads to premature leaf senescence, slowing of metabolic processes, and cell death (Isayenkov and Maathuis 2019: Roy et al. 2014).

Although modern rice varieties are sensitive to salt stress, rice is the predominant and preferred crop in many areas, owing to its inherent capacity to grow in stagnant water. Further, salinity rarely occurs in isolation and is often associated with deficiencies of certain ions such as phosphorus and zinc or toxicities of other ions such as aluminum and iron. Rice has an edge over other crops for growing in salt-affected agroecosystems because of its ability to grow in waterlogged conditions. For sodic soil reclamation, rice cultivation is recommended to start with because of water ponding for the leaching of salts during the first year. Similarly, in coastal saline areas, rice can tolerate both excessive rains and waterlogging conditions, which are inimical to other field crops. Since rice puddling requires stagnant water, the dilution of salts takes place in the process, thus decreasing the adverse effects of salts on plant growth and development.

However, in such a scenario, the use of salt-tolerant rice varieties plays an important role, which has been widely demonstrated and appreciated for harnessing the productivity potential of salt-affected lands, especially in India as well as in other South Asian countries. Nevertheless, this has to be augmented in conjunction with appropriate management practices. Therefore, exploiting salt tolerance cannot be considered as a complete answer and needs to be seen as part of the total package inclusive of management options. Although the Green Revolution provides testimony to the power of conventional plant breeding, in favorable ecosystems, it is insufficient to produce more grains from fragile environments such as salty soils. The pace of progress in breeding salt-tolerant rice has been slow owing to varied factors, including the lack of salt-tolerant germplasm, poor understanding of the mechanisms underlying tolerance, complexity of the trait needed for efficient selection criteria, absence of a reliable and repeatable screening methodology, field heterogeneity for salt distribution, and variation of tolerance with ontogeny.

However, parallel developments in understanding the physiological/biochemical bases of salt tolerance and mapping salt tolerance genes using molecular markers paved the way for faster and more precise designing of salt-tolerant germplasm. The complexity of the trait is manifested by its polygenic nature (quantitative), large interactions with the environment (soil-water-temperature-relative humidity), and ontogeny of the rice plant. Past experimental evidence indicates that salt tolerance is a multigenic trait and the components of salt tolerance (tissue tolerance, Na:K ratio, K accumulators, early vigor, and water use efficiency), in turn, are determined by many genes. Several breeding programs have worked on developing saline-tolerant genotypes. Extensive research has been redirected toward the advancement of salt tolerance at the seedling stage of rice, but little effort has been made in reproductive-stage salinity effects. QTL mapping is mostly based merely on phenotypic performance under stress alone. Therefore, the use of the relative performance of various genotypes under stress and nonstress conditions will be helpful in identifying and mapping the appropriate QTLs. The introgression of QTLs into high-yielding salt-sensitive mega-varieties is a short-term quick solution to increase productivity with the help of molecular breeding programs. Marker-assisted selection (MAS), therefore, offers immense scope for improving salt tolerance in rice, and QTL mapping studies are a prerequisite for MAS, which can be used to enhance salinity tolerance through pyramiding of several responsible QTLs.

Rice has one of the biggest collections of germplasm in the world. This crop is also considered as salt-sensitive because of its lower degree of threshold salinity

(3 dS $m^{-1}$). But this does not mean that all the germplasm is salt-sensitive because vast intra-crop variability exists in germplasm for salt tolerance. Farmers and researchers are equally trying to find out the most tolerant and acceptable local and improved varieties that can grow well under marginal soils (due to salt stress) as a best-bet solution that is both economically viable and environment friendly. In addition to salinity/sodicity stress and water stagnation in the rhizosphere, numerous other factors directly or indirectly affect plant growth under stress environments: climatic conditions, soil texture, inherent soil fertility, and available water quality and quantity. These factors also determine the suitability of target-specific rice varieties. Because of such a vast array of diverse factors, this represents "one size doesn't fit all." Therefore, the variety best suited for a specific location should be selected to maximize the benefits to farmers from a particular cropping system.

## 1.2 Breeding as a Key Tool for Augmenting Salinity Tolerance in Rice

Breeding techniques form an efficient and cost-effective way of generating rice cultivars to grow successfully in saline environments. The primary goal of rice breeding programs for salinity tolerance is to enhance the underpinning alleles involved in salinity tolerance in rice. Assessing the differences in physiological traits among varieties is the best way to define and refine salinity-tolerant genotypes. However, physiological traits are polygenic and difficult to improve through breeding. Salinity stress affects flowering and yield traits, including spikelet number per panicle, panicle length, and grain yield, and delays panicle emergence. Rice is not uniformly sensitive or tolerant at different growth stages but exhibits a differential response to diverse concentrations of salinity at various growth stages due to stage-specific tolerance mechanisms (Singh et al. 2010: Fita et al. 2015). Screening for salinity tolerance at the seedling stage is simple and straightforward because it is duration dependent as all the genotypes follow the same seedling stage pattern irrespective of early-, medium-, or late-duration genotypes. However, this is not the case for the reproductive stage as it is growth stage dependent, which exhibits at a different time depending upon the duration of the genotype (Singh et al. 2021). Hence, most salinity breeding programs focused on screening segregating populations in the seedling stage and later assessed the potential of these selected salinity-tolerant genotypes at the reproductive stage under the target field conditions (Li and Xu 2007).

By exploring conventional breeding methods, considerable success has been achieved in enhancing salinity tolerance. Several tolerant cultivars were developed and commercialized in various countries by exploiting the existing salt tolerance conferred by prominent donor sources such as Pokkali, Kalarata, and Nona-Bokra and other salt-tolerant wild relatives (Mangrauthia et al. 2014). Of these, Pokkali is highly tolerant compared with other tolerant genotypes, and therefore, it is used as a promising donor for several salinity tolerance breeding programs (Muthuramalingam et al. 2022).

Marker-assisted breeding (MAB) in rice has become more usable, and many SSR and SNP markers are available for rice breeding programs. Traditional map-based cloning, GWAS (genome-wide association studies), MABC (marker-aided backcrossing), and MARS (marker-aided recurrent selection) are the various marker-based strategies now followed for the mapping and introgression of QTLs/genes for salinity tolerance in rice. Multi-parent advanced generation intercross (MAGIC) populations are more useful for harnessing multiple QTLs/genes (Samantara et al. 2021). GWAS of a MAGIC population revealed a novel QTL for salinity tolerance in rice (Krishnamurthy et al. 2022). Moreover, recent advances, including gene editing, high-throughput genotyping, genomic selection (GS), and marker-assisted selection, along with speed breeding (SB) have accelerated crop improvement (Samantara et al. 2022).

## 1.3 Stress-Tolerant Rice for Africa and South Asia

Considering all these factors, it was conceptualized to combine and synergize the efforts of several rice researchers from different institutions from India and Bangladesh under the auspices of the Salinity-Tolerant Breeding Network (STBN), which was coordinated by the International Rice Research Institute (IRRI), Philippines. The Bill & Melinda Gates Foundation (BMGF), through the Stress-Tolerant Rice for Africa and South Asia (STRASA) project, financially supported this important and extremely successful initiative. Sixteen institutions from India and two from Bangladesh formed part of the STBN. These partner institutions and IRRI contributed to the improved germplasm to be evaluated across the target sites. Based on the performance of the germplasm contributed for evaluation across locations and over the years, the best entries were selected for specific as well as general adaptability and overall stability. This marked exemplary cooperation among partners in which they could obtain improved germplasm and use it at their location or recommend it to national or state-coordinated trials to initiate the varietal release process for commercial cultivation. STRASA has played a catalytic role in the testing and dissemination of tolerant varieties by various salinity research centers in India and Bangladesh. Especially, the components of farmers' participatory varietal selection and seed production have further reinforced the uptake and spread of salt-tolerant rice varieties in the target niche areas.

From 2008 to 2019, the STRASA project was operational in India, Bangladesh, and Nepal in South Asia; in Nigeria, Benin, and Senegal in sub-Saharan Africa; in Burkina Faso, Ghana, Guinea, The Gambia, Mali, Côte d'Ivoire, Guinea Conakry, and Sierra Leone in West Africa; and in Mozambique, Tanzania, Uganda, Ethiopia, Madagascar, Rwanda, Burundi, and Kenya in East and Southern Africa. STRASA Phase 3 (2014–2019) made good progress in releasing and disseminating *71* stress-tolerant rice varieties (STRVs) in India (58), Bangladesh (6), and Nepal (7), of which 27 are drought-tolerant, 23 are submergence-tolerant, 12 are salinity-tolerant, and 9 are tolerant of multiple stresses [drought and heat (2), drought and submergence (6), and submergence and salinity (1)]. The nine multiple-stress-tolerant rice varieties included five from India (DRR 52, DRR 47, Sukhadhan 5, CR dhan 801, and CR dhan 802), one from Bangladesh (BRRI dhan78), and three from Nepal (Sukhadhan 6, Bahuguni dhan-1, and Bahuguni dhan-2), which were also released and disseminated in South Asia. The coverage of these rice varieties is progressively increasing every year because of their tolerance of multiple stresses, early maturity, suitability to local cropping systems, good taste, and good market price. Twenty-two more stress-tolerant rice varieties had been released by the final year of the project completion in 2018–2019 (https://strasa.irri.org/home). Therefore, under the three phases of the STRASA project, more than **150 climate-smart rice varieties** that are tolerant of flood, drought, and salinity (including multiple tolerance in a variety) were released in various countries in South Asia and in sub-Saharan Africa. Many more varieties are also in the pipeline for release. In addition to improving varieties and distributing seeds, the STRASA project was able to train **74,000 farmers and scientists (including 19,400 women farmers)** in producing good-quality seeds through its capacity-building programs (https://strasa.irri.org/home).

## 1.4 Conclusions and Future Perspectives

For abiotic stresses, salinity stress is one of the predominant stressors, which dauntingly affects the overall plant growth and yield of rice plants across the world. In addition, saline stress is a highly complex process wherein its components are still far away from clarity and understanding because several enzymes and metabolites have overlapping cellular compartments owing to possible divergence in tolerance. Therefore, to enhance rice yield during salinity stress, new tools and methods are needed. The primary methods for crop improvement in modern agriculture are cross-breeding, mutation breeding, and transgenic breeding. However, these time-consuming and labor- and cost-intensive breeding approaches cannot meet the increasing expectations and global food demand. To deal with this challenging situation and to enhance crop selection efficiency, marker-assisted breeding and transgenic approaches have been adopted, thus generating the desired traits via exogenous transformation into elite varieties. It is also important to fortify the salt-tolerant varieties with biotic stress resistance genes through novel molecular breeding and pyramiding tools for sustaining crop productivity under salt stress. In addition, genome editing systems are excellent tools that provide rapid and targeted mutagenesis and can identify specific plant molecular mechanisms for crop improvement. Crop breeding was revolutionized by the development of next-generation breeding techniques. Genome editing technologies have many advantages over traditional agricultural methods, given their simplicity, efficiency, high specificity, and amenability to multiplexing. To conclude, speed breeding, combined with genetic tools and resources, will enable plant biologists to scale up their research in the field of crop improvement.

This book has endeavored to compile and document the research and development work carried out at salinity research centers in the past, current status, and future perspectives, including some success stories. We hope that this book will be quite useful for policymakers, development agencies, and researchers to harness the advantages of the genetic attributes of salt tolerance for successfully growing crops in salt-stressed areas.

## References

Fita A, Rodríguez-Burruezo A, Boscaiu M, Prohens J, Vicente O (2015) Breeding and domesticating crops adapted to drought and salinity: a new paradigm for increasing food production. Front Plant Sci 6:978.

Isayenkov SV, Maathuis FJM (2019) Plant salinity stress: many unanswered questions remain. Front Plant Sci 10:80

Krishnamurthy SL, Sharma PC, Dewan D, Lokeshkumar BM, Rathor S, Warraich AS, Vinaykumar NM, Leung H, Singh RK (2022) Genome wide association study of MAGIC population reveals a novel QTL for salinity and sodicity tolerance in rice. Physiol Mol Biol Plants 8:819–835

Li Z-K, Xu J-L (2007) Breeding for drought and salt tolerant rice (*Oryza sativa* L.): Progress and perspectives. In: Jenks MA, Hasegawa PM, Jain SM (eds) Advances in molecular breeding toward drought and salt tolerant crops. Springer, Dordrecht, pp 531–564

Mangrauthia SK, Revathi P, Agarwal S, Singh AK, Bhadana V (2014) Breeding and transgenic approaches for development of abiotic stress tolerance in rice. In: Improvement of crops in the era of climatic changes. Springer, Berlin/Heidelberg, pp 153–190

Muthuramalingam P, Jeyasri R, Rakkammal K, Satish L, Shamili S, Karthikeyan A, Ramesh M (2022) Multi-omics and integrative approach towards understanding salinity tolerance in rice: a review. Biology 11(7):1022

Roy SJ, Negrão S, Tester M (2014) Salt resistant crop plants. Curr Opin Biotechnol 26:115–124

Samantara K, Bohra A, Mohapatra SR, Prihatini R, Asibe F, Singh L, Reyes VP, Tiwari A, Maurya AK, Croser JS (2022) Breeding more crops in less time: a perspective on speed breeding. Biology 11:275

Samantara K, Reyes VP, Agrawal N, Mohapatra SR, Jena KK (2021) Advances and trends on the utilization of multi-parent advanced generation intercross (MAGIC) for crop improvement. Euphytica 217:189

Singh RK, Kota S, Flowers TJ (2021) Salt tolerance in rice: seedling and reproductive stage QTL mapping come of age. Theor Appl Genet 134:3495–3533. https://doi.org/10.1007/s00122-021-03890-3

Singh RK, Flowers TJ (2010) The physiology and molecular biology of the effects of salinity on rice. In: Pessarakli M (ed) Third edition of "handbook of plant and crop stress". Taylor and Francis, Florida, pp 899–939

Singh RK, Redoña ED, Refuerzo L (2010) Varietal improvement for abiotic stress tolerance in crop plants: special reference to salinity in rice. In: Pareek A, Sopory SK, Bohnert HJ, Govindjee (eds) Abiotic stress adaptation in plants: physiological, molecular and genomic foundation. Springer, New York, NY, pp 387–415. 119

Wicke B, Smeets E, Dornburg V, Vashev B, Gaiser T, Turkenburg W, Faaij A (2011) The global technical and economic potential of bioenergy from salt-affected soils. Energy Environ Sci 4(8):2669–2681

# Facilitating Role of ICAR-Indian Institute of Rice Research, Hyderabad in Multilocation Evaluation and Rice Varietal Release for Salinity and Alkalinity Tolerance in India

**2**

G. Padmavathi, P. Senguttuvel, K. Suneetha, R. A. Fiyaz, and L. V. Subbarao

**Abstract**

Rice is the principal staple food crop for more than half of the world's population. India ranks first in the area and second in rice production worldwide. The ICAR-Indian Institute of Rice Research (ICAR-IIRR), Hyderabad, was established in1965 as the All India Coordinated Rice Improvement Project, now known as All India Coordinated Research Project on Rice (AICRPR) under the aegis of ICAR, and continues to play a pivotal role in conducting nationwide interdisciplinary research for maximizing rice production. The mandate of AICRPR is to coordinate a rice research program in India, basic and strategic research for enhancing rice productivity by developing location-specific varieties and technologies for various ecosystems. Besides, it is involved in the dissemination of technologies, capacity building, and establishing linkages with rice research centers.

Under changing climatic conditions, salt stress is becoming devastating abiotic stress next to drought, affecting rice production. Therefore, special focus is given by AICRPR to salt tolerance, and salinity/alkalinity varietal trials are composed to screen elite lines in hotspots to identify the climate-resilient varieties suitable for salt-affected soils. Since its inception, AICRPR has shown significant progress and contributed 63 saline/alkaline tolerant rice varieties, which were released and notified for cultivation in problem soils. This chapter discusses the genesis and role of ICAR-Indian Institute of Rice Research under the coordinated multilocation testing program, significant achievements made in salt tolerance

G. Padmavathi (✉) · K. Suneetha · R. A. Fiyaz · L. V. Subbarao
Plant Breeding Department, ICAR-IIRR, Hyderabad, Telangana, India

P. Senguttuvel
Hybrid Rice Department, ICAR-IIRR, Hyderabad, Telangana, India



R. K. Singh et al. (eds.), *Genetic Improvement of Rice for Salt Tolerance*,
https://doi.org/10.1007/978-981-99-3830-8_2

breeding under AICRPR, and future directions for enriching rice productivity in salt-affected soils of India.

**Keywords**

Rice · ICAR-IIRR · AICRPR · Rice growing ecosystems · Salt tolerance · Varietal improvement

## 2.1 Introduction

Rice *(Oryza sativa* L.*)* is a major cereal crop and a primary food source for more than half of the global population. Nearly 50% of the world population depends on rice, which provides >70% of human dietary caloric intake. Rice is grown in almost all continents except Antarctica, while Asia is the major continent producing and consuming nearly 90% of the world's rice. Globally, rice, the third largest cereal, is grown in 162 million hectares (mha), with an annual production of 755 million tons (mt) (FAO, 2020). India is the leading producer of rice, ranking first in area and second in production next to China in the world. Rice is cultivated in an area of about 44 mha with an estimated production of 127.93 mt during 2021–2022 under varying climatic and soil conditions (Agricultural Statistics at a Glance, 2021).

In the past seven decades, rice production in India increased nearly six times from 20.58 mt in 1950–1951 to 122.27 mt in 2020–2021. There has been a fourfold increase in rice yield wherein the yield of rice was just 668 kg/ha during 1950–1951, and it gradually increased up to 2713 kg/ha during 2020–2021 (Agricultural Statistics at a Glance, 2021). The country, which used to import rice in the past, now become the world's largest rice exporter wherein it has exported 20 mt of rice worth more than $ 9.6 billion in 2021–2022 to around 150 countries. The adoption of semidwarf high-yielding varieties along with superior crop production practices originated from the All India Coordinated Rice Improvement Project (AICRPR) of the Indian Council of Agricultural Research (ICAR), which was started way back in 1965, and is mainly responsible for this remarkable achievement.

It is projected that India's population will reach 1.4 billion by 2025, and India has to produce a minimum of 160 mt by 2050 to meet the food requirement of its growing population (Subba Rao et al., 2019). The enhanced production must result from increased rice production. In this context, AICRPR is playing a pivotal role in providing the right stewardship to rice research programs in network mode linking state agricultural universities, national institutions, and private research organizations involved in varietal development, evaluation, and release. This chapter discusses about AICRPR on rice and its role in coordinating nationwide multidisciplinary and multilocation testing of varietal improvement for increasing rice production in India in general and progress in rice varietal development for salt-affected areas in particular, including evaluation, identification, release, and notification of varieties and future breeding strategies to fast track the varietal improvement under salt-prone areas.

## 2.2 Major Rice Production Ecosystems

Rice is the only cereal food crop that grows in diverse *agroecological conditions. It is grown in* irrigated (24.3 m ha), rainfed upland (4.2 m ha), rainfed lowland (14.3 m ha), and flood-prone (1.2 m ha) ecologies as in India (Shobha rani et al., 2010). Further, rice is grown from below sea level, for instance, in Kuttanadu district of Kerala up to an altitude as high as 2155 m above mean sea level in high elevated hills of Larnoo in Uttarakhand. It also grows in areas where the water level reaches 5 meters or more. It can also be cultivated in problem soils (coastal and inland salinity, alkalinity, acidity, etc.). India is the only country in the world with such a variety of rice ecosystems, as mentioned below (Fig. 2.1).

### 2.2.1 Irrigated Ecosystem

The irrigated ecosystem is the most productive one, which occupies 24.3 mha, accounting for 50% of the total area in the country. Irrigated rice (70.5 mt) contributes to nearly 75.5% of total rice production, with average productivity of 2.87 t/ha. Much of the irrigated rice is grown in Punjab, Haryana, Uttar Pradesh, Jammu and Kashmir, Andhra Pradesh, Tamil Nadu, Telangana, Karnataka, Kerala, Himachal Pradesh, and Gujarat states.

### 2.2.2 Rainfed Upland Ecosystem

Rice is predominantly grown under rainfed upland conditions in poor soils, limiting yield, and is generally cultivated by small and marginal farmers. Eastern India has the maximum area of upland rice covering Assam, Bihar, Chhattisgarh, Orissa, Eastern U.P, West Bengal, Madhya Pradesh, and Odisha and the North Eastern hill region. This ecosystem is spread across the country, occupying 6 mha with a maximum area in Eastern India and mostly affected by drought.

### 2.2.3 Rainfed Lowland Ecosystem

Rice is grown in about 14 mha of rainfed lowlands. Most of the country's rainfed lowland areas are in the eastern region with low soil fertility leading to yield reduction. It is often affected by drought in later stages of crop growth /flood in early stages due to unpredictable rains. This ecosystem is categorized according to the water depth and duration of water logging into i) rainfed shallow lowland (up to 50 cm) occupying an area of 14 mha; ii) semi-deep (up to 100 cm) and iii) deep (> 100 cm) water ecosystems constituting 1.3 mha of flood-hit areas of eastern Uttar Pradesh, Assam, Bihar, West Bengal, and Odisha.

**Fig. 2.1** Diverse ecologies of rice

## 2.2.4 Flood-Prone Ecosystem

Flood-prone rice is distinguished by the submergence of water with a depth of 50–400 cm in the field for an extended period of time, subject to monsoon rains. It occupies 2.5 mha in the country's eastern states of Assam, Bihar, West Bengal, and

Uttar Pradesh. The average yield is low (1 to 1.5 tons/ha) and is variable. Submergence is a major challenge for rice production in the country. Under such harsh environmental conditions, varietal improvement has very limited potential to increase yield.

## 2.3 Genesis of All India Coordinated Research Project on Rice (AICRPR)

The diversity of ecosystems in which rice is grown in India poses enormous challenges to rice farmers regarding abiotic and biotic stresses affecting rice production. Realizing the ecosystem complexities, destabilizing pest and disease issues, food and nutritional demand of burgeoning population, and the need to attain self-sufficiency in food grain production, the ICAR, New Delhi, launched AICRPR, a landmark in the history of rice research in 1965 with its headquarters at Hyderabad. At that time, the objective was to have a national network for multilocation evaluation of rice lines to expedite breeding efforts with semidwarf varieties.

AICRPR is the largest coordinated research network compared to other crops in India, mainly to generate new rice technologies by coordinating multidisciplinary and multilocation testing of varietal improvement, crop protection, and crop management technologies in different ecosystems of the country and identifying the most stable superior genotypes possessing broad-spectrum resistance to the targeted insect pest and diseases thereby increasing and stabilizing rice production in India. It also strengthens the network of collaboration between various ICAR institutes, SAUs, and the private sector.

It is the first nationally Coordinated Rice Improvement Project having a full-time coordinator with a mandate (1) to conduct basic and strategic research for enhancing rice productivity under irrigated ecosystems, (2) to coordinate multilocation testing to develop location-specific varieties and technologies for various ecosystems, and (3) to disseminate technologies, capacity building, and to establish linkages with national and international institutions of rice research.

Until 1970s, Rockefeller and Ford Foundations funded the AICRPR and supported the system in terms of technical and scientific manpower. Year-round screening against insect pests and diseases was made possible with the construction of glasshouses through financial support extended by Ford Foundation. However, the foreign funding support dwindled and gradually faded over time. In 1966, a big *breakthrough* came in *rice production* with the release of a miracle variety, IR8, from IRRI, Philippines. It more or less coincided with the release in India of another semidwarf high-yielding miracle rice variety, Jaya by AICRPR in 1968. It brought a phenomenal increase in rice production paving the way for green revolution. Such a significant yield improvement enabled the country to move from a state of starvation to exportation by the nineties earning foreign exchange (Brainerd and Menon, 2014).

AICRPR has served over five and a half decades since its inception, with 64 researchers working at the nodal center and around 500 researchers, including scientists, technical, administrative, and supporting staff working at 131 centers, i.e.,

45 funded and 86 voluntary research centers (Table 2.1) spread over seven zones of the country associated with State Agricultural Universities, State Departments of Agriculture, ICAR Institutions, and corporate sector institutions carrying out the planned activities of Breeding, Entomology, Pathology, Agronomy, Soil Science, and Plant Physiology disciplines.

The AICRPR was upgraded to the **Directorate of Rice Research** (DRR) in 1983. The name of the project coordinator was changed to project director. Later, DRR was elevated to a full-pledged national institute and renamed as the Indian Institute of Rice Research (IIRR), a single umbrella to coordinate rice research and development in India in 2014. The position of project director was redesignated as director. The institute improved its infrastructure and human resources over time to meet domestic and global challenges.

AICRPR network also facilitates access to the evaluation of international rice nurseries comprising global exotic germplasm under the International Network for Genetic Evaluation of Rice (INGER) program, which facilitated the introduction of improved varieties such as IR8, IR20, and IR64 from IRRI and brought phenomenal growth in rice production during the 1970s and 1980s.

### 2.3.1 Major Rice Production Zones

Currently, AICRPR trials are evaluated in seven diverse agroecological zones in India based on homogeneity in climatic conditions and soil types. They are designated as Zone 1 (Hills), Zone 2 (northern), Zone 3 (eastern), Zone 4 (northeastern), Zone 5 (central), Zone 6 (western), and Zone 7 (southern). There are 45 funded and around 86 voluntary centers located across 27 states and four union territories (Andaman and Nicobar Islands, Delhi, Puducherry, and Jammu and Kashmir).

Karnataka state in the southern zone holds the maximum number (5) of funded centers while Haryana (northern) and Maharashtra (western) contribute a major share of six and seven centers, respectively, assisting in the assessment of varietal trials across India. The list of zones, states, and their respective funded and voluntary centers are furnished in Table 2.1 and Figs. 2.2 and 2.3.

## 2.4 Chronology of Activities of AICRPR in the Varietal Development Process

The release of a miracle variety, Jaya, from the cross T(N)1 x T-141 by AICRPR, a short-statured high-yielding variety during 1968 through AICRPR system, has revolutionized rice research in India, ushering in green revolution. During the 1970s, a major focus was laid on enhancing rice production by formulating need-based programs or AICRPR trials to detect superior genotypes mainly for irrigated ecology. With this, the country underwent a huge transformation, and by the mid-80 s, India rose to the level of self-sufficiency and began exporting rice to around 150 countries earning foreign exchange.

**Table 2.1** Classification of rice production zones of India

| Name of the zone/state/UT | Funded centers | Voluntary centers |
|---|---|---|
| Zone 1-Hills | | |
| *Northwestern hills* | | |
| Jammu Kashmir | Khudwani, Chatha | Rajouri, Pombay, Wadura, Bandipore, Shalimar, Larnoo |
| Himachal Pradesh | Malan | Dhaula Kuan, Palampur, Sundernagar, Barjura |
| Uttarakhand | | Almora, Bageshwar |
| *Northeastern hills* | | |
| Nagaland | Kohima | Medziphema |
| Meghalaya | Upper Shillong | Umiam-ICAR, Umiam-CAU |
| Manipur | Wangbal | Lamphelpat-ICAR, Imphal-CAU |
| Mizoram | – | - |
| Sikkim | | Gangtok |
| West Bengal | | Kalimpong |
| *Southern hills* | | |
| Karnataka | – | Sirsi |
| Tamil Nadu | – | Gudalur |
| Andhra Pradesh | – | Chintapalli |
| *Zone II-Northern* | | |
| Uttarakhand | Pantnagar | |
| Punjab | Ludhiana | Gurdaspur, Rauni, Kapurthala |
| Haryana | Kaul | Karnal, Rohtak, Jind, Kurukshetra, Panipat, Gautam Buddha Nagar |
| Uttar Pradesh | Nagina, Kanpur | Modipuram |
| New Delhi | | IARI-New Delhi |
| Rajasthan | Kota | Banswara |
| *Zone III-Eastern* | | |
| Odisha | Jeypore, Chiplima | Bhubaneswar, Cuttack, Motto, |
| | | Ranital |
| Bihar | Bikramganj (Patna), Pusa | Patna-ICAR, Sabour |
| Jharkhand | Ranchi | Hazaribagh |
| West Bengal | Bankura, Chinsurah | Canning, Pundibari, Chakdaha, Hathwara, Gosaba |
| Uttar Pradesh | Masodha, Ghaghra Ghat, Varanasi | Lucknow, Gorakhpur |
| *Zone IV-Northeastern* | | |
| Assam | Titabar | Gerua, Shillongoni |
| Tripura | Arundhutinagar | Lembucherra |
| Manipur | Wangbal | Lamphelpat, Imphal (CAU) |
| *Zone V- Central* | | |
| Madhya Pradesh | Rewa | Waraseoni, Jabalpur |

(continued)

**Table 2.1** (continued)

| Name of the zone/state/UT | Funded centers | Voluntary centers |
|---|---|---|
| Chhattisgarh | Raipur, Jagdalpur | Bilaspur, Ambikapur |
| Maharashtra | Sakoli | Sindewahi |
| *Zone VI- Western* | | |
| Maharashtra | Karjat | Panvel, Radhanagari, Shirgaon, Vadgaon, Phondaghat, Parbhani, Palghar |
| Gujarat | Nawagam, Navsari | Danti, Derol, Dabhoi, Vyra, Baruch |
| Goa | – | Goa |
| *Zone VII-Southern* | | |
| Andhra Pradesh | Maruteru | Machilipatnam, Ragolu, Bapatla, Nellore |
| Telangana | Rajendranagar, Warangal | Kunaram, Jagtial, Rudrur, Kampasagar |
| Tamil Nadu | Aduthurai, Coimbatore | Trichy, Annamalai Nagar, Tirur |
| Kerala | Moncompu, Pattambi | Vyttila |
| Karnataka | Mandya, Mugad, Brahmavar, Ponnampet, Gangavati | Sirsi, Malagi |
| Puducherry | Kurumbapet | Karaikal |
| Andaman and Nicobar | – | Port Blair |
| **Total** | **45** | **86** |

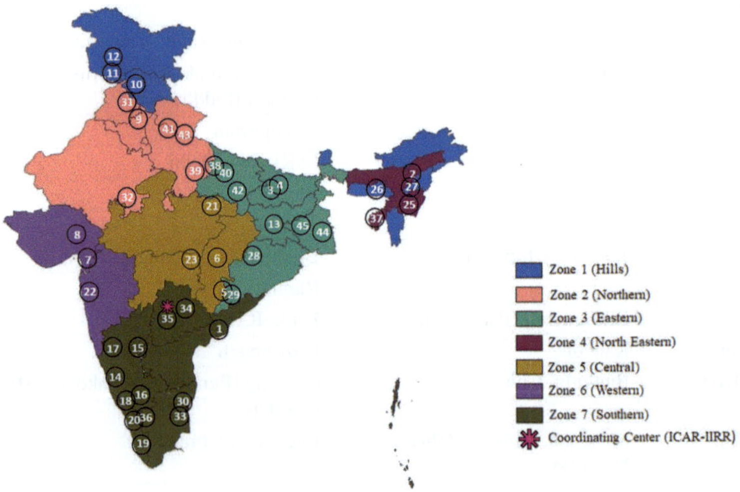


**Fig. 2.2** Funded centers of AICRPR

There have been consistent strides in developing and testing suitable varietal technologies for rainfed uplands, lowlands, semi-deep and deep water, hills, and salt-affected ecologies. During the 1980s, having witnessed the rapid dynamics in insect

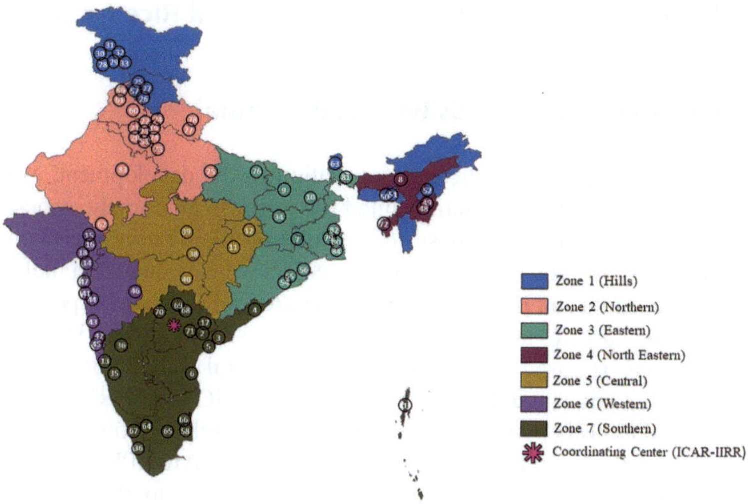


**Fig. 2.3** Voluntary centers of AICRPR

pests and pathogens, apart from major pests (Planthoppers, Stem borer) and diseases (Blast, BLB), resistance to minor pests such as gall midge, green leafhoppers, hispa, leaf folder, and gundhi bug and minor diseases like RTV, sheath blight, false smut, brown spot, and sheath rot received attention. Rice grain quality improvement of both non-basmati and basmati varieties was given priority to sustain yield increase.

The 1990s witnessed the generation of rice hybrids with a minimum of 10% higher magnitude of heterosis over varietal checks and 5% over hybrid checks in different duration groups. Priority was also given to the development of semidwarf aromatic short-grain rice varieties with exquisite aroma and quality attributes by constituting rice aromatic short-grain trials targeting the export markets.

During the 2000s, the soil stress trials, drought trials, and hill trials received much attention. Considering the accelerated variety development through marker-assisted selection (MAS), the trials on near-isogenic lines (NILs) introgressed with genes/ QTLs for biotic (blast, bacterial blight, gall midge) and abiotic (drought, salt and submergence) stresses were constituted.

The enhancement of nutritional quality with high proteins (≥10%) and zinc (≥ 24 ppm) content was addressed through biofortification trials in 2013. While the focus on Boro rice improvement continues, aerobic trials were constituted to address water scarcity. As there is a national demand for medium slender grain varieties fetching good price, trials on MS grain rice lines were constituted in 2015. Trials were formulated on nutrient use efficiency (NUE), particularly nitrogen and phosphorous in 2017 to reduce the cost incurred on nitrogen and phosphorous fertilizers.

## 2.5 Achievements of All India Coordinated Rice Improvement Project

### 2.5.1 Varieties and Hybrids Released in India

The AICRPR has made a significant contribution to the development, evaluation, and release of improved rice varieties suited for various agroclimatic zones. During 1966 to 2021, under the AICRPR system, a total of 32,342 elite lines contributed from different funded centers, and voluntary centers were evaluated in multilocation trials across the length and breadth of the country. AICRPR played a spectacular role in releasing 1495 varieties and 134 hybrids till 2021. Among these, 335 varieties and 114 hybrids were released and notified by the Central Sub-Committee on Crop Standards, Notification and Release of Varieties (CSCS & NRV), while 1160 varieties and 20 hybrids by the respective state variety release committees (SVRC) for commercial cultivation (Fig. 2.4). Thus, the impact of AICRPR on rice improvement is remarkable, which has contributed substantially to the country's food security. Besides releasing varieties and hybrids, India could become the world's largest rice exporter earning $9.6 billion during 2021–2022, and this transformation

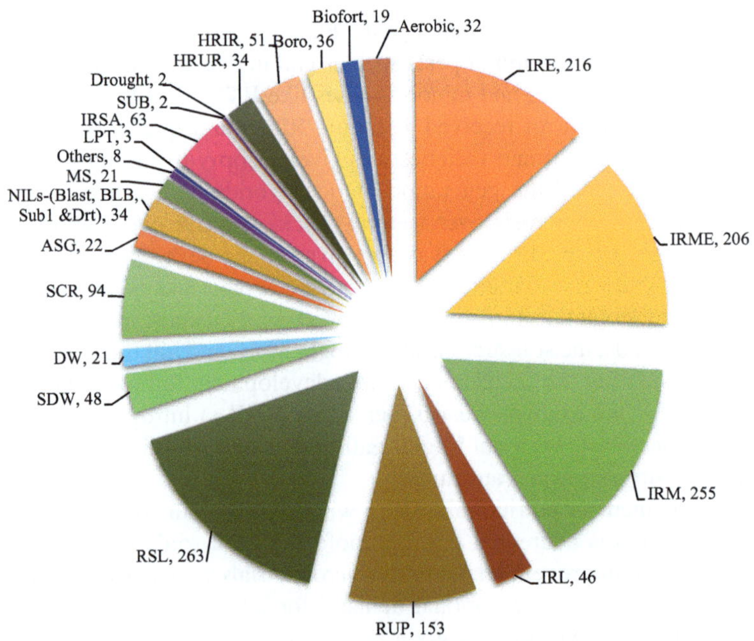


**Fig. 2.4** Rice varieties and hybrids released for different ecosystems in India. *IRE* Irrigated early, *IRME* Irrigated mid-early, *IRM* Irrigated medium, *IRL* Irrigated late, *RUP* Rainfed upland, *RSL* Rainfed shallow lowland, *SDW* Semi-deep water, *DW* Deep water, *SCR* Scented rice, *ASG* Aromatic short grain, *NILs* Near-isogenic lines, *MS* Medium slender, *LPT* Low phosphorus, *IRSA* Irrigated saline and alkaline, *SUB* Submergenece, *HRUR* Hill rice-upland rice, *HRIR* Hill rice-irrigated rice

of the country from importer to exporter happened mainly due to significant progress made by AICRPR.

## 2.6 Role of AICRPR in Coordinating and Developing Tolerant Varieties for Salt-Affected Areas

### 2.6.1 Salt Tolerance in Rice

Soil salinity/alkalinity is the second most devastating abiotic stress after drought, limiting irrigated rice production worldwide and threatening the livelihoods of paddy farmers. Salt-affected regions are characterized by seawater intrusion, low-lying waterlogged areas, flood-prone and ill-drained soils. It is estimated that soil salinization causes annual global economic losses of US$27.3 billion (Qadir et al., 2014). In India, 6.72 mha of land is affected by salt, including 2.95 by salinity and 3.77 mha by sodicity/alkalinity (Arora and Sharma, 2017). In the coming years, soil salinization is expected to increase under changing climate scenarios, such as a rise in sea level causing coastal areas to become more saline and rise in temperature, thereby increasing evaporation and salt deposition. By 2050, it is projected that there will be a threefold rise in the salt-affected area from 6.74 to 16.2 million hectares unless ameliorative steps are taken up (Sharma et al., 2014).

Broadly, the salt-affected problem soils can be classified into two types, viz., saline and alkaline. Carbonates and bicarbonates of sodium and magnesium are the dominating anions in alkaline soils, whereas chlorides and sulfates of sodium and magnesium are most commonly found in saline soils. Salinity is a soil condition characterized by a high concentration of soluble salts and is generally measured in units of electrical conductivity of a saturated soil paste extract (ECe). Saline soil is defined as the soil having ECe of $>4$ $dSm^{-1}$ ($\sim$ 40 mM NaCl), pH of $<8.5$, sodium absorption ratio (SAR) of $<15$, and exchangeable sodium percentage (ESP) of $<15\%$. Again saline soils are grouped into 1. slightly saline (2 to 4 $dSm^{-1}$), 2. moderately saline (4 to 8 $dSm^{-1}$), 3. strongly saline (9 to 16 $dSm^{-1}$), and 4. very strongly saline ($>16$ $dSm^{-1}$). The alkaline/sodic soils are those soils with ECe below 4 $dSm^{-1}$, ESP $>15\%$, SAR $\geq 15$, and pH is usually $>8.5$ or between 8.5 and 10 (Eynard et al. 2005). Further, in rice crops, a soil pH of 8.8–9.2 is considered non-stress, while 9.3 to 9.7 as moderate alkaline stress and $>9.7$ as high alkaline stress.

Salinity in turn is divided into two classes, viz., (i) in land salinity which is caused by heavy waterlogging and evaporation caused by the progressive rise in groundwater level over time in command areas leading to salt deposition and (ii) coastal salinity consequent to seawater intrusion, which introduces salts into agricultural lands in the coastal areas. The coastal saline soils in India occur along the coastlines of India in the states of Odisha, West Bengal, Andhra Pradesh, Pondicherry, Tamil Nadu, Kerala, Karnataka, Maharashtra, Gujarat, Goa and Andaman, and Nicobar Island. The majority of alkaline or sodic soils in India are found in in Uttar Pradesh and Haryana.

Excess salt in soil adversely affects plant growth, development, and productivity when osmotic stress reduces water uptake by roots (Munns and Tester 2008). Direct accumulation of salts affects not only metabolic processes but also yield by affecting physiological and yield-related traits, including tiller number, panicle length, spikelet number per panicle (Khatun et al., 1995, Zeng and Shannon 2000a, b), grain filling (Rao et al., 2013), flowering time (Saade et al., 2016), plant biomass (Zeng et al., 2007), harvest index (Gholizadeh et al., 2014), and photosynthesis (Ismail et al., 2007).

The common field symptoms of salt stress in rice are white leaf tip, leaf-tip burning (salinity), leaf browning and necrosis (sodicity), leaf rolling, drying of older leaves, stunted plant growth, reduced tillering, spikelet sterility, altered flowering duration, reduced root growth, grain sterility, patchy appearance of plants, and ultimately >50% of yield loss occurs (Zeng and Shannon 2000a). Generally, the seedling and reproductive stages are more sensitive to salinity (Waziri et al., 2016), wherein seedling stage salt sensitivity leads to the poor establishment of crop, and reproductive stage sensitivity causes yield reduction.

Rice is moderately tolerant to salinity with a threshold salinity of 3 dS $m^{-1}$. Rice yields decline by 12% for every unit (dS/m) increase in ECe beyond 3.0 dS/m (Grieve et al., 2012). Rice production in India's coastal regions is also affected by abiotic stresses such as submergence, flash and stationary flooding by salt water, drought, high temperatures, and cyclones, apart from salinity stress at distinct growth periods. Improving salt tolerance without yield penalty and grain quality when grown in problem soils is a challenge for rice breeders. Hence, in national rice breeding programs, the deployment of salt tolerance is given high priority to increase rice production in salty areas.

### 2.6.2 AICRPR Trials on Salt Tolerance

Salt-tolerant rice breeding program is coordinated by ICAR-IIRR, Hyderabad, through AICRPR in India. Special trials on salt stress were constituted for screening elite lines developed at various research centers working on genetic improvement of rice for salt-affected areas. There are about 31 hotspot centers consisting of 14 alkaline, five inland saline, and 12 coastal saline centers spread over four zones: northern, eastern, western, and southern (Tables 2.2, 2.3, and 2.4).

The superior advanced breeding lines identified in station yield were initially nominated to a salt-specific nursery, namely, the National Saline and Alkaline Screening Nursery (NSASN). CSR13 and CSR 10 (alkalinity and inland salinity), CST 7–1 (coastal salinity), and Jaya (biological yield check) served as checks along with a local high-yielding stress tolerant popular variety. At the time of nominating elite test lines to the varietal trials, it is vital to provide information to the coordiating centre on parentage and pedigree by the contributory center. The technical program prepared by the coordinating center for conduct of trial is followed uniformly at all rice centers.

**Table 2.2** List of AICRPR funded and voluntary test centers under alkalinity

| SN | Name of the institution | State | Location | Zone | Status |
|---|---|---|---|---|---|
| 1 | ICAR-Central Soil Salinity Research Institute | Haryana | Karnal | Zone 2 | Voluntary |
| 2 | ICAR-Central Soil Salinity Research Institute | Haryana | Kurukshetra | Zone 2 | Voluntary |
| 3 | ICAR-Central Soil Salinity Research Institute | Haryana | Jind | Zone 2 | Voluntary |
| 4 | ICAR-Central Soil Salinity Research Institute | Haryana | Rohtak | Zone 2 | Voluntary |
| 5 | ICAR-Central Soil Salinity Research Institute | Haryana | Gautam Buddha Nagar | Zone 2 | Voluntary |
| 6 | Punjab Agricultural University | Punjab | Kapurthala | Zone 2 | Voluntary |
| 7 | ICAR-, CSSRI Regional Research Station, Shivri Farm | Uttar Pradesh | Lucknow | Zone 3 | Voluntary |
| 8 | C.S. Azad. University of Agriculture and Technology | Uttar Pradesh | Kanpur | Zone 3 | Funded |
| 9 | Crop Research Station, NDUA & T | Uttar Pradesh | Masodha | Zone 3 | Funded |
| 10 | Main Rice Research Station, Anand Agril. University | Gujarat | Nawagam | Zone 6 | Funded |
| 11 | Annamalai University | Tamil Nadu | Annamalai Nagar | Zone 7 | Voluntary |
| 12 | Pandit Jawaharlal Nehru College of Agriculture and Research | Puducherry | Karaikal | Zone 7 | Voluntary |
| 13 | Anbil Dharmalingam Agrl. College & Res. Institute | Tamil Nadu | Trichy | Zone 7 | Voluntary |
| 14 | Agricultural Research Station | Telangana | Kampasagar | Zone 7 | Voluntary |

**Table 2.3** List of AICRPR funded and voluntary test centers under inland salinity

| SN | Name of the institution | State | Location | Zone | Status |
|---|---|---|---|---|---|
| 1 | ICAR-Central Soil Salinity Research Institute | Haryana | Karnal | Zone 2 | Voluntary |
| 2 | ICAR-Central Soil Salinity Research Institute | Haryana | Panipat | Zone 2 | Voluntary |
| 3 | Main Rice Research Station, Anand Agril. University | Haryana | Palwal | Zone 2 | Voluntary |
| 4 | CSSRS Soil and Water Management Research Institute, (Navsari Agril. University), | Gujarat | Danti | Zone 6 | Voluntary |
| 5 | Agricultural Research Station | Karnataka | Gangavati | Zone 7 | Funded |

**Table 2.4** List of AICRPR funded and voluntary test centers under coastal salinity

| SN | Name of the institution | State | Location | Zone | Status |
|---|---|---|---|---|---|
| 1 | ICAR-National Rice Research Institute | Odisha | Cuttack | Zone 3 | Voluntary |
| 2 | ICAR-Central Soil Salinity Institute Regional Research Station | West Bengal | Canning | Zone 3 | Voluntary |
| 3 | Rice Research Station, Govt. of West Bengal | West Bengal | Chinsurah | Zone 3 | Funded |
| 4 | Salt & Flood-Resistant Paddy Research Station | West Bengal | Gosaba | Zone 3 | Voluntary |
| 5 | Main Rice Research Centre, Navsari Agril. University | Gujarat | Navsari | Zone 6 | Funded |
| 6 | Khar Land Research Station | Maharashtra | Panvel | Zone 6 | Voluntary |
| 7 | ICAR-Central Coastal Agricultural Research Institute | Goa | Goa | Zone 6 | Voluntary |
| 8 | Agricultural Research Station | Andhra Pradesh | Machilipatnam | Zone 7 | Voluntary |
| 9 | ICAR-Central Island Agricultural Research Institute | Andaman & Nicobar | Port Blair | Zone 7 | Voluntary |
| 10 | Rice Research Station, (KAU) | Kerala | Vyttila | Zone 7 | Voluntary |
| 11 | Zonal Agril. & Hort. Research Station | Karnataka | Brahmavar | Zone 7 | Funded |
| 12 | Agricultural Research Station | Karnataka | Kumta | Zone 7 | Voluntary |

The coordinating center codes the nominated entries for an unbiased appraisal. NSASN accommodates a large number of lines and facilitates shortlisting the best tolerant lines stress-wise. They were advanced for screening in the respective saline and alkaline experiments of the Saline and Alkaline Tolerant Variety Trial (SATVT) for 3 years under diverse agroclimatic conditions. In 2000, in place of CSR 10, two new checks, i.e., CSR 13 (alkaline) and CSR 27 (inland saline), were included for judging the performance of test entries. In 2007, the SATVT trial was bifurcated into two separate stress trials, i.e., the Alkaline and Inland Saline Tolerant Variety trial (AL & ISTVT) and Coastal Saline Tolerant Variety trial (CSTVT). In 2013, the screening nursery, NSASN, was discontinued for testing in AICRPR. Consequent to the release of new varieties, CSR 13 check was replaced by CSR 36 (Al) and CSR 27 by CSR 23 (IS) and included short-duration check, CSR 10 in both AL & ISTVT and CSTVT trials for comparison of short-duration test entries during 2014. Each trial of AL & ISTVT and CSTVT is subdivided into IVT, AVT-1, and AVT-2 trials. During 2015, FL 478 (tolerant check) and Samba Mahsuri (sensitive check) were added in the trials for effective screening of test lines. Samba Mahsuri was eventually replaced in the trials by Pusa 44, a more salt sensitive check.

### 2.6.3 Testing Procedure of Salinity/Alkalinity Trials Under AICRPR

The varietal testing program follows a three-tier system involving 1 year of initial varietal trial (IVT), namely, IVT-AL & ISTVT or IVT-CSTVT, which constitutes the first level of testing after assigning the initial evaluation trial (IET) number to every nominated entry developed by different cooperating centers. It is followed by 2 years of testing in advance varietal trials (AVT-1 and AVT-2), namely, AVT-1-AL & ISTVT/AVT-1-CSTVT or AVT-2-AL & ISTVT/AVT-2-CSTVT to identify the best elite lines with consistent yield superiority over the best stress specific checks. Simultaneously, the elite breeding lines are screened at hotspot locations as well as under artificial conditions for insect pest and disease resistance. Grain quality components and agronomic performance is evaluated for all the promising entries.

#### 2.6.3.1 Initial Variety Trial (IVT): first Year of Testing

The nominated entries are introduced directly into IVT-AL & ISTVT or IVT-CSTVT, the first year of testing and grown in a complete randomized block design in a 10 $m^2$ plot size following the appropriate crop and water management practices as per the guidelines. The trial for sodic soils (alkaline) could be conducted in soil with pH >8.5, ECe <4 $dSm - ^{1}$, and ESP more than 15%, whereas inland salinity trial will be conducted in soil with pH ≤ 8.0 and ECe >4 $dSm^{-1}$. In coastal saline soils, the trial could be conducted in soils with electrical conductivity >4 dSm $^{-1}$ and pH <8. The observations on days to 50% flowering, plant height, lodging, panicles/$m^2$, reaction to disease and pests, grain yield, and soil stress parameters have to be recorded by the cooperating scientists conducting the trial. The entries at the IVT stage are tested for their reaction to pests and diseases in National Screening Nursery-2 (NSN-2) across hotspots. The superior lines with ≥5% yield gain over the best stress-tolerant check would be advanced to the next stage, AVT-1.

#### 2.6.3.2 Advance Variety Trial (AVT-1): second Year of Testing

The promoted entries from IVT are constituted as a separate trial, AVT-1-AL & ISTVT or AVT-1- CSTVT, for the second year of testing in the same manner as that of IVT. Care is taken to conduct the trial in larger plots of 15 $m^2$ to reduce the measurement errors inherent in small plot size. Data recording and promotion criteria of test entries are similar to the norms specified under IVT. The AVT-1 cultures are simultaneously analyzed for grain quality and response to pests and diseases in National Screening Nursery-1 (NSN -1). As per the recommendations of the 53rd Annual Rice Group Meeting, 2018, the entries with the following grain quality parameters are promoted.

Entries with Head Rice Recovery (HRR): >45% (Basmati) and 55% (non-basmati and hybrids).

Entries with AC: 20 to 25%, GC: 20 to 80 mm and ASV: 4–7.

Entries with AC: 25 to 27%, GC: >40 mm and ASV: 5–7.

Entries with AC: > 27%, GC: > 60 mm and ASV >5.

Entries with AC <20% are dropped. However, in the Northeastern region, where sticky rice is preferred for consumption, entries with AC < 20%, GC: <60 mm, and ASV < 5 are promoted.

Along with desirable grain quality traits, the superior entries exhibiting ≥5% yield improvement over the respective stress-tolerant check are promoted to the next stage of testing, i.e., AVT-2.

#### 2.6.3.3 Advance Variety Trial (AVT-2): third Year of Testing

The entries qualified for advancement from AVT-1 are further evaluated in AVT-2-AL & ISTVT or AVT-2-CSTVT in the third year of testing following similar plot size ($15m^2$), experimental design, and agronomic practices as that of AVT-1. The procedure of recording data, screening against pests and diseases, and analysis for cooking and eating quality attributes is the same as that of the AVT-1 trial. AVT-2 trials are simultaneously evaluated for their response to two recommended doses of fertilizer (50% and 100% RDF) at specific centers by agronomists. The criteria specified under IVT and AVT-1 for yield and quality evaluation shall be implemented during this stage in identifying the best entries at the zone level.

At each year of testing during the crop season, multidisciplinary monitoring teams are constituted by the director, ICAR-IIRR, Hyderabad involving scientists from the sections of crop improvement, crop production and crop protection etc. The team pays a visit to trial sites during flowering to maturity stage. They will inspect the conduct of trial according to the layout, management as per the specified norms, plot size adopted, and uniformity of crop stand and record observations on days to flowering, maturity, plant height, lodging, threshability, grain type, pest and disease pressure at hotspots, and controlled conditions and damage due to birds and yield estimates. While monitoring the trials on near-isogenic lines, a stringent inspection of the entries is made for the presence of the trait introgressed and similarity with the recurrent parent. The report of the monitoring team is crucial as it forms the basis for acceptance or denial of the yield data for inclusion in the analysis.

### 2.6.4 Recording Data

A prescribed datasheet is furnished to test centers for recording data as per the guidelines. The replication-wise data on grain yield (kg/plot) based on net plot size, and the ancillary traits such as plant height, days to 50% flowering as well as panicles/$m^2$, seedling survival percentage, salt injury scores, phenotypic acceptability, and reaction to the insects and diseases prevalent in the location including weather parameters and occurrence of abiotic stresses (submergence, drought, etc.) are recorded during crop growth. Additionally, the data on soil stress parameters such as soil pH, ECe, and ESP are recorded by drawing soil samples at a depth of 0–15cm and 15–30 cm in three stages of crop growth, i.e., (1) before transplanting, (2) maximum tillering, (3) flowering stage and sent to the coordinating center within the time frame. The yield data would not be considered for analysis in the absence of data on stress parameters, as it is impossible to determine the degree of stress the crop has experienced.

The data will be critically checked before inclusion in statistical analysis at the coordinating center. The data from any location is rejected in case the trial is not conducted under appropriate stress, or the sensitive check is performing better than the tolerant checks, or the trial mean yield is lower than the recommended yield level (2 t/ha) or high, unrealistic yield value or CV exceeding the set norms (5 to 30%). Of late, various AICRPR partners are advised to submit data online through the AICRPR-intranet program.

### 2.6.5 Criteria for Promoting the Entries During Testing

In the AL & ISTVT trial, the performance of entries is compared for >5% yield superiority over the best check between CSR 36 and local check under alkalinity while against the best check between CSR 23 and local check under inland salinity. The short-duration entries are compared with the short duration check CSR 10 under alkalinity and inland salinity. Similarly, in CSTVT under coastal salinity, the entries are compared with the best check between Bhuthnath and local check for >5% yield superiority, while the short-duration entries are compared with the short duration check CSR 10. The entries outperforming the best check are promoted in zones if they show >5% yield advantage across locations in that zone or 10% yield gain in case of a single location to the second year of testing (AVT-1 AL & ISTVT or AVT-1-CSTVT). The same procedure is followed to promote entries from AVT-1 to AVT-2-AL & ISTVT or AVT-1-CSTVT to AVT-2-CSTVT.

### 2.6.6 Method for Identification of Varieties

After completion of evaluation in AVT-2 stage, a few entries would be identified with superior yield performance, adaptability, disease and pest resistance, tolerance to salt stress, and acceptable grain quality. The respective plant breeder submits a proposal for varietal identification in a variety identification proforma specified by ICAR to the coordinating center before the commencement of the annual rice workshop. It consists of compiled data on grain yield across 3 years (IVT, AVT-1, and AVT-2), a minimum of 2 years' data on grain quality traits and disease and pest reaction for 2 years (AVT-1 and AVT-2) and one-year agronomy data (AVT-2) on the response of entry to recommended fertilizer doses. While submitting the variety proposals to the variety identification committee (VIC), the yield superiority of the entry in the third year is obligatory. Hence, the entries exhibiting yield advantage over the check in the first year and third year; or the second year and third year; or the first, second, and third years are considered for submission to VIC. DNA fingerprinting of the candidate variety along with parents; and hybrid and of its parental lines in case of rice hybrids with molecular markers and good photographs of field, of a single plant and seed is required during submission. DUS characterization of the candidate variety has to be done for descriptors of DUS traits (Rani et al., 2006) to ensure that it is distinct from the released varieties, the expression of morphological characteristics is uniform and characteristics are stable over a period of time. A Variety Identification Committee established by ICAR meets during every annual rice workshop headed by the Deputy Director General (Crop science) consisting of ADG (FFC), ADG (seeds), director, director of research of SAU, agricultural commissioner, one representative of seed organization, ICAR. The director serves as a member secretary and principal investigators of various disciplines act as resource persons for scrutinizing the proposals. The superior candidate varieties shall be identified and recommended for release in the states of the respective zones.

#### 2.6.6.1 Release and Notification of Varieties

The proposals approved and recommended by the VIC are then submitted in another proforma for the release of crop varieties/hybrids to the Central Sub-Committee on Crop Standards, Notification and Release of Varieties (CCSNRC). Submission of a certificate from the National Bureau of Plant Genetic Resources with the IC number is essential for releasing a variety (Tandon et al., 2015). The members of CSCSNRC of the Government of India scrutinize the data on proposed cultivars for release (ICAR 1992).

Upon acceptance of the proposal by the Central Sub-Committee, the variety/hybrid is released for the concerned state(s) as a central variety, and it is simultaneously notified in the official Government Gazette for seed production in the country. Departments of Agriculture of various states and seed producers across the country, with the Directorate of Seeds, Ministry of Agriculture, GOI, would place the indents for breeder seed. GOI allot breeder seed production targets in cultivars through AICRPR to the respective originating breeders. Certified seed production will be taken up by either public or private seed organizations using the breeder seed of notified inbred cultivars for distribution to farmers (GOI 1996–2020).

Any variety/hybrid which is dropped at IVT stage in the first year or found superior continuously for 2/3 years of evaluation in any one of the states could further be subjected to multilocation testing of that particular state for 2–3 additional years. If found superior, the release proposal could be submitted by the State Agricultural University to the State Variety Release Committee for release in that state.

### 2.6.7 Breeding Efforts at AICRPR Centers for Genetic Enhancement of Salt Tolerance

Tolerance to salt is a complex quantitative trait controlled by polygenes present at different loci in the rice genome (Flowers, 2004). Basically, three mechanisms that control salinity tolerance were proposed (Munns and Tester, 2008; Roy et al. 2014), namely, (1) exclusion of toxic ions from the shoot by adjusting transport pathways in the roots, preventing the excess accumulation of $Na^+$ and $Cl^-$ in the leaves and including the retrieval of $Na^+$ from the xylem, and efflux of ions back to the soil; (2) tissue tolerance, by compartmentalizing toxic ions, namely $Na^+$, in specific tissues, cells, and subcellular organelle, in addition to the synthesis of compatible solutes and production of enzymes catalyzing detoxification of reactive oxygen species; and (3) osmotic tolerance, by regulating long-distance signals and reducing shoot growth, independently of the extent of $Na^+$ accumulation in the shoot.

In India, traditional breeding methods such as introduction and selection of landraces, pedigree, modified bulk pedigree, bulk method, mutation, and shuttle breeding were followed to breed new salt-tolerant varieties (Singh et al., 2009). Initially, at different AICRPR centers, the improvement was focused on the selection of existing germplasm. Several landraces were widely utilized to generate locally adapted and improved cultivars that were cultivated on a large scale in farmers' fields. To name a few, Pokkali, Nona Bokra, and Kalarata originated in the coastal areas of India were extensively utilized in breeding programs in the mid-1970s. Among them, Pokkali is the most frequently used donor of salt tolerance traits. It maintains low $Na^+$ to $K^+$ ratio in the shoot and grows relatively better under the saline situation. However, the salinity tolerance potential of newly bred varieties was lower than that of donors. Due to linkage drag, the undesirable genes associated with tallness, photosensitivity, lateness, poor grain quality, and low yield of donors were often brought from the donors and could not be completely eliminated during the selection process. Thus, early breeding efforts were aimed at the genetic improvement of locally domesticated land races having remarkable tolerance to salinity by pure line selections. Some of the locally adapted popular varieties selected from local landraces are Arya 33 and Ankola in Karnataka; CSR1(Damodar), CSR 2 (Dasal), CSR 3 (Getu), Matla, Hamilton, CSR 6, Bhaluki, Rupsail, Patnai 23, and Patnai 298 in West Bengal; Kalar samba, and Kallimadayam in Tamil Nadu; MCM 2 in Andhra Pradesh; Pokkali, Vyttila-1, Vyttila-2, Vyttila-3, Vyttila-4, Vyttila, MO1, MO2, and MO 3 in Kerala,; Jhona 349 in Punjab; T 21 and Lakra or T 22A in Uttar Pradesh; SR 8, SR 26B, and Raspanjore Orissa; Korgut and Azgo in Goa; Kalarata 1–24, Bhura Rata 4–10, MK 47–22, and SR 3–9 in Maharashtra.

With the development of more efficient screening techniques (Gregorio et al., 1997), a large number of germplasm accessions were systematically evaluated to identify donors with considerable yield potential. Hence, in the late 1980s, emphasis was shifted to selection and recombination breeding at different centers. The identified donors were exploited in hybridization programs, and tolerance genes were transferred to improved genetic backgrounds through the pedigree method or bulk method of selection under salt stress, and subsequently, varieties were released after subjecting them to multilocation testing in AICRPR. Heterosis breeding and mutation breeding were also executed at partner centers.

#### 2.6.7.1 Marker-Assisted Back Cross-Breeding to Introgress *Saltol* in Rice

Despite the progress made in salt tolerance through traditional breeding methods, the longer breeding time and possible transfer of undesirable traits due to linkage drag affect yield and slow down breeding success. Consequent to the remarkable efforts made in the identification of a major *Saltol* QTL maintaining shoot $Na^+/K^+$

homeostasis under salinity, its fine mapping and detection of tightly linked markers within the *Saltol* QTL region (AP3206, RM8094, and RM3412), flanking markers, *i.e.,* RM1287 and RM10694, RM493 and RM10793 (Thomson et al., 2010), it was successfully transferred into many popular varieties in India within the shortest time through marker-assisted back cross-breeding. Several researchers have successfully introduced the *Saltol* QTL into various rice varieties to breed new, improved salt-tolerant lines in the gentic background of ADT 43 (Geetha et al., 2017), PB 1121 (Babu et al., 2017), Pusa Basmati 1 (Singh et al., 2018), Pusa 44 and Sarjoo 52 (Krishnamurthy et al., 2020), Pusa Basmati 1509 (Yadav et al., 2020), Aiswarya (Nair and Shylaraj 2021), and improved Samba Mahsuri (Rekha et al., 2022) in India.

### 2.6.8 Progress on Rice Variety Development Under Salt-Affected Soils in India Through AICRPR

Significant progress has been made in breeding salt-tolerant high-yielding rice varieties for various inland saline, coastal saline, and alkaline soils under AICRPR. Different breeding approaches were used to alter the genetic makeup of rice to increase its salt tolerance at partner centers. Diverse parents were crossed to broaden the genetic base for salt tolerance. Single crosses, double-crosses, three-way crosses, and poly crosses were used during hybridization. The vast number of germplasm accessions with varying tolerance levels were exploited in the breeding program. The comprehensive efforts in rice improvement made in the AICRPR program in collaboration with key partners resulted in the development and release of around 63 salt-tolerant high-yielding rice varieties with improved genetic background for cultivation suited for salt-affected regions of the country.

Some of the notable varieties released over the years across states through conventional breeding approaches at the national level by NARCs include CSR10, CSR13, CSR 23, CSR 27, CSR30, CSR 36, CSR43, CSR46, CSR49, CSR52, CSR56, CSR60, and CSR76 developed from CSSRI, Karnal; Bhuthnath, CST 7-1, Sumati, and Amal Mana from CSSRI regional station Canning Town; Lunishree, Luna suvarna, Luna sampada, and Luna Ambika from ICAR-Cuttack; and Vikas, DRR Dhan 39, and Jarava from Hyderabad. At the state level, the popular varieties released include Sagara, VTL 1 to 9 (Pokkali varieties), Ezhome-1, Ezhome-3 Lavanya, and Jyotsna in Kerala; Panvel 1, Panvel 2, and Panvel 3 from Maharashtra; MCM 1 from AP, AU 2, PVR1, TRY 1, and TRY 2 from Tamil Nadu; Mohan from West Bengal; CSR 22 from Karnataka; type 100, Usar Dhan 1, Narendra Usar Dhan 2, Narendra Usar Dhan 3, and Narendra 2 from UP; Dandi from Gujarat; Luna Suvarna and Luna Sankhi from Odisha; and Jhona 349 from Punjab (Table 2.5).

Attempts were also made to exploit wild species of rice, namely, *Oryza rufipogon*, and succeeded in the generation of Jarava variety at ICAR-IIRR, Hyderabad. Two salt-tolerant mutant varieties, Lunishree and Vyttila-5, derived from Nonasail and Mahsuri, were developed through mutation breeding at Cuttack and Vyttila, respectively. Through heterosis breeding, a salt-tolerant hybrid, namely,

**Table 2.5** Salt-tolerant rice varieties released in India till 2022

| S N | Variety name | IET No | Designation | Cross combination | 50% Flowering duration | Grain type | Year of notification | Notification number | Notification date | States recommended |
|---|---|---|---|---|---|---|---|---|---|---|
| 1 | Jhona 349 | – | Jhona 349 | Pure line selection | 96 | LB | 1933 | 716 (E) | 20-2-1970 | Haryana |
| 2 | PVR 1 | – | PVR 1 | SR 26-B/MTU 1 | 115 | SB | 1968 | – | – | Tamil Nadu |
| 3 | Type 100 | – | – | Sel. from Bhanslot | 120 | SB | 1978 | 13 (E) | 19-12-1978 | Uttar Pradesh |
| 4 | Narendra 2 | 4555 | IR 2071-625-1-252 | IR 8/Tadukan// TKM 6/TN 1 // IR 8 × IR 24 | 85 | LS | 1982 | 165 (E) | 03-06-1987 | Uttar Pradesh |
| 5 | Vikas | 3116 | RP 516-31-6 | IR 8/TKM 6 | 84 | MS | 1983 | 499 (E) | 07-08-1983 | Andhra Pradesh, Bihar, Madhya Pradesh, West Bengal, Kerala |
| 6 | Mohan | 12,494 | CSR 4 | Mutant of IR 8 | 85 | MS | 1984 | – | – | West Bengal |
| 7 | Panvel 1 | 7337 | PNL 5-30 | BR 4–10/IR 8 | 95 | SB | 1984 | 540 (E) | 24-7-1985 | Maharashtra |
| 8 | Usar 1 | – | – | Jaya/Getu | 100 | MS | 1984 | 540 (E) | 24-7-1985 | Uttar Pradesh |
| 9 | Panvel 2 | 8118 | PNL 32-10-1-1 | BR 4–10/IR 8 | 90 | LS | 1987 | 386 (E) | 15-5-1990 | Maharashtra |
| 10 | CSR 10 | 10,349 | 81-H 21-2-4 | M 40–431–24–114/Jaya | 95 | SB | 1989 | 915 (E) | 11-06-1989 | Uttar Pradesh, Haryana, Punjab, Gujarat |
| 11 | AU 2 | – | IR 2071-586-5-6-3 (IR 42) | IR 1561–228–1-2/IR 1737 | 102 | LB | 1989 | – | – | Tamil Nadu |
| 12 | CST 7–1 | 9341 | CST 7-1 | Damodar/ IR 24 | 115 | LB | 1991 | 793 (E) | 22–11–1991 | Andhra Pradesh, Odisha, West Bengal |
| 13 | Vyttila-4 | 13,418 | KAU 906 | Chettivirippu/ IR 4630–22–2-17 | 93 | LB | 1991 | – | – | Kerala |
| 14 | Lunishree | 10,678 | CRM 30 | Nonasail mutant | 115 | LS | 1992 | 814 (E) | 11–04–1992 | Odisha, West Bengal |
| 15 | Sagara | – | – | Selection from Orumundakan | 155 | SB | 1993 | 599 (E) | 25-4-2006 | Kerala |

(continued)

**Table 2.5** (continued)

| S N | Variety name | IET No | Designation | Cross combination | 50% Flowering duration | Grain type | Year of notification | Notification number | Notification date | States recommended |
|---|---|---|---|---|---|---|---|---|---|---|
| 16 | Vyttila-5 | 14,527 | KAU 655 | Mutant of Mahsuri | 90 | SB | 1995 | – | – | Kerala |
| 17 | Narendra Usar Dhan 2 | 13,556 | NDRK 5020 | IRRI Line F2 | 95 | LB | 1998 | 401 (E) | 15-5-1998 | Uttar Pradesh |
| 18 | CSR 13 | 10,348 | 80-H-3-13 | CSR I / Basmati 370 // CSR 5 | 95 | LS | 1999 | 425 (E) | 06-08-1999 | Uttar Pradesh, Haryana, Bihar, Gujarat, Maharashtra |
| 19 | CSR 27 | 13,765 | CSR 88-IR-6 | Nona Bokra/ IR 5657-33–2 | 110 | LB | 1999 | 1050 (E) | 26-10-1999 | Uttar Pradesh, Haryana, Maharashtra, Bihar, Gujarat, Kerala, Tamil Nadu, West Bengal |
| 20 | Narendra Usar Dhan 3 | 14,659 | IR 46330 (NDRK 14659) | Lung Y AI 148/IR 9125–209–2-2-2-1// IR1872–27–3-1 | 105 | LS | 2000 | 92 (E) | 02–02–2001 | Uttar Pradesh |
| 21 | Panvel 3 | 15,368 | PNL 18–5-H 7–2 | Damodar/ Pankaj | 100 | LB | 2000 | – | – | Maharashtra |
| 22 | TRY 1 | 16,643 | TRY 1 | BR 153–2B-10-1-3 | 110 | SB | 2000 | 92 (E) | 02–02–2001 | Tamil Nadu |
| 23 | TRY (R) 2 | 12,863 | RP 2597–14-250 | IET 6238/IR 36 | 85 | LS | 2001 | 1134 (E) | 15–11-2001 | Tamil Nadu |
| 24 | Dandi | 14,906 | PNL 2–58–1-1 | PNL 2/IET 8320 | 100 | LB | 2002 | 283 (E) | 03–12–2003 | Gujarat |
| 25 | CSR 23 | 13,769 | CSR 89-IR 5 | IR 64// IR 4630–22–2-5-1-3/IR 9764-45-2-2 | 103 | MS | 2003 | 161 (E) | 02-04-2004 | Uttar Pradesh, Haryana, Maharashtra, Gujarat, Kerala, Tamil Nadu |
| 26 | Sumati | 13,428 | CSRC(S)2–1-7 | Pankaj/NC 678 | 105 | SB | 2004 | 161 (E) | 02-04-2004 | West Bengal, Odisha, Kerala |

| | | | | | | | | | | |
|---|---|---|---|---|---|---|---|---|---|---|
| 27 | Jarava | 15,420 | B 90–15 | B 32 Sel.4/ *O. rufipogon*// B 29–6 | 113 | SB | 2005 | 1566 (E) | 11-05-2005 | West Bengal, Andaman & Nicobar, Puducherry |
| 28 | Naina | 17,340 | CSR 36 | CSR 13/Panvel 2// IR 36 | 115 | LS | 2005 | 1566 (E) | 11-05-2005 | Haryana, Puducherry |
| 29 | Narendra Usar Sankar Dhan 3 | 16,651 | NDURH 3 | IR 58025 A/NDRK 5026–1 R | 105 | SB | 2005 | – | – | Uttar Pradesh |
| 30 | Vyttila 6 | – | – | Chiriviruppu/ IR 5// Jaya | 80 | LB | 2005 | 1566 (E) | 11-05-2005 | Kerala |
| 31 | Bhuthnath | 12,855 | CSRC(S) 5–2–2-5 | SR 26B/Pankaj | 105 | LS | 2006 | 1572 (E) | 20-9-2006 | West Bengal, Odisha, Maharashtra |
| 32 | CSR 22 | 15,942 | CSR 22 | IR 64/IR 4630–22–2-5-1-3/IR 7969-45-2-2 | 107 | LS | 2008 | 2187 (E) | 27-8-2009 | Karnataka |
| 33 | Amal Mana | 18,250 | CSRC(S) 7–1-4 | Pankaj/SR 26 B | 117 | ELS | 2009 | 454 (E) | 02-11-2009 | West Bengal, Odisha, Andhra Pradesh |
| 34 | Narendra Usar Dhan 2008 | 18,699 | NDRK 5088 | TCCP 266–249-B-B-3/IR 262–43–8-1 | 100 | LB | 2009 | 2187 (E) | 27-8-2009 | Odisha, Andhra Pradesh, West Bengal |
| 35 | GNR-2 | 21,640 | NVSR-6029 | GR-103 X POKKALI | 95–100 | MS | 2009 | 312(E) | 02-01-2013 | Gujarat |
| 36 | DRR Dhan 39 | 19,487 | RP 4631-46-6-5-1-1-1 | CSR 3/Kasturi | 103 | LS | 2010 | 211 (E) | 29-1-2010 | Odisha, Kerala and Gujarat |
| 37 | Vyttila-8 | – | – | IR 47310–94–4-3-1/CSR 10 | 85 | MB | 2010 | 733 (E) | 04-01-2010 | Kerala |
| 38 | Luna Sampada | 19,470 | CR2095–181–1 | Mashuri/ Chakrakonda | 113 | MB | 2011 | 632-E | 25-03–2011 | Odisha |
| 39 | Luna Suvarna | 18,697 | CR2096-71-2 | Mashuri/ Orumundakan | 108 | LS | 2011 | 632-E | 30-03–2011 | Odisha |

(continued)

**Table 2.5** (continued)

| S N | Variety name | IET No | Designation | Cross combination | 50% Flowering duration | Grain type | Year of notification | Notification number | Notification date | States recommended |
|---|---|---|---|---|---|---|---|---|---|---|
| 40 | CSR 43 | 18,259 | CSR89-IR 8 | KDML 105/IR 4630–22–2-5-1-3/IR 20925–33–3-1-28 | 85 | LS | 2012 | 244(E) | 24-01-2014 | Uttar Pradesh |
| 41 | Luna Barial (CR Dhan 406) | 19,472 | CR 2092–158-3 | Jaya/ Lunishree | 125 | SB | 2012 | 312 (E) | 01–02–2013 | Odisha |
| 42 | Luna Sankhi (CR Dhan 405) | 21,237 | CR 2577–1 | IR 31142–14–1-1-3-2/IR 71350 | 85 | MS | 2012 | 312 (E) | 01–02–2013 | Odisha |
| 43 | Binadhan 10[a] | – | PBRC-37 | IR42598-B-B-B-12/Nona Bokra | 95–100 | MS | 2015 | 921(E) | 01-04-2015 | West Bengal, Tripura, Assam |
| 44 | Binadhan 8[a] | – | IR66946–3R-149-1-1 | IR 29/POKKALI | 10–105 | Bold | 2015 | 921(E) | 01-04-2015 | West Bengal, Tripura, Assam |
| 45 | Chinsurah Nona-I | 23,403 | Gosaba 5 (IR 551793B-11–3) | IR 4630–22–2-5-1-3/Nona Bokra | 110 | SB | 2016 | 3540(E) | 22-11-2016 | West Bengal |
| 46 | CARI Dhan-5 | 16,885 | BTS 24 | Pure line selection from Pokkali somaclones | 114 | SB | 2017 | 1007(E) | 30-03–2017 | Andaman and Nicobar |
| 47 | CSR 60 | 25,378 | CSR 2013 MI-10 | IR4630–22–2-5-1-3/ IR05N204 | 100 | LS | 2018 | 6318 (E) | 26–12–2018 | Uttar Pradesh, Puducherry |
| 48 | CSR 56 | 24,537 | CSR 56 | CSR 21/CSR 10 | 120 | LB | 2018 | 6318 (E) | 26–12–2018 | Uttar Pradesh, Haryana |
| 49 | GNR-5 | 25,075 | NVSR-6137 | Jaya/GR-6 | 105 | LS | 2018 | 1379(E) | 27–03–2018 | Gujarat |
| 50 | Pandu Ranga | 22,611 | MCM 100 | MTU 1042/ MTU 1061 | 110 | MS | 2019 | 3220 (E) | 05-09-2019 | Andhra Pradesh |

| | | | | | | | | | | |
|---|---|---|---|---|---|---|---|---|---|---|
| 51 | GNR 7 | 25,093 | NVSR 6128 | GR-103/GR-11 | 100 | SS | 2019 | 3220 (E) | 05-09-2019 | Gujarat |
| 52 | Chinsurah Nona 2 (Gosaba 6) | 21,943 | RP 4919-50-13 | KMR 3/*Oryza rufipogon* | 104 | MB | 2019 | 3220 (E) | 05-09-2019 | West Bengal |
| 53 | CSR 76 | 27,070 | CSR 2748–4441–195 | CSR 27/MI48 | 100 | LS | 2021 | 8 (E) | 24–12–2021 | Uttar Pradesh |
| 54 | TRY 4 | 21,730 | TR 05031 | ADT 39/CO 45 | 100 | MB | 2021 | 8 (E) | 24–12–2021 | Tamil Nadu |
| 55 | DRR Dhan 58 | 28,784 | RP 6287–188–45–12-88 | Improved Samba Mahsuri[a]3/// FL 478 | 100 | MS | 2021 | 8 (E) | 24–12–2021 | Andhra Pradesh, Telangana, Tamil Nadu, Karnataka, Jharkhand, Odisha, Chhattisgarh, Maharashtra, Gujarat, Bihar |
| 56 | Lavanya (Kau-Vyttila 10) | 25,083 | KAU VTL – 51-5 | VTL-3 mutant | 83 | SB | 2021 | 2986 (E) | 20-07-2021 | Kerala |
| 57 | Jyotsna (KAU, BIL4) | 26,057 | KAU VTL-11 | Jyothi/FL 478/Jyothi// Jyothi | 75 | LS | 2021 | 2986 (E) | 20-07-2021 | Kerala |
| 58 | Auranga | 27,043 | GR 19 | Dandi/IET 15429 | 100 | SB | 2021 | 8(E) | 24.12.2021 | Gujarat |
| 59 | NICRA Dhan, (Luna Ambiki) | 27,852 | CR Dhan 412 (CR 2851-S-1-6-2B-4-1) | Gayatri/SR 26B | 101 | MS | 2021 | 8(E) | 24.12.2021 | Odisha |
| 60 | Rajendranagar Vari-1 | 27,077 | RNR 11718 | MTU1010/ NLR 34449 | 100 | MS | 2022 | 4065(E) | 31-08-2022 | Haryana, Gujarat, Karnataka, Punjab |
| 61 | CR Dhan 414 | 27,051 | CR2851-S-1-B-4-1-4-1-1 | Gayatri/SR 26B | 113 | MS | 2022 | 4065(E) | 31-08-2022 | Andhra Pradesh, Odisha, West Bengal |
| 62 | MTU Rice 1293 | 28,010 | MTU Rice 1293 | MTU 1010/FL 478//[a]3 MTU 1010 | 86–88 | LS | 2022 | 4065(E) | 31-08-2022 | Andhra Pradesh, Telangana, Tamil Nadu, Kerala, Karnataka |
| 63 | TRY 5 | 26,068 | TR 09030 | Mutant of TRY 2 | 75–82 | LS | 2022 | 4065(E) | 31-08-2022 | Tamil Nadu |

[a]Released by the Philippines, Bangladesh, and India under a tripartite agreement

Narendra Usar Sankar Dhan 3 was developed in Faizabad in UP. These varieties/hybrids can be cultivated in soils with pH and ECe ranging from 9.4 to 9.8 and 6–11 $dSm^{-1}$, respectively.

Apart from the traditional breeding approach, concerted efforts were made to improve the salt tolerance of otherwise well-adapted varieties by introgression of *Saltol* QTL through marker-assisted selection. Recently, a couple of varieties, namely, DRR Dhan 58 and IET 28010 (MTU Rice 1293) were generated using the MABB approach. DRR Dhan 58 variety was developed at IICAR-IRR, Hyderabad, by introgressing the *Saltol* QTL locus into the genetic background of Improved Samba Mahsuri suitable for coastal saline soils and was released in 2021. Along with *Saltol* QTL, it possesses three bacterial blight resistance genes (*Xa21, xa13* and *xa5*). Similarly, another variety, IET 28010 (MTU Rice 1293), a MAS-derived NIL with seedling stage salinity tolerance (*Saltol* QTL) in the background of MTU 1010 has been developed at RARS, Maruteru in ANGRAU, and released during 2022 for cultivation in the states of Andhra Pradesh and Telangana.

### 2.6.9 Zonal and State Level Contribution of Test Centers and Test Lines in AICRPR Salt Stress Trials

Currently, the AICRP on rice has a network of 31 test centers, including seven funded and 24 voluntary centers (Figs. 2.5, 2.6, and 2.7) located in all the major rice-

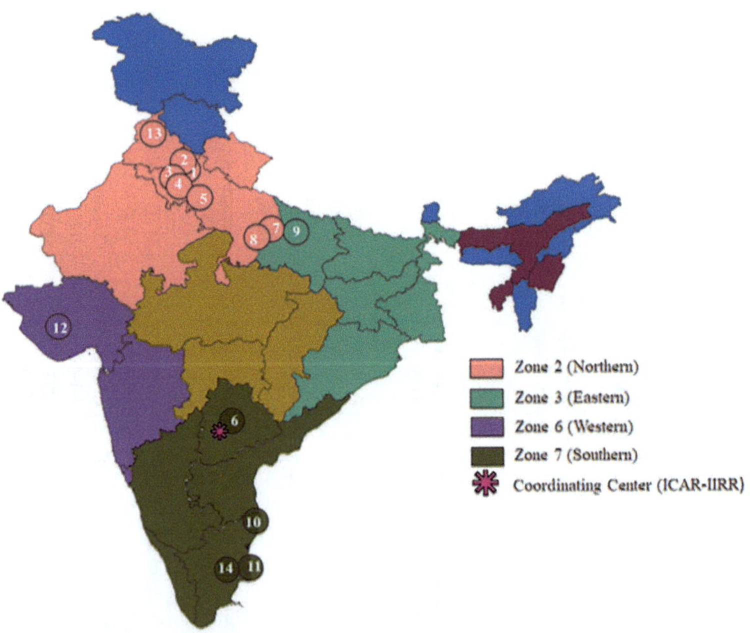


**Fig. 2.5** Location of AICRPR zones and alkaline test centers

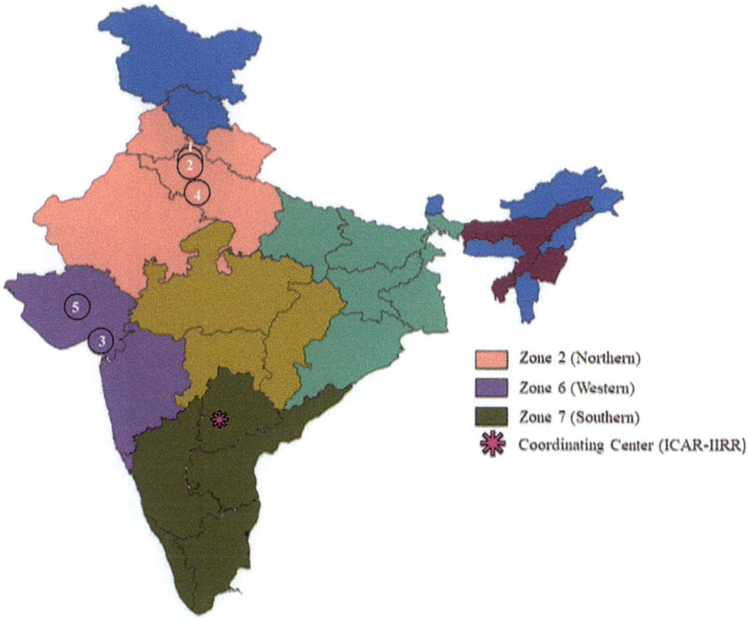


**Fig. 2.6** Location of AICRPR zones and inland saline test centers

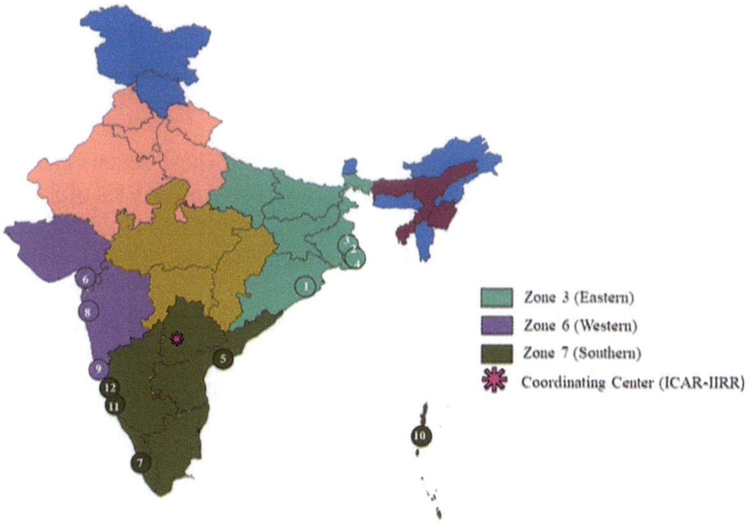


**Fig. 2.7** Location of AICRPR zones and coastal saline test centers

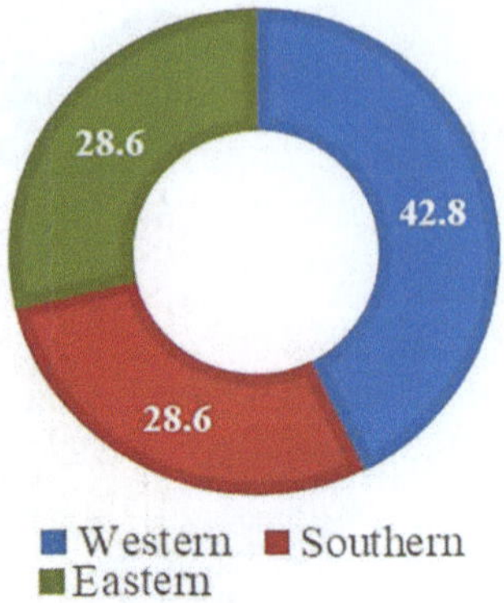


**Fig. 2.8** Funded centers of AICRPR conducting salt stress trials

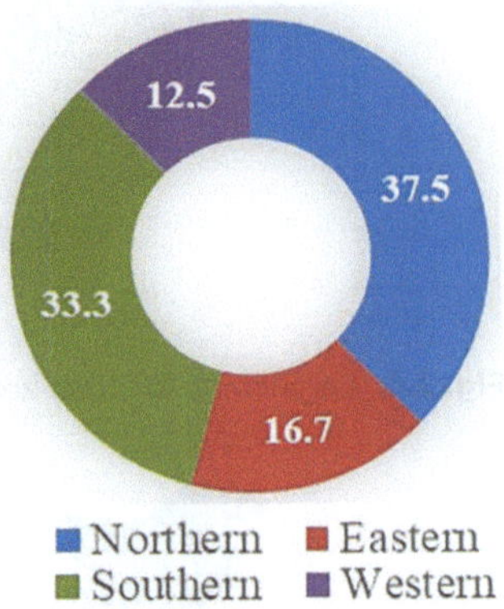


**Fig. 2.9** Voluntary centers of AICRPR conducting salt stress trials

growing zones except hills, northeast, and central zones. These centers are situated in ICAR institutions, regional agricultural research stations, and KVKs of SAUs across different agroclimatic zones of the country pursuing mandated salt tolerance research activities of rice crop. Seven funded centers conduct AL & ISTVT and CSTVT trials which are situated in three zones covering four states, i.e., eastern (Uttar Pradesh & West Bengal), western (Gujarat), and southern (Karnataka). The western zone had a maximum share of locations (42.8%), followed by the southern (28.6%) and eastern (28.6%) zones in the funded center category for testing the genotypes (Fig. 2.8). There are 24 voluntary centers located across four zones spreading over 13 states and two union territories, i.e., Haryana and Punjab (northern); Tamil Nadu, Andhra Pradesh, Kerala, Karnataka, Telangana, Puducherry, Andaman, and Nicobar (southern); Uttar Pradesh, Odisha, and West Bengal (eastern); and Maharashtra, Gujarat, and Goa (western) conducting salt trials. Among them, the northern zone reflected the maximum share (37.5%) of locations, followed by the southern (33.3%), eastern (16.7%), and western (12.5%) zones (Fig. 2.9).

A total of 1740 elite cultures were nominated to soil stress trials of AICRPR so far by 15 states and two union territories. Statewise comparison revealed that Uttar Pradesh (299) and Odisha (298) contributed the highest number of elite entries,

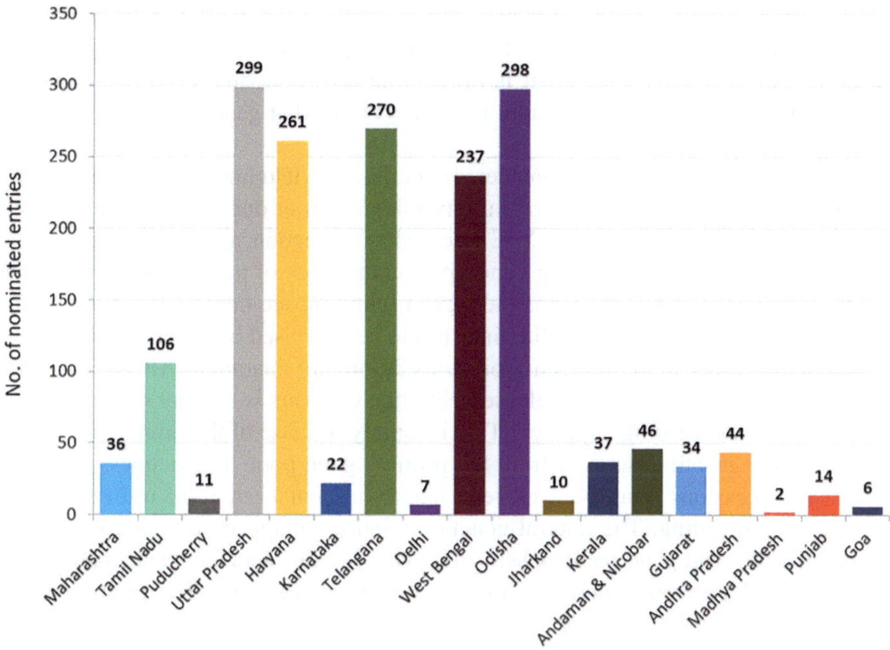


**Fig. 2.10** Frequency distribution of statewise nominated entries in salinity/alkalinity varietal trials in AICRPR

followed by Telangana (270), Haryana (261), West Bengal (237), and Tamil Nadu (106). These six states together accounted for 85% of total entries (1740) nominated in the last three and a half decades. The remaining 15% of the entries were contributed jointly by two union territories, i.e., Andaman and Nicobar (46) and Delhi (7); ten states, namely, Andhra Pradesh (44), Kerala (37), Maharashtra (36), Gujarat (34), Karnataka (22), Punjab (14), Puducherry (11), Jharkhand (10), Goa (6), and Madhya Pradesh (2) (Fig. 2.10). This indicates the need for other states to step up breeding efforts for increasing rice production under salty soils significantly. The pedigree database reveals that only 16% of the total nominations had the landraces (253) and accessions (25) of wild species (*O. rufipogon* and *O. nivara*) as one of the parents. This shows the limited utilization of genetic resources in the primary genetic pool warranting broadening the genetic base by utilizing the genetically diverse sources from secondary and tertiary gene pools possessing various adaptive salt tolerance mechanisms.

## 2.7 Challenges and Future Directions for Enhancing Yield Under Salt-Affected Areas

Soil salinization is a severe problem in rice production. The continued expansion of salt-affected soils is one of the serious constraints limiting rice yield. Nevertheless, AICRPR played a major role in the evaluation and release of numerous superior salt-tolerant rice varieties. In this direction, the success of AICRPR is reflected in the release of 63 tolerant rice varieties having the potential to yield reasonably in salty areas of the country. However, it is hoped that the following strategies could be implemented in the AICRPR system for achieving further genetic gain.

- A major challenge for plant breeders in increasing yield under salt stress appears to be the genetic complexity of salinity tolerance trait due to polygenic control, low heritability, and high G X E interaction. Selection and nomination to the AICRPR system of promising genotypes on the basis of performance in a single environment where it is bred is the lacuna found in rice improvement programs. Hence, selection of tolerant elite lines has to be exercised based on yield stability and adaptability in multiple environments before nominating to AICRPR testing.
- The narrow genetic base of released elite rice cultivars is another concern in the conventional breeding approach. This is mainly because of the limited utilization of diverse genetic resources from the primary gene pool. The richness of germplasm accessions conserved in gene banks has not been fully utilized in salt tolerance breeding. The natural genetic variation untapped and underutilized in compatible wild gene pools possessing adaptive salt tolerance mechanisms must be fully characterized and exploited in breeding programs. To circumvent the cross-incompatibility barriers and linkage drags associated with wild species could be tackled by strengthening prebreeding, which enables the introgression of new and novel genes from wild sources in the elite genetic backgrounds to withstand salt stress.
- The currently cultivated salt-tolerant varieties have the potential to tolerate pH and ECe up to ~10 and 11 $dS/m^{-1}$, respectively, producing an average yield of 3 to 4 t/ha under salt stress. But in the future, the genetic gain for grain yield under salt stress is expected to be low because of the rapid buildup of soil salinity associated with climate change affecting rice productivity. Therefore, it is necessary to breed new cultivars possessing higher genetic tolerance levels ($pH > 10$ and ECe up to $>11 dS/m^{-1}$) than the existing ones with yield potential of >4 t/ha as against present yield level of 3 to 4 t/ha under salt stress. For this, it is essential to study genetic diversity and categorize mechanisms of salt tolerance functioning in the donor sources, which will help in the accumulation of the favorable alleles from different donors into elite rice varieties to elevate the magnitude of salt tolerance.
- Salinity/alkalinity coexists with other challenges, such as drought, submergence, and heat. Therefore, breeding climate-resilient varieties with inbuilt tolerance to multiple stresses in superior rice cultivars could be the solution to have greater adaptation. Initiation of development of MAGIC (multiparental advanced generation intercross) populations involving intercrossing among multiple founder lines having diverse traits such as adaptability to salinity, submergence, drought, heat, high grain yield, and grain quality is required instead of routine classical biparental matings. This method facilitates the pooling of several beneficial traits and the recovery of a genetically potential MAGIC population having diverse agromorphological traits with a broader genetic base without yield penalty. At

various AICRPR centers, on priority basis, MAGIC approach has to be taken up to address salt tolerance as this method is not yet widely adopted at various centers.

- Periodic replacement of tolerant check varieties is essential under AICRPR trials on alkalinity/inland salinity/coastal salinity tolerance in case they show large GXE and poor stability across multilocations with recently released stable and best performing rice varieties against which the mean yield performance of test entries is compared.
- Increasing the number of testing sites or environments for screening is the need of the hour, which enables the selection of the best genotypes giving stable and consistent yield performance across environments.
- Yield is the main criterion while promoting a superior entry in AICRPR over the best check. The trials with a mean yield less than national average (2 t/ha) in salinity/alkalinity trials are not considered in data analysis while determining promotion and release of entries. This practice has to be continued stringently in AICRPR system.
- Even though the integration of genomic technologies in conventional breeding, particularly introgression of seedling stage *Saltol* QTL, has considerably accelerated breeding for salt-tolerant elite lines, emphasis should be laid on reproductive stage salinity tolerance, which is more critical in determining the grain yield. It is imperative to focus on the identification of novel and major QTLs at the reproductive stage other than *Saltol* conferring vegetative tolerance with consistent expression in different genetic backgrounds for successful introgression into high-yielding backgrounds.
- The Constitution of the shuttle breeding network involving all the partner centers could be an important strategy for screening and evaluating breeding material in cooperation with the scientists working in target areas. This approach facilitates the exchange of genetically diverse potential donors and early generation breeding material and provides an opportunity to screen and select desirable genotypes as per the local requirement.
- Collaborative funded research programs between NARS partners can accelerate the joint development of new products
- Organization of training and capacity building of cooperators is essential to strengthen their capabilities for the effective conduct of salinity/alkalinity trials till reporting data.

**Conflict of interest** The authors declare no conflict of interest.

## References

Agricultural Statistics at a Glance (2021) Directorate of Economics & Statistics, Department of Agriculture, Cooperation and Farmers Welfare, Ministry of Agriculture & Farmers Welfare, Govt of India, New Delhi, p 99

Arora S, Sharma V (2017) Reclamation and management of salt-affected soils for safeguarding agricultural productivity. J Safeguarding Agric 1:1–10

Babu, N.N., Krishnan, S.G., Vinod, K.K., Krishnamurthy, S.L, Singh, V.K., Singh, M.P., Singh, R., Ellur, R.K., Rai, V., Bollinedi, H., Bhowmick, P.K., Yadav, A.K., Nagarajan, M., Singh, N.K., Prabhu, K.V., & Singh, A.K. (2017). Marker aided Incorporation of *Saltol*, a major QTL associated with seedling stage salt tolerance, into *Oryza sativa* 'Pusa Basmati 1121'. Front Plant Sci, 8, 41

Brainerd E, Menon N (2014) Seasonal effects of water quality: the hidden costs of the Green Revolution to infant and child health in India. J Dev Econ 107:49–64

Eynard A, Lal R, Wiebe K (2005) Crop response in salt-affected soils. J Sustain Agric 27:5–50

Flowers TJ (2004) Improving crop salt tolerance. J Exp Bot 55(396):307–319

Food and Agriculture Organization (FAO) (2020) FAOSTAT

Geetha S, Vasuki A, Jagadeesh SP, Saraswathi R, Krishnamurthy SL, Manikandan P (2017) Development of sodicity tolerant rice varieties through marker assisted backcross breeding. Electronic J Plant Breeding 8(4):1013–1021

Gholizadeh A, Dehghania H, Dvorakb J (2014) Determination of the most effective traits on wheat yield under saline stress. Agric Adv 3(4):103–110

GOI (1996–2020) Agricultural statistics at a glance. Ministry of Agriculture and Farmers Welfare, Department of Agriculture Cooperation and Farmers Welfare, Directorate of Economics and Statistics, Government of India, New Delhi

Gregorio GB, Senadhira D, Mendoza RD (1997) Screening rice for salinity tolerance. In: IRRI discussion paper series, vol 22, pp 1–30

ICAR (1992) Management of change in all India coordinated crop improvement projects. 24–26 February 1992. In: Muralidharan K, Prasad GSV, Mauriya S, Rai M, Siddiq EA (eds) Proceedings of the group discussion. Indian Council of Agricultural Research, New Delhi, p 216

Ismail AM, Heuer S, Thomson MJ, Wissuwa M (2007) Genetic and genomic approaches to develop rice germplasm for problem soils. Plant Mol Biol 65(4):547–570

Khatun S, Rizzo CA, Flowers TJ (1995) Genotypic variation in the effect of salinity on fertility in rice. Plant Soil 173:239–250

Krishnamurthy SL, Pundir P, Warraich AS, Rathor S, Lokeshkumar BM, Singh NK, Sharma PC (2020) Introgressed Saltol QTL lines improves the salinity tolerance in rice at seedling stage. Front Plant Sci 11:833

Munns R, Tester M (2008) Mechanisms of salinity tolerance. Annu Rev Plant Biol 59:651–668

Nair MM, Shylaraj KS (2021) Introgression of dual abiotic stress tolerance QTLs (Saltol QTL and Sub1 gene) into rice (*Oryza sativa* L.) variety Aiswarya through marker assisted backcross breeding. Physiol Mol Biol Plants 27(3):497–514

Qadir M, Quillerou E, Nangia V, Murtaza G, Singh M, Thomas R, Drechsel J, Nobel AD (2014) Economics of salt-induced land degradation and restoration. Nat Res Forum 38(4):282–295

Rani NS, Rao LVS, Viraktamath BC (2006) National guidelines for the conduct of distinctness, uniformity and stability: rice. ICAR-Indian Institute of Rice Research (Directorate of Rice Research), Hyderabad, pp 1–39

Rao PS, Mishra B, Gupta SR (2013) Effects of soil salinity and alkalinity on grain quality of tolerant, semi-tolerant and sensitive rice genotypes. Rice Sci 20(4):284–291

Rekha G, Abhilash Kumar V, Gokulan CG, Koushik MBVN, Laxmi Prasanna B, Kulkarni S, Aleena D, Harika G, Hajira SK, Pranathi K, Punniakoti E, Kale RR, Dilip Kumar T, Ayyappa D, Anila M, Sinha P, Manohara KK, Padmavathi G, Subba Rao LV, Laha GS, Srinivas Prasad MS, Fiyaz RA, Suneetha K, Balachandran SM, Patel HK, Ramesh V, Sonti Senguttuvel P, Sundaram RM (2022) DRR Dhan 58, a seedling stage salinity tolerant NIL of improved samba Mahsuri Shows superior performance in multi-location trials. Rice 15:45

Roy SJ, Negrao S, Tester M (2014) Salt resistant crop plants. Curr Opin Biotechnol 26:115–124

Saade S, Maurer A, Shahid M, Oakey H, Schmockel SM, Negrao S, Pillen K, Tester M (2016) Yield-related salinity tolerance traits identified in a nested association mapping (NAM) population of wild barley. Sci Rep 6:32586

Sharma DK, Chaudhari SK, Singh A (2014) CSSRI Vision 2050. Central Soil Salinity Research Institute, Karnal

Rani S, Prasad GSV, Prasad ASR, Sailaja B, Muthuraman P, Meera SN, Viraktamath BC (2010) Rice Almanac. DRR Technical Bulletin No. 50/2010, Directorate of Rice Research, Rajendranagar, Hyderabad, pp 1–307

Singh RK, Mishra B, Ismail AM, Gregoria GB (2009) Breeding rice for salt affected areas of India. In: Hossain M, Bennet J, Mackill D, Hardy B (eds) Progress in crop improvement research. IRRI, Los Banos, pp 78–90

Singh VK, Singh BD, Kumar A, Maurya S, Gopala Krishnan S, Vinod KK, Singh MP, Ellur RK, Bhowmick PK, Singh AK (2018) Marker-Assisted Introgression of Saltol QTL enhances seedling stage salt tolerance in the rice variety Pusa Basmati 1. Int J Genomic 8319879:1–12

Subba Rao LV, Fiyaz RA, Jukanti AK, Padmavathi G, Badri J, Anantha MS, Nagarjuna E, Voleti SR (2019) Coordinated Rice Improvement Project in India: Its Significant Achievements and Future prospects. Oryza 56:82–91

Tandon JP, Sharma SP, Sandhu JS, Yadava DK, Prabhu KV, Yadav OP (2015) Guidelines for Testing Crop varieties under the All India Coordinated Crop Improvement Projects. Indian Council of Agricultural Research, New Delhi, p 250

Thomson MJ, de Ocampo M, Egdane J, Rahman MA, Sajise AG, Adorada DL, Tumimbang-Raiz E, Blumwald E, Seraj ZI, Singh RK, Gregorio GB, Ismail AM (2010) Characterizing the *Saltol* quantitative trait locus for salinity tolerance in rice. Rice 3:148–160

Waziri A, Kumar P, Purty RS (2016) Saltol QTL and their role in salinity tolerance in rice. Austin J Biotechnol Bioeng 3(3):1067

Yadav, A.K., Kumar, A., Grover, N., Ranjith Kumar, E., Gopala Krishnan, S., Haritha, B., Bhowmick, P.K., Vinod, K.K, Nagarajan, M., Krishnamurthy, S.L., Kumar & Singh, A.K. (2020). Marker aided introgression of *Saltol*, a major QTL for seedling stage salinity tolerance into an elite Basmati rice variety 'Pusa Basmati 1509'. Sci Rep, 10, 13877

Zeng L, Shannon MC (2000a) Effects of salinity on grain yield and yield components of rice at different seeding densities. Agron J 92(3):418–423

Zeng L, Shannon MC (2000b) Salinity effects on seedling growth and yield components of rice. Crop Sci 40(4):996–1003

Zeng Y, Zhang H, Li Z, Shen S, Sun J, Wang M, Liao D, Liu X, Wang X, Xiao F, Wen G (2007) Evaluation of genetic diversity of rice landraces (*Oryza sativa* L.) in Yunnan, China. Breed Sci 57(2):91–99

# Genetic Improvement of Rice for Salt Tolerance in Andaman and Nicobar Islands, India

# 3

R. K. Gautam, P. K. Singh, K. Venkatesan, and K. Sakthivel

**Abstract**

Paddy is the mainstay food crop in Andaman and Nicobar Islands covering about 17% of the total cultivable land. The islands are characterized by high rainfall (3000 mm/year) coupled with waterlogging and coastal salinity. To address the island-specific variety development, ICAR-Central Island Agricultural Research Institute (CIARI), Port Blair, has developed efficient phenotyping or screening facilities/infrastructures such as micro-plots, glasshouses and/ or lab facilities for breeding for different biotic and abiotic stress tolerance. Consequent to the research work carried out under seven rice projects across many years, a total of nine improved rice varieties have been developed and recommended for cultivation for normal and saline soils to alleviate the problem of low productivity in the islands. These varieties are derived from exotic breeding lines, somaclonal variation induced in traditional saline-tolerant paddy Pokkali, and purification of highly popular rice landrace C14–8. In addition, few salt-tolerant and high-yielding varieties developed in Indian mainland research organizations have also been found superior in the islands. About 150.7 quintal of truthfully labeled (TFL) seeds of salt-tolerant varieties were produced and disseminated to island farmers during 2011–2018 covering about 498 ha of area. Further, MAGIC (multi-parent advance generation inter-cross) populations involving eight diverse parents having economic traits and adapted to the islands' conditions have been developed by our group for creating high genetic variability in rice, thus leading to high-yielding and climate-resilient varieties for islands.

R. K. Gautam (✉)
ICAR-Central Island Agricultural Research Institute, Port Blair, Andaman & Nicobar Islands, India
e-mail: raj.gautam@icar.gov.in

P. K. Singh · K. Venkatesan · K. Sakthivel
ICAR-Central Island Agricultural Research Institute, Port Blair, Andaman & Nicobar Islands, India



R. K. Singh et al. (eds.), *Genetic Improvement of Rice for Salt Tolerance*,
https://doi.org/10.1007/978-981-99-3830-8_3

**Keywords**

Andaman · Climate resilience · Genetic variability · Rice · Salt tolerance

## 3.1 Introduction

In India, the Andaman and Nicobar (A&N) group of islands is an archipelago in the Bay of Bengal consisting of 306 named islands and 206 rocks and rocky outcrops with outstanding natural beauty and ecological diversity. It is situated approximately 1200 km away from the Indian mainland, between 6°45'N to 13°41'N longitude and 92°12′E to 93°57′E latitude. These islands are bestowed with a maritime climate, influenced by both southwest and northeast monsoons. The average rainfall is about 3100 mm, distributed over 8 months, between May and December. However, the rainfall pattern is irregular and thereby causes flooding and water stagnation in low-lying areas and drought in the hilly uplands. No perennial water resources and assured irrigation systems are available in these islands; thus, the field crops including rice are grown under rainfed conditions.

Out of the total geographical area of Andaman and Nicobar islands (8249 sq. km), only 6% (50,000 ha) is available for agriculture, whereas the remaining land is covered by forests and urban settlements. Of the total cultivable land, rice occupies a maximum area of about 8390 ha followed by spices (1661 ha), pulses (2610 ha), and vegetables (5150 ha) (Gautam et al. 2013). The net area sown (12,755.74 ha) of the Andaman group of islands forms only 1.6% of the total geographical area, whereas it is only 0.1% for the Nicobar group of Islands. Paddy is the main crop during Kharif (rainy) season covering about 17% of the total cultivable land of islands, and pulses and oilseeds are grown during Rabi (dry) season.

The tsunami that struck the Indian Ocean in 2004 has affected a large area of cultivable land owing to the intrusion of seawater. Since then, about 9% (4207 ha) of pre-tsunami agricultural land is under partial/permanent submergence. Presently, rice is grown in about 5700 ha area of cultivated land (DES 2019–20). Among the three districts in Andaman and Nicobar Islands, the North and Middle Andaman district has the major share of rice-growing area (5630.47 ha), followed by the South Andaman district (67.55 ha) and Nicobar Islands (3.10 ha). Annual rice production of these islands is around 17,981 t with productivity of 3.15 t $ha^{-1}$ (DES 2019–20), whereas the annual rice requirement of island population is around 60,000 t (Singh et al. 2020). Salinity, a serious problem affecting 33% of the irrigated agricultural land in the world (Shrivastava and Kumar 2015) and 4207 ha in Bay Islands, impairs normal growth and limits the realization of yield potential of modern cultivars. Salinity is one of the major obstacles to increase production in rice-growing areas. A large area available for the field crops (>4000 ha) is affected by excess salt and water stagnation of varying levels during the monsoon season. Besides salinity stress, Fe and Al toxicities and acid sulfate soils are also major constraints in these coastal areas. Thus, the crop often encounters dual soil stress of moderate salinity and acid sulfate soils.

Rice is generally considered as salt-sensitive crop (Joseph et al. 2010). According to the classification of salt tolerance to salinity, the rice crop is within the sensitive range from 1.9 $dSm^{-1}$ to 8 $dSm^{-1}$ with reductions in growth and number of filled grains per panicle (Zeng et al. 2001; Farshid and Hassan 2012). Rice is affected by salinity at different stages. Rice plant responses to salt stress are complex and depend on duration and type of salt stress, development stage of rice, day length and other factors (Bernardo et al. 2000; Cramer et al. 2001). Rice is relatively tolerant to both salinity and alkalinity during the germination stage. The rice areas adjoining to sea are often prone to frequent high tides and seepage of saline water of sea in the water table, thus affecting the root zone to varying degrees. As rice is grown only during Kharif season, there is no assured irrigation in dry season. However, the high rainfall during Kharif season (about 3000 mm/year) contributes to some salt dilution, thus resulting in moderate salinity stress in the islands.

## 3.2 Historical Perspectives

ICAR-Central Island Agricultural Research Institute (CIARI)-Port Blair came into being in 1978 with the amalgamation of different research stations of ICAR located in the islands, which later formed different Divisions of CIARI. Historically, rice is the only feasible crop in low-lying areas where nothing can be grown in the wet season in Andaman and Nicobar Islands. Since its inception, ICAR-CIARI researchers collected, documented, characterized, and evaluated rice germplasm originating from exotic and native sources for the islands' conditions. The traditional rice varieties commonly grown in different regions of Andaman and Nicobar Islands are C14–8 (popularly called *Aath Number Dhan*), Black Burma, Gol Dhan, Jeera Dhan, Nata Khushbu, Jaya, Swarna Jaya, Nata Jaya, Pagla Jaya, Jagannath, etc. These diverse rice varieties were preferred and originally brought by different ethnic settlers in the islands, which got coevolved and adapted in the rice-growing system. However, cultivation of these cultivars in the islands is in few patches except for C14–8 that occupies a sizeable area of about 50% of rice cultivation in the islands. Therefore, development of suitable rice varieties has been recognized as a potential approach for increasing rice yields under island conditions (Ahmed et al. 2014).

The concerted efforts have been historically made by ICAR-Central Island Agricultural Research Institute (CIARI) for collecting, identifying and characterizing various exotic and indigenous germplasm and developing improved rice varieties for normal and saline soils (Mandal et al. 2004; Abraham et al. 2008). Consequently, these concerted works led to the development of nine improved rice varieties, viz., CARI Dhan 1, CARI Dhan 2, CARI Dhan 3, CARI Dhan 4, CARI Dhan 5, CARI Dhan 6, CARI Dhan 7, CARI Dhan 8, and CARI Dhan 9 (Gautam et al. 2016). Of these, CARI Dhan 1, CARI Dhan 2, CARI Dhan 3, CARI Dhan 4, and CARI Dhan 5 are recommended by A&N State Variety Release Committee for the islands. CARI Dhan 1, CARI Dhan 2, and CARI Dhan 3 are recommended for normal soils of islands, whereas CARI Dhan 4 and CARI Dhan 5 developed from somaclonal variation induced in Pokkali landrace are suitable for saline and acid sulfate coastal soils of the island region. Two rice varieties CARI Dhan 6 and CARI

Dhan 7 with medium duration and bacterial blight resistance have also been developed. In addition, two new varieties CARI Dhan 8 (yellow grains) and CARI Dhan 9 (brown grains) have been developed through pure line selection from a native popular variety C14–8. Besides, the other high-yielding rice varieties developed by other rice centers like Savitri, Gayatri, and Ranjeet under normal soil conditions, whereas CSR 36 under coastal saline conditions have also shown better performance in the islands. Farmers' participatory variety selection has been suggested and practiced as an integral component of varietal development and dissemination due to the diverse preferences of cultivators and consumers (Singh et al. 2014; Witcombe et al. 2001; Witcombe 2002). Our recent farmers' participatory variety selection trials revealed the better appreciation and adoption of CARI Dhan 5 and CSR 36 in salt-affected conditions while for CARI Dhan 6, CARI Dhan 7, CARI Dhan 8, and CARI Dhan 9 in normal soil conditions by islands' farmers.

## 3.3 Constraints in Breeding Work

Rice being the main cereal crop of the islands is cultivated as a single crop in a year wherein the popular C14–8 (Aath Number Dhan) variety historically dominates the scene (Mandal et al. 2004). Though C14–8 coevolved and is adaptive to the rice culture over time, availability of its true-to-type seed was the main constraint faced by the native farmers. Interaction with few elderly farmers revealed that this variety has reduced its yield potential over time perhaps due to admixture. Thus, it was considered imperative to purify this cultivar to restore its identity and potential not only for the islands but also for the conservation and utilization of resilient traits for other ecosystems as well.

The main reasons for low productivity of the rice crop in the islands are as follows: (1) cultivation of traditional photosensitive, tall, low-yielding, and disease-susceptible varieties under majority of the cultivated area (about 40% of the area); (2) poor availability of quality seed of island-specific, high-yielding, and stress-tolerant varieties; (3) prevailing climatic conditions of high rainfall (3000 mm/year), warm conditions, high humidity and cloudiness, all conducive for disease and pest development and infestation; (4) coastal salinity affecting about 3000 ha land, which got accentuated due to the 2004 tsunami and (5) recommendations of the islands' administration for adopting organic agriculture, etc. Further, to meet the per capita annual rice requirement of islands (64.1 kg/year against the present 14.64 kg/year), yield enhancement through varietal improvement is also a very crucial step (Singh et al. 2020).

## 3.4 Need of Efficient Phenotyping

To address the above mentioned breeding constraints in the islands, there is a need for the development of efficient phenotyping or screening facilities/infrastructures, which may be micro-plot, glasshouse, and/ or lab facilities for different biotic and abiotic stress tolerance in rice breeding.

## 3.5 Infrastructure Developed

### 3.5.1 Establishment of Micro-Plot Facility at ICAR-CIARI, Port Blair

Often, the results of natural field screening for salinity stress tolerance are vitiated by variable distribution of salts and unexpected rainfall conditions. To obviate such problems, the facility of 18 micro-plots with covered rainout shelter has been recently established at ICAR-CIARI, Port Blair, costing more than 1.00 crore rupees. The facility inaugurated in 2018 is now functional and is being used for phenotyping varieties and germplasm for salt tolerance under island conditions (Plates 3.1 and 3.2). Development of island-specific salt-tolerant varieties will be accelerated in the future with the establishment of this facility.

Inauguration of micro-plot facility on November 15, 2018, at ICAR-CIARI, Port Blair

**Plate 3.1** Visit of micro-plot facility by QRT Chairman and members on November 29, 2018, at ICAR-CIARI, Port Blair

**Plate 3.2** Screening of rice genotypes for salinity tolerance in micro-plots at ICAR-CIARI, Port Blair

### 3.5.2 Glasshouses and Biotechnology Laboratory

During the early years of the 1990s and later, glasshouse facilities were established for conducting saline hydroponic, pot and tank experiments at ICAR-CIARI, Port Blair. Similarly, tissue culture and genetic transformation laboratories were well established through funding from institutional and external grants for conducting in vitro studies resulting in many scientific findings and conclusions. Recently, marker-assisted breeding laboratory has been established under a DBT-funded project "Marker assisted introgression of bacterial blight resistance in popular rice cultivars of Andaman and Nicobar Islands" to harness the strengths of molecular marker-assisted technology for crop improvement.

## 3.6 Breeding Work

### 3.6.1 Germplasm Evaluation and Utilization

The islands are considered more vulnerable to the effects of climate change, sea level rise, and coastal salinity as witnessed during the 2004 tsunami. Systematic screening and identification of salt-tolerant varieties/germplasm is very important for such conditions. A total of 308 salt-tolerant rice lines received over the years (2014–2018) under IRRI-STRASA project were evaluated under Salinity Tolerance Breeding Network (STBN) trials along with national checks in coastal saline and acid sulfate soils of Andaman and Nicobar Islands (Krishnamurthy et al. 2016, 2017).

### 3.6.2 Creation of Variability

#### 3.6.2.1 Hybridization

A project on "Marker assisted introgression of bacterial blight resistance in popular rice cultivars of Andaman and Nicobar Islands" funded by the Department of

Biotechnology, Ministry of Science and Technology, Government of India, was started in 2012. Under this project, four bacterial blight resistance genes, viz., *Xa4, xa5, xa13, and Xa21* (donor IRBB60) have been attempted for transfer in the background of CARI Dhan 5, which is a salt-tolerant popular rice variety developed for Andaman and Nicobar Islands by ICAR-Central Island Agricultural Research Institute. Under this project, 41 backcross-derived lines (BILs) have been derived ($BC_3F_5$ stage) from a cross between CARI Dhan 5 and bacterial blight-resistant line IRBB 60 (CARI Dhan 5 x IRBB 60). These lines were evaluated under natural field as well as controlled salinity micro-plots during Kharif 2020 for salinity tolerance (ECe ~5.5 $dSm^{-1}$) and bacterial blight resistance. Out of 41 rice lines, only 17-1-69-34 was found tolerant for salinity with a score of "1," while seven lines (17-1-69-43, 17-1-69-55, 17-1-69-215, 17-1-69-316, 17-1-69-324, 17-1-69-334 and 17-1-69-337) showed moderate tolerance with a score of "3." For BLB resistance scoring, lines were artificially inoculated by bacterial leaf blight pathogen, and scoring was done using a 1–9 scale. A total of 10 lines, viz., 17-1-69-55, 17-1-69-72, 17-1-69-179, 17-1-69-215, 17-1-69-334, 17-1-69-375, 46-3-95-655, 46-3-5-659, 46-3-95-717, 131-2-175-1196 and 131-2-175-1123 were found resistant for BB, whereas seven lines (17-1-69-43, 17-1-69-179, 17-1-69-181, 17-1-69-324, 46-3-95-683, 131-2-190-1197,131-2-175-1209 and 131-2-175-1239) were found moderately resistant. Five lines viz., 17-1-69-43, 17-1-69-55, 17-1-69-215, 17-1-69-324 and 17-1-69-334, showed both BLB resistance and salinity tolerance. These BILs along with recurrent parent CARI Dhan 5 have also been characterized for DUS traits. Further, MAGIC (multi-parent advance generation inter-cross) populations, involving eight diverse parents having economic traits and adapted to the islands' conditions, have been developed by our group for creating high genetic variability in rice, thus leading to high-yielding and climate-resilient varieties for islands in the near future.

#### 3.6.2.2 Tissue Culture-Derived Lines

##### Hydroponic and Tissue Culture

Tissue culture-derived salt-tolerant lines and fiberglass tanks based study with excess salt (8–12 $dSm^{-1}$) in the form of sodium chloride-calcium chloride (1:1) revealed five (IR 21820–154–3-2-2-3, IR 9884–54–3-IE-PI, ITA 249, MTU 7633, and CR 1009) out of ten varieties as the most promising to salinity stress (Mandal et al. 2004). Conventional approaches are not only cumbersome but also time-consuming. To reduce the breeding cycle, a total of 2310 calli have been screened in which those regenerates with higher salt tolerance were recovered from calli derived from varieties B 116 and C14–8 using plant cell and tissue culture techniques. For this, the two strategies adopted were as follows: (1) direct callus induction on salt-stressed medium followed by regeneration on salt-stressed and salt-free medium and (2) soaking mature seed for 24 h in NaCl solution and subsequent callus induction on stress-free medium. Subsequently, in vitro regeneration study on two rice varieties, Pokkali (salt-tolerant line) and Basmati 370 (salt-susceptible scented line) revealed a maximum callus induction of 92.7% with Pokkali and

88.25% with Basmati 370 on MS with 2 mg/l 2,4-D and 5 g/l NaCl. In another study, it was found that more plants could be regenerated from salt-tolerant varieties SR26-B, Nona Bokra and C14-8 than the high yielding varieties suitable for mass cultivation in Bay islands (IR 18351-229-3 and IR 185-6-3-3-2) on CIM (MS +2 mg/l 2,4-D + 0.5 mg/l kinetin) stressed with 1,2 and 4 g/l of NaCl.

Somatic embryogenesis and plantlet regeneration of a salt-tolerant tall traditional indica variety Pokkali was also compared with a salt-susceptible variety PTB 10 following four culture routes. The MS, LS, and N6 culture media were used for callus induction and plantlet regeneration. Plantlet regeneration as high as 68.9% for PTB 10 on N6 was followed by 66.5% on MS and 55.8% on LS. Sucrose supplement at 6% resulted in the maximum plantlet regeneration (55.6%) for Pokkali and 63.5% for PTB 10 on N6. Calli from individual seeds of PTB 10 produced an average of 6.65 plants, while Pokkali produced 3.83 plantlets across successful treatments in MS, LS, and N6. Somatic embryo-derived plants produced fertile healthy seeds, while in vitro developed plantlets showed more spikelet sterility than control and were found to be true to type. These findings indicated that genotypes and hormones played a crucial role in callus induction and plantlet regeneration under in vitro culture system, while 2,4-D alone was found to be sufficient for callus induction. The *Agrobacterium*-mediated genetic transformation of the osmotin gene that confers salt and drought tolerance to the plants was incorporated into high-yielding indica rice variety Annada. Under in vitro assessment, the seeds of $T_1$ progenies survival percent under salt stress (8 $dSm^{-1}$) ranged from 38.25 to 88.2. Considering tolerance values across three phases, T46–1 was found to be the most tolerant line toward excess NaCl among the progenies tested.

Anther culture (AC) is a kind of breeding method, which is proved effective in expeditious introgression of diverse traits in breeding populations in rice. This method is considered more reliable with high selection efficiency (Morrison and Evans 1988; Foroughi-Wehr and Wenzel 1989) due to its unique form of gamete selection (Kasperbauer et al. 1980). This method is also cost-effective by its reduced space and labor requirements during varietal development because homozygosity can be achieved in one generation itself (Sanint et al. 1996).

Mandal and Bandyopadhyay (1997) subjected 30 hybrids ($F_1$) developed by involving different salt-tolerant and susceptible varieties to in vitro anther culture (AC) on $N_6$, $B_5$, and MS media with varying concentrations of auxins, cytokinins, and 2,4-D (mostly 2 $mg^{-1}$). This study concluded that AC response is largely genotype dependent as 13 hybrids did not produce calli. Across the 17 responsive crosses, an average of 3.87% callus induction was observed. Being a callus-inducing medium, $N_6$ proved more suitable (14 hybrids with best callus development) than MS (2 hybrids superior) or $B_5$ media (1 hybrid superior). Anthers from ITA 249 x Jhona 5867 showed maximum callus induction of 8.40%, whereas IR 52713-8-8-8-1-2 x C018 produced a maximum of 4.5 calli per responding anther. In their study, MS media with different concentrations of kinetin and BAP were found superior in plantlet regeneration in the cross ITA 249 x Jhona 5867 (72.60%) followed by B 42 x SR 26-B (70.5%) with the least in the cross Pokkali x IR 8.

**In Vitro Screening for Salt Stress**

Maximum green plantlets were found to be NaCl tolerant in Nona Bokra, C14–8, and SR 26–8. On the other hand, no plants from calli stressed with 5.0 g $dm^{-3}$ of NaCl were regenerated in salt-susceptible varieties. The genotypes IR 18351-229-3 and IR 3185-6-3-3-2 (salt susceptible), Nona Bokra, and SR26-B (salt tolerant) and a traditional cultivar C14–8 were subjected to in vitro screening to develop NaCl-tolerant rice genotypes grown on MS medium supplemented with 2 mg $dm^{-3}$ 2, 4-D, 30 g $dm^{-3}$ sucrose, and three levels – low, medium, and high (1.0, 3.0, and 5.0 g $dm^{-3}$) – of NaCl stress along with the control set for comparison of callus induction (Sarangi et al. 2002). In this study, two genotypes Nona Bokra and C14–8 showed better callus induction even at higher concentrations of NaCl in comparison to salt-susceptible genotypes (IR 18351-229-3, IR 3185-6-3-3-2), indicating the presence of built-in salt-tolerant genes in these varieties at the cellular level. Embryogenic callus percentage and plantlet regeneration varied substantially and decreased with increased NaCl concentrations across different genotypes.

**Isolated Microspore Culture**

The approach of isolated microspore culture has the benefit of anther culture by facilitating the production of a large number of haploid single-celled microspores capable of producing homozygous doubled haploid Iines in a single step (Chen 1986). A salt-susceptible x salt-tolerant indica rice $F_1$ hybrid involving IR 24 and CRM 30 was subjected to shed microspore embryogenesis, and intact anther culture revealed higher regeneration of green plantlets in liquid medium containing the synthetic polymer Ficoll 400 than from intact anthers cultured on a semisolid system (Chowdhury and Mandal 2001).

**Somaclones for Salt Tolerance**

During the process of culturing the ex-plants in synthetic media with different levels of phytohormones, somaclonal variations are generated under in vitro conditions (Larkin and Scowcroft 1981). This variation arises de novo during the process of dedifferentiation and redifferentiation as of ex-plants into unorganized cell masses (calli) and calli into plantlets, respectively (Chopra et al. 1989). The variation has been found to be widespread and heritable (Karp 1995). While mutagenesis causes dramatic changes in the genome, somaclonal variation has precise effects (Chopra et al. 1989).

Pokkali cultivation is mainly restricted to the hot humid tropics of the state of Kerala, India. Obviously, Pokkali has undergone more than 100 years of natural selection under salinity stress and insect/disease pressures as it is developed as a system of integrated farming, which does not require any external inputs and also not affected by the natural ecological processes (Thomas 2002; Kerala State Biodiversity Board 2009). Performance of selected Pokkali somaclones with respect to major agronomic traits including insect and disease tolerances grain quality, and salt-soluble grain polypeptide profile studied at $SC_6$ generation at ICAR-CIARI, Port Blair, revealed a great deal of variation in the majority of the traits among the somaclones and substantial deviation from their parents (Mandal et al. 1999).

The electrophoretic analysis of seed polypeptides under this study also revealed the existence of somaclonal variations at salt-soluble grain polypeptide level. Among the somaclones, BTS 23 and BTS 24 had produced maximum yield under both normal and saline soil conditions in addition to their tolerance to major insect pests and diseases (Mandal et al. 1999). These somaclonal variations were found to be widespread and heritable in nature with no dramatic change in genetic architecture compared to conventional mutagenesis. Therefore, these two somaclones were tested under AICRIP on rice and were subsequently named as CARI Dhan 4 and CARI Dhan 5, respectively. Further, they were recommended for cultivation by farmers under normal soil conditions in general and under saline soils in particular in Bay Islands.

## 3.7 STRASA Work

The islands are considered more vulnerable to the effects of climate change, sea level rise and coastal salinity as witnessed during the 2004 tsunami. Systematic screening and identification of salt-tolerant varieties/germplasm are very important for such conditions. A total of 308 salt-tolerant rice lines received over the years (2014–2018) under the IRRI-STRASA project were evaluated under Salinity Tolerance Breeding Network (STBN) trials along with national checks in coastal saline and acid sulfate soils of Andaman and Nicobar Islands (Krishnamurthy et al. 2016, 2017). The objectives of the of the IRRI-BMGF project "Stress Tolerant Rice for Poor Farmers of Africa and South Asia" at CARI, Port Blair, were as follows:

- To identify and disseminate salt-tolerant rice varieties in the target stress areas.
- To involve farmers in varietal uptake through mother and baby trials approach.
- To meet the seed requirements of salt-tolerant rice varieties.

Traits recorded under STBN trials were plant height (PH in cm), days to 50% flowering (DF), number of effective tillers per plant (NTPP), panicle length (PL in cm), and grain yield (GY in t/ha) as shown in Table 3.1. The range of variation observed for important traits such as plant height, days to flowering, number of productive tillers/plant, panicle length, and grain yield on rice entries evaluated under Salinity Tolerance Breeding Network (STBN) trials during 2012–2013 to 2018–2019 were presented in Table 3.1. The EC ($dSm^{-1}$) and pH of the experimental field ranged across the years as 4.0 to 5.5 and 5.2 to 7.8, respectively. Genetic variation on grain yield in rice entries over the years under study was observed as 8.5% to 32%, and it also ranged from 0.21 to 5.03 t/ha. The year-wise details on genotype with better grain yield (t/ha) are described below: STBN 12-4 (3.0) in 2012–2013; STBN11–9 (2.35) and STBN 11-11 (2.32) in 2013–2014; SIT 4446 (>4.0) in 2014–2015; SIT 5168 (>4.0); NDRK 11-20 (4.92) and PAU7114–3480-1-1-1-0 (4.44) in 2016–2017; CSR2016 18–15 (2.97 t/ha) and CSR2016-IR18–2 (2.69 t/ha) in 2017–2018; and CSR 2016 IR18-1 (3.2) in 2018–2019.

**Table 3.1** Evaluation of rice entries in coastal saline soils of A&N Islands over the years

| Year (entries + checks) | Mean | Range | CD | CV |
|---|---|---|---|---|
| 2012–2013 (31 + 4) | $EC_{iw}$ ~ dS/m = 4.0 | | pH ~ 5.2 | |
| Plant height (cm) | 97 | 71–147 | 14.74 | 7.39 |
| Days to flowering | 123 | 114–136 | 3.81 | 1.51 |
| Productive tillers/plant | 6 | 5–8 | N/A | 15.21 |
| Panicle length (cm) | 23 | 19–29 | N/A | 10.98 |
| Grain yield (t/ha) | 1.14 | 0.21–3.00 | 0.75 | 32 |
| 2013–2014 (30 + 4) | $EC_{iw}$ ~ dS/m = 4.5 | | pH ~ 5.5 | |
| Plant height (cm) | 102 | 79–140 | 8.76 | 4.2 |
| Days to flowering | 114 | 106–143 | 3.18 | 1.36 |
| Productive tillers/plant | 8 | 5–10 | | 24.5 |
| Panicle length (cm) | 25 | 22.5–30.5 | 1.99 | 3.84 |
| Grain yield (t/ha) | 2 | 1.1–2.3 | 0.55 | 15.01 |
| 2015–2016 (25 + 5) | $EC_{iw}$ ~ dS/m = 5.0 | | pH ~ 5.5 | |
| Plant height (cm) | 110.3 | 78–137 | 1.05 | 0.58 |
| Days to flowering | 101.8 | 74–121 | 17.67 | 10.59 |
| Productive tillers/plant | 6.3 | 5–7 | 0.79 | 7.6 |
| Panicle length (cm) | 24.2 | 17.5–27.6 | 1.01 | 2.56 |
| Grain yield (t/ha) | 3.23 | 0.63–5.03 | 0.46 | 8.78 |
| 2017–2018 (102 + 5) | $EC_{iw}$ ~ dS/m = 4.3 | | pH ~ 7.7 | |
| Plant height (cm) | 98 | 74–154 | – | – |
| Days to flowering | 104.6 | 77–128 | – | – |
| Productive tillers/plant | 7.7 | 4–11 | – | – |
| Panicle length (cm) | 22.9 | 15–28 | – | – |
| Grain yield (t/ha) | 1.62 | 0.4–3.0 | – | – |
| 2018–2019 (96 + 6) | $EC_{iw}$ ~ dS/m = 5.5 | | pH ~ 7.8 | |
| Plant height (cm) | 95.9 | 46–119 | – | – |
| Days to flowering | 109.2 | 81–133 | – | – |
| Productive tillers/plant | 7.7 | 4–12 | – | – |
| Panicle length (cm) | 22.7 | 17–29 | – | – |
| Grain yield (t/ha) | 1.83 | 0.9–3.2 | – | – |

The most promising salt-tolerant varieties filtered through agronomic performance and farmers' preferences scores across 8 years (2011–2018) under Salinity Tolerance Breeding Network (STBN) trials are presented in Table 3.2. Therefore, three salt-tolerant rice varieties (CSR 36, CARI Dhan 4, CARI Dhan 5) were targeted for seed dissemination in the islands and thus have replaced the existing rice varieties, viz., C14–8, Jaya, Jaganath, Nata Jaya, and Ranchi Dhan etc in the islands over the period.

**Table 3.2** Promising salt-tolerant rice varieties for Andaman and Nicobar Islands

| Year | Salinity ($EC_{iw}$ ~ dS/m) | pH | Farmers' preferred salt-tolerant varieties |
|---|---|---|---|
| 2011 | 4.0 | 5.3 | CARI Dhan 5, CSRC(S) 21-25 |
| 2012 | 4.0 | 5.2 | CARI Dhan 5, CSR36 |
| 2013 | 4.5 | 5.5 | CST71, CSRC(D)-2-17-5 |
| 2014 | 6.8 | 5.5 | CARI Dhan 6, NDRK1114 |
| 2015 | 4.0 | 5.5 | IR84645-305-6-1-1-1, IR87938-1-1-2-1-3-B |
| 2016 | 5.0 | 5.5 | RP 4353-MSC-38-43-6-2-4-3 and CARI Dhan 5 |
| 2017 | 4.3 | 7.8 | CSR2016 18-15 and CSR2016-IR18-2 |
| 2018 | 5.5 | 7.8 | CSR2016 IR18-1 and KS-12 |

## 3.8 Lines Developed, Nominated, Released as Varieties and Genetic Stocks Registered

Based on the multilocation/year's performance, ICAR-CIARI, Port Blair, has identified three rice varieties CSR 36, CARI Dhan 5, and CST 7–1 with high yield range of 2.5–3.5 t/ha under coastal saline soils of Andaman and Nicobar Islands. In addition, 11 out of 308 salt-tolerant rice lines evaluated under Salinity Tolerance Breeding Network Trials along with national checks in rainfed lowland saline soils of Andaman and Nicobar Islands were also identified as promising cultures for the benefit of island farmers. The main salt-tolerant rice varieties, which have been recommended for cultivation in these islands based on their proven superiority and farmers' preferences, are CARI Dhan 4, CARI Dhan 5, CSR36, and CSR23 (Table 3.3). Therefore, based on our yield performance trials across the years and island farmers' preferences, three salt-tolerant rice varieties (CSR 36, CARI Dhan 4, CARI Dhan 5) have considerably replaced the low-yielding local varieties like C14–8, Jaya, Jaganath, Nata Jaya, Ranchi Dhan, etc., in the islands through their large-scale seed production and dissemination (Table 3.4).

### 3.8.1 Seed Dissemination and Probable Impact

About 150.7 quintal truthfully labeled (TFL) seeds of three salt-tolerant varieties (CARI Dhan 4, CARI Dhan 5, and CSR 36) were produced and disseminated to island farmers to fulfill their needs during 2011–2018, which covered an area of about 498 ha (Singh et al. 2020). The maximum area was covered by the variety CSR 36 (215 ha), followed by CARI Dhan 5 (155 ha) and CARI Dhan 4 (128 ha). These TFL seeds of salt-tolerant rice varieties were disseminated through complementation and synergy of projects including STRASA, Tribal Sub Plan (TSP), ICAR Seed Project, and NABARD supported outreach center at ICAR-CIARI, Port Blair. Besides, seed was also given to the Department of Agriculture, Andaman and Nicobar Islands, and some NGOs. Among the three popular salt-tolerant rice

**Table 3.3** Salient traits of high-yielding rice varieties in A&N Islands

| Name of variety | CARI Dhan 4 (SVRC)[a] | CARI Dhan 5 (SVRC) | CSR36 (CVRC)[a] | CSR23 (CVRC) |
|---|---|---|---|---|
| Parentage | Somaclonal variant of Pokkali | Somaclonal variant of Pokkali | CSR13/ Panvel 2// IR36 | IR64//IR4630-22-2-5-1-3/IR964-45-2-2 |
| IET no. | 16,904 | 16,885 | 17,340 | 13,769 |
| Year of release | 2009 | 2009 | 2005 | 2004 |
| Plant height (cm) | 120 | 100 | 110 | 115 |
| Maturity days | 140 | 140 | 140 | 130 |
| Yield (t/ha) | | | | |
| Nonstress | >4.5 | >5.0 | >5.5 | >5.0 |
| Salt stress | >3.2 | >3.6 | >4.0 | >4.0 |
| Grain type | Medium bold | Medium bold | Long slender | Long slender |
| Recommended ecology | Rainfed lowland and coastal saline soils | Rainfed lowland and coastal saline soils | Sodic and saline soils | Sodic and saline soils |
| Developed by | ICAR-CIARI, Port Blair | ICAR-CIARI, Port Blair | ICAR-CSSRI, Karnal | ICAR-CSSRI, Karnal |

[a]*SVRC* Andaman and Nicobar State Variety Release Committee; *CVRC* Central Variety Release Committee

**Table 3.4** Year-wise seed distribution (quintal) of salt-tolerant rice varieties and their area coverage (ha) in Andaman and Nicobar Islands during 2011–2018

| | CSR 36 | | CARI Dhan 4 | | CARI Dhan 5 | | Total | |
|---|---|---|---|---|---|---|---|---|
| Years | SD[a] | AC[a] | SD | AC | SD | AC | SD | AC |
| 2011 | 8.1 | 26 | 00 | 0 | 11.86 | 39 | 19.92 | 65 |
| 2012 | 16.8 | 56 | 1.50 | 5 | 7.00 | 23 | 25.3 | 84 |
| 2013 | 5.0 | 16 | 4.80 | 16 | 7.00 | 23 | 16.8 | 55 |
| 2014 | 5.0 | 16 | 4.1 | 13 | 7.0 | 23 | 16.1 | 52 |
| 2015 | 7.0 | 23 | 5.5 | 18 | 1.1 | 4 | 13.6 | 45 |
| 2016 | 11.0 | 37 | 10.7 | 36 | 1.0 | 3 | 22.7 | 76 |
| 2017 | 6.3 | 21 | 7.0 | 23 | 7.0 | 23 | 20.3 | 67 |
| 2018 | 6.0 | 20 | 5.0 | 17 | 5.0 | 17 | 16 | 54 |
| Total | **65.2** | **215** | **38.6** | **128** | **47.0** | **155** | **150.7** | **498** |

[a]*SD* Seeds distributed (quintals), *AC* Area covered (ha)

varieties, CSR 36 was the most preferred variety by the farmers due to its long slender grain and high-yielding ability (3.5 t/ha) followed by CARI Dhan 4 and CARI Dhan 5, which are having bold grain and gave 2.8–3.2 t/ha grain yield under coastal saline and acid sulfate conditions of Andaman and Nicobar Islands.

## 3.9 Acknowledgment with Due Credit to Funding Agencies

A total of seven research projects were implemented over different years for rice improvement through conventional and biotechnological approaches. These projects were funded by ICAR-CIARI, Port Blair [breeding for mineral stress (PI: N.D. Mazumdar); stress tolerance by biotechnological approaches (PI: A.B. Mandal); biotic and abiotic stress tolerance by genetic manipulation (PI: A.B. Mandal) or by external grants [salt tolerance by molecular approach (AP CESS funded- PI: A.B. Mandal), stress-tolerant rice for poor farmers of Africa and South Asia (Bill and Melinda Gates Foundation and IRRI (PI: R.K Gautam), marker-assisted introgression of bacterial blight resistance in popular rice cultivars of Andaman and Nicobar Islands (DBT-PI: R.K. Gautam), and Seed Production of Agricultural Crops (ICAR Seed Project-PI: P.K. Singh)].

## 3.10 Future Outlook

Rice improvement along with matching agrotechnological management (Gautam et al. 2013) offers attractive option to harness the productivity potential of salinity-affected coastal areas of tropical islands. Varietal development for higher degree of salt stress along with tolerance to acid sulfate soils and multiple stresses of mineral toxicities and deficiencies occurring in coastal saline areas is required to attain higher genetic gain and sustainable productivity. Tropical island ecosystem being conducive to biotic pressures requires continuous monitoring of the emergence of new pests and pathogens for planning and deploying resistance genes in breeding programs. Therefore, it is also important to upgrade the salt-tolerant varieties with island-specific resistance genes for major diseases like bacterial blight (Gautam et al. 2015) to get higher and sustained yields from such agroecologies. Precisely locating the island-specific QTLs/genes governing salinity stress tolerance along with other edaphic stresses through the development of mapping populations followed by gene pyramiding with other traits through conventional and genomic technologies is needed to plan and prepare for future breeding programs. Farmers' participatory variety selection and seed production and variety release process through administrative and policy supports will accelerate the adoption and spread of desirable varieties. Post harvest interventions to minimize production losses to minimize milling losses along with handholding of farmers through market trends and better price support are required for harnessing and sustaining rice production system in these marginal environments.

## 3.11 Conclusion

While the tropical islands are vulnerable to the effects of sea level rise and climate change, there are limited options for livelihood sustenance and employment. Rice is the only major feasible food crop in the coastal line due to high rainfall and waterlogging in these areas. The development and adoption of appropriate high-yielding and stress-resilient rice varieties offer attractive, simple, and cost-effective options for attaining higher and sustainable yields from the coastal saline areas of geographically isolated Andaman and Nicobar Islands. This chapter documents the research and development works on breeding high-yielding salt-tolerant rice varieties suitable for Andaman and Nicobar Islands. Future areas for research and development are also highlighted to attain incremental gains in this direction.

**Acknowledgments** We are grateful to the Director of ICAR-CIARI, Port Blair, for the support and the various funding agencies for the financial assistance.

## References

Abraham Z, Senthilkumar R, John KJ, Sharma TVRS, Nair NV, Unnikrishnan M, Pareek SK (2008) Collection of plant genetic resources from Andaman and Nicobar Islands. Genet Resour Crop Evol 55(8):1279–1289

Ahmed SZ, Singh PK, Gautam RK, Roy SD (2014) Yield gap analysis of rice through front line demonstration in tropical Andaman Islands. J Indian Soc Coastal Agric Res 32(2):42–46

Bandyopadhyay AK (1999) Coastal areas and their management. In: Singh GB, Sharma BR (eds) 50 years of natural resource management research. Division of Natural Resources Manangement, ICAR, New Delhipp, pp 635–644

Bernardo MA, Dieguez ET, Cortes AL, Ojanguren CLT, Jones HG, Chairez FA (2000) Path analysis of cowpea early seedling growth under saline conditions. Int J Exp Botany 67:85–92

Chen Y (1986) Haploids of higher plants in vitro. In: Chen (ed) The inheritance of rice pollen plant and its application in crop improvement, pp 118–136

Chopra VL, Narasimhulu SB, Kirti PB, Prakash S, Anuradha G (1989) Studies of somaclonal variation in brassica spp. and its relevance to improvement of stress tolerance and yield. In: Review of advances in plant biotechnology, pp 1985–1988

Chowdhury B, Mandal AB (2001) Microspore embryogenesis and fertile plantlet regeneration in a salt susceptible× salt tolerant rice hybrid. Plant Cell Tissue Organ Cult 65(2):141–147

Cramer GR, Schmidt CL, Bidart C (2001) Analysis of cell wall hardening and cell wall enzymes of salt-stressed maize (*Zea mays*) leaves. Funct Plant Biol 28(2):101–109

DES (2019–20). Basic statistics, Directorate of Economics and Statistics (DES), Andaman and Nicobar Administration., Andaman and Nicobar Islands, India. http://andssw1.and.nic.in

Farshid A, Hassan ER (2012) Physiological characterization of rice under salinity stress during vegetative and reproductive stages. Indian J Sci Technol 5(4):2578–2586

Foroughi-Wehr B, Wenzel G (1989) Androgenetic haploid production. IAPTC News Letter 58:11–18

Gautam RK, Roy SD, Singh PK, Singh S, Awnindra K, Singh SK, Ahmed SZ, Jaisankar S, K. (2016) Improved varieties of field and horticultural crops for Andaman & Nicobar Islands, India, ICAR-CIARI Port Blair, PP 36

Gautam RK, Singh PK, Birah A, Kumar K, Singh AK, Kumar N, Swain S, Ahmed SZ, Singh A, Ravisankar N, Roy SD (2013) Agro-technology for higher rice productivity in Andaman & Nicobar Islands, p 42

Gautam RK, Singh PK, Sakthivel K, Srikumar M, Kumar N, Kumar K, Singh AK, Roy SD (2015) Analysis of pathogenic diversity of the rice bacterial blight pathogen (*Xanthomonas oryzae* pv. *Oryzae*) in the Andaman Islands and identification of effective resistance genes. J Phytopathol 163(6):423–432

Joseph B, Jini D, Sujatha S (2010) Biological and physiological perspectives of specificity in abiotic salt stress response from various rice plants. Asian J Agric Sci 2(3):99–105

Karp A (1995) Somaclonal variation as a tool for crop improvement. Euphytica 85(1):295–302

Kasperbauer MJ, Buckner RC, Springer WD (1980) Haploid plants by anther-panicle culture of tall fescue 1. Crop Sci 20(1):103–107

Kerala State Biodiversity Board. (2009). Available at http://www.keralabiodiversity.org/pdf/policy.pdf. Accessed on 11 Feb 2009

Krishnamurthy SL, Sharma PC, Sharma DK, Ravikiran KT, Singh YP, Mishra VK, Burman D, Maji B, Mandal S, Sarangi SK, Singh RK (2017) Identification of mega-environments and rice genotypes for general and specific adaptation to saline and alkaline stresses in India. Sci Rep 7(1):1–14

Krishnamurthy SL, Sharma SK, Sharma DK, Sharma PC, Singh YP, Mishra VK, Burman D, Maji B, Bandyopadyay BK, Mandal S, Sarangi SK (2016) Analysis of stability and G× E interaction of rice genotypes across saline and alkaline environments in India. Cereal Res Commun 44(2):349–360

Larkin PJ, Scowcroft WR (1981) Somaclonal variation—a novel source of variability from cell cultures for plant improvement. Theor Appl Genet 60(4):197–214

Mandal AB, Bandyopadhyay AK (1997) In vitro anther culture response in indica rice hybrids. Cereal Res Commun 25(4):891–896

Mandal AB, Pramanik SC, Chowdhury B, Bandyopadhyay AK (1999) Salt-tolerant Pokkali somaclones: performance under normal and saline soils in Bay Islands. Field Crop Res 61(1): 13–21

Mandal, A. B., Basu, A. K., Roy, B., Sheeja, T. E., & Roy, T. (2004). Genetic management for increased tolerance to aluminium and iron toxicities in rice—A review, 106

Morrison RA, Evans DA (1988) Haploid plants from tissue culture: new plant varieties in a shortened time frame. Bio/Technology 6(6):684–690

Sanint LR, Martinez CP, Lentini Z (1996) Another culture as a rice breeding tool: a profitable investment. In: Rice Genetics III: (In 2 Parts), pp 511–518

Sarangi S, Chowdhury B, Battacharya A, Mandal AB (2002) In vitro screening for increased tolerance to excess sodium chloride in rice. Asia-Pac J Mol Biol Biotechnol 10(1):49–58

Shrivastava P, Kumar R (2015) Soil salinity: a serious environmental issue and plant growth promoting bacteria as one of the tools for its alleviation. Saudi J Biol Sci 22(2):123–131

Singh PK, Gautam RK, Sakthivel K, Ahmed SZ, Venkatesan K, Varadan RJ (2020) Augmenting rice productivity in Andaman and Nicobar Islands through improved varieties and seed replacement rate. J Soci Coast Agric Res 38(1):1–10

Singh YP, Nayak AK, Sharma DK, Gautam RK, Singh RK, Singh R, Ismail AM (2014) Farmers' participatory varietal selection: a sustainable crop improvement approach for the 21st century. Agroecol Sustain Food Syst 38(4):427–444

Thomas, P. M. (2002). Problems and prospects of paddy cultivation in Kuttanad region. Thiruvananthapuram: Kerala research Programme on local level development, draft report. Available at http://www.krpcds.org/reportleft.htm. Accessed on 25 Sept 2010

Witcombe JR (2002) A mother and baby trial system. In: Breeding rainfed rice for drought-prone environments: integrating conventional and participatory plant breeding in south and Southeast Asia. Proceedings of a DFID Plant Sciences Research Programme/IRRI Conference. IRRI, Los Baños, pp 79–89

Witcombe JR, Joshi KD, Rana RB, Virk DS (2001) Increasing genetic diversity by participatory varietal selection in high potential production systems in Nepal and India. Euphytica 122(3): 575–588

Zeng L, Shannon MC (2000) Salinity effects onseedling growth and yield components of rice. Crop Sci 40:996–1003

Zeng L, Shannon MC, Lesch SM (2001) Timing of salinity stress affects rice growth and yield components. Agric Water Manag 48(3):191–206

# Recent Approaches in Breeding for Salt Tolerance in Rice in Bangladesh

4

M. Akhlasur Rahman, Hasina Khatun, Hosneara Hossain, K. M. Iftekharuddaula, and R. K. Singh

**Abstract**

Salinity is one of the most important abiotic stresses which limit agricultural productivity and pose threats to food security in Bangladesh as rice is the main staple food but sensitive to salt stress. As of now, farmers grow local varieties in the southern coastal region during T. Aman (wet) season which is moderately tolerant to salt stress with low yielding potential, and large areas remain fallow during boro (dry) season because of high salinity (12–16 $dSm^{-1}$). Breeding high yielding salt-tolerant rice is a sustainable way to develop varieties for this highly saline zone and brings these highly saline areas under rice cultivation that leads to maintaining food security for achieving the Sustainable Development Goals (SDGs). The physiological, genetic, and molecular pathways of salt tolerance have extensively been studied, but these pathways need to be addressed from the salt tolerance enhancement perspective, and these should appropriately be combined with the breeding approaches. There is an adequate variation for salt tolerance in rice germplasm, and salt tolerance is a complex quantitative trait. Hence, attempts are being focused on using this variation to develop high yielding salt-tolerant varieties through applying holistic approaches of breeding including conventional, molecular, and modern breeding methods. Understanding the gene action for salt tolerance in rice is important for use in the future breeding program. Selection for salinity tolerance with high and low heritable traits may be made in the early and later generations, respectively. A number of

M. Akhlasur Rahman (✉) · H. Khatun · H. Hossain · K. M. Iftekharuddaula
Plant Breeding Division, Bangladesh Rice Research Institute, Gazipur, Bangladesh
e-mail: akhlas08@gmail.com; akhlasur.breeding@brri.gov.bd

R. K. Singh (✉)
International Center for Biosaline Agriculture (ICBA), Dubai, UAE
e-mail: r.singh@biosaline.org.ae





R. K. Singh et al. (eds.), *Genetic Improvement of Rice for Salt Tolerance*,
https://doi.org/10.1007/978-981-99-3830-8_4

QTLs and gene-based markers for salt tolerance have been reported which have an immense potential for use during genomic assisted breeding. Seventeen salt-tolerant rice varieties have been developed and released in Bangladesh, and a series of salt-tolerant breeding lines were developed based on region-specific product profiles for the target region of Bangladesh.

**Keywords**

Breeding · QTL · Gene · Product profile · Salt stress · Rice

## 4.1 Introduction

Salinity is one of the major threats in the southern coastal region of Bangladesh which encompasses around 30% of the total crop production areas. Seventeen districts such as Cox's Bazar, Chittagong, Noakhali, Feni, Laxmipur, Bhola, Barisal, Jhalakathi, Pirojpur, Patuakhali, Borguna, Bagerhat, Khulna, Satkhira, Jashore, Narail, and Gopalganj are affected by different levels of salinity. In Bangladesh, salinity is caused by (1) presence of salts in the soils as ions are released through mineral weathering in the soil, (2) agricultural land affected by seawater flooding/storm surges, (3) entry of the saltwater into the shrimp *gher* (shallow pond for shrimp cultivation) in an unplanned way, (4) use of saline groundwater (EC >3.0 $dSm^{-1}$) for irrigation, and (5) non-leaching of ions (Na) from the soil because of insufficient rainfall during November–May consequently accumulating salts in the soil and salinity buildup in Bangladesh. Recently, the Bangladesh Climate Change Resilience Fund (BCCRF) Management Committee emphasized salinity problem in coastal Bangladesh as a crucial issue of climate change adaptation. Moreover, climate change may have adverse effects in soil and water salinity in different southern coastal districts of Bangladesh during the dry season (DasGupta et al. 2015) agriculture in different crops including rice.

Salinity buildup is a special issue for rice based agriculture in Bangladesh. Several reports revealed the damaging effects and appropriate tolerance mechanisms of salt stress during different susceptible stages of seedling and reproductive growth including early seedling establishment in field, panicle initiation, booting, heading, spikelet fertility-sterility/seed set, yield, and other salt tolerance indices (Khatun and Flowers 1995; Grattan et al. 2002; Singh et al. 2007; Hasanuzzaman et al. 2009; Singh and Flowers 2010; Hossain et al. 2015; Munns and Gilliham 2015; Rahman et al. 2016; Rahman et al. 2021). In this paper, we discuss the recent progresses in the physiological, breeding/varietal development and molecular basis of salinity tolerance, and status of salt-tolerant rice product development in Bangladesh.

## 4.2 Review

### 4.2.1 Physiological Basis of Salt Tolerance

Salt stress entails changes in different physiological and metabolic pathways, based on severity and duration of the stress and eventually decreased rice productivity (Munns 2005; Rozema and Flowers 2008; Rahnama et al. 2010; James et al. 2011; Rahman et al. 2016). The salt stress tolerant genotypes use the mechanisms mostly in sensing and signaling of the salt stress, transcription regulation, ionic homeostasis, and reactive oxygen species (ROS) scavenging in response to salt stress (Mirdar Mansuri et al. 2020).

When rice plants are continuously exposed to salt stress with NaCl concentration of 100 mM and 200 mM, then the genes (*OsHSP23.7*, *OsHSP71.1*, *OsHSP80.2*- Heat-shock proteins, molecular chaperones, folding, assembling, and transporting proteins) and proteins (*OsHsp17.0*, *OsHsp23.7*- Heat-shock proteins, molecular chaperones, and folding, assembling, and transporting proteins) are upregulated and involved in response to salt stress in rice (Zou et al. 2009; Zou et al. 2012). The combined approach is necessary for determining the key mechanisms or pathways responsible for salt tolerance. Moreover, the significant progress has so far been made to understand plant responses to salt stress (Gupta and Huang 2014). The future study should be focused on the intercellular and intracellular molecular interactions that are involved in plant response toward salt stress (Gupta and Huang 2014; Mirdar Mansuri et al. 2020).

### 4.2.2 Genetic Basis of Salt Tolerance

Genetic characterization of salt tolerance related traits in 24 rice landraces and 6 modern varieties has shown significant phenotypic coefficient of variation (PCV), genotypic coefficient of variation (GCV), and broad-sense heritability and genetic advance (GA) as shown in Table 4.1. PCV and GCV for salt tolerance related traits varied from 21.98% to 46.51% and from 19.95% to 45.25% respectively, while heritability ranged from 80.24% to 94.66% and GA varied from 0.17% to 34.71%. The highest PCV and GCV values were found in root dry weight trait, while the lowest in SES. Heritability and genetic advance were highest in root dry weight and survival rate, respectively. The results revealed a sufficient variation for salt tolerance related traits with high heritability and genetic advance. Consequently, there is possibility to improve the traits related with salt tolerance in rice by selecting and exploiting salt-tolerant landraces/germplasm in the breeding programs.

Genetic component analysis (GCA) study made by Gregorio and Senadhira (1993) showed that both additive and dominance gene effects controlled low Na-K ratio. The combining ability analysis shows that both general combing ability (GCA) and specific combining ability (SCA) effects are also important for deciphering the genetics of salt tolerance. They also revealed that selection may be made in later

**Table 4.1** Genetic parameters for salt tolerance related traits in 30 rice genotypes

| Traits | PCV (%) | GCV (%) | Heritability (%) | GA (% mean) | Reference |
|---|---|---|---|---|---|
| Survival rate | 34.23 | 30.66 | 80.24 | 34.72 | Huda et al. (2017) |
| Shoot fresh weight | 43.83 | 42.43 | 93.70 | 7.74 | |
| Shoot dry weight | 46.15 | 44.82 | 94.33 | 1.64 | |
| Root length | 25.07 | 24.10 | 92.42 | 4.01 | |
| Root fresh weight | 30.84 | 28.29 | 84.15 | 3.86 | |
| Root dry weight | 46.51 | 45.25 | 94.66 | 0.166 | |
| SES | 21.98 | 19.95 | 82.38 | 1.17 | |

*PCV* phenotypic coefficient of variation, *GCV* genotypic coefficient of variation, *GA* genetic advance, *SES* standard evaluation score

generations under controlled conditions for minimizing environmental effects for low heritable traits like Na-K ratio.

Varieties such as Nona Bokra and Pokkali (tolerant), Damodar and Jhona 349 (moderately tolerant), and IR28 and IR35657–33–2 (sensitive) were evaluated at EC 12 $dSm^{-1}$ for genetic analysis of salt tolerance in both seedling and reproductive stages (Akbar et al. 1986). Both additive and dominance effects were important in the inheritance of all the traits studied. Shoot length, Na content in the shoots, and shoot and root dry weight showed high heritability and significant additive effects. Genes controlling Na level in shoots exhibited partial dominance. Three types of gene effects were observed in the inheritance of Na content during seedling stage. Also highly significant additive effects with high heritability were found for plant height and yield per plant. Predominance of additive effects with high heritability can be useful for salt-tolerant variety development, while selection was made based on the plant height, shoot Na level, shoot length, shoot and root dry weight, and yield per plant (Akbar et al. 1986).

Ray and Islam (2008) estimated the genetic parameters for salt tolerance. They observed nonallelic interactions (epistasis) in the cross between moderately tolerant and sensitive parents. There was no evidence of nonallelic interactions in the cross between tolerant and sensitive parents such as Nona Bokra/BR29 and Pokkali/BR29. Additive effects of genes contributed more in the inheritance of salinity tolerance in Pokkali/BR29, while Nona Bokra/BR29 had predominantly additive and dominant gene actions, and there was no epistatic gene action. The additive effects could enhance fixation of the desirable combination of genes, and also additive gene action was correlated to narrow sense heritability (Mohammadi et al. 2014). Thus, selection for salinity tolerance may be made in the early generation. Understanding the gene action for salt tolerance in rice will be useful in the future breeding program (Souleymane et al. 2017).

### 4.2.3 Molecular Basis of Complex Salt Tolerance

The molecular dissection of salt tolerance has considerably enhanced using the molecular platforms for identifying quantitative trait loci (QTL) controlling related genetic variation in crops including rice (Tanksley 1993; Ribaut and Hoisington 1998; Flowers et al. 2000; Koyama et al. 2001; Munns 2005; Tuberosa and Salvi 2006; Thomson et al. 2010; Hossain et al. 2015; Ismail and Horie 2017; Rahman et al. 2017; Rahman et al. 2019a, 2019b; Haque et al. 2020; Maniruzzaman et al. 2022).

Moreover, several QTLs related with physiological, agronomic traits conferring salinity tolerance at seedling and reproductive stage have been reported (Thomson et al. 2010; Hossain et al. 2015; Rahman et al. 2017, Mondal et al. 2019) including major QTLs for salinity tolerance such as *SKC1* (Lin et al. 2004); a sodium transporter *OsHKT1; 5* in the *SKC1* locus (Ren et al. 2005) and *Saltol* (Bonilla et al. 2002; Thomson et al. 2010) on chromosome 1 (Table 4.2). Recently molecular basis of salt tolerance has been unraveled in various rice landraces such as Pokkali (Bonilla et al. 2002; Thomson et al. 2010), Nona Bokra (Ren et al. 2005), Hasawi (Rahman et al. 2017), Capsule (Rahman et al. 2019a, 2019b), Changmaogu (Sun et al. 2019), and Horkuch (Haque et al. 2020) which can withstand different levels of salt stress at various sensitive growth stages of rice.

### 4.2.4 Breeding Informatics Aspects for Salt Stress

Breeding informatics helps plant breeder to increase breeding efficacy using modern biometric tools including data management system (Breeding4Results-B4R) and selecting elite breeding lines with higher *breeding* value for using as parents in the subsequent *breeding* cycle as a part of cyclic breeding (Ye et al. 2013; Rahman et al. 2019a, 2019b).

Widely used tools for quantitative genetics such as genomic estimated breeding value (GEBV) and best linear unbiased predictions (BLUPs) are applied to evaluate the performance to construct high throughput new breeding populations for selecting the superior breeding lines when combined with genetic relatedness or kinship matrix/information (i.e., coefficient of coancestry) using pedigree information to calculate estimated breeding values (EBVs). This is a key parameter for selecting complex traits like salt tolerance and yield through estimating parents' genetic potential to produce better descendants on the basis of parent's own performance, pedigree information, and progeny data. EBVs play important role to select parent for higher rate of genetic gain (Bernardo 2010; Quddus et al. 2019).

Theory of genetic gain in breeding and classical biometrical genetics indicates the improvement of performance compared to a standard or baseline. It is generally evaluated after advancing one generation to the subsequent generation, and artificial selection is done (Falconer et al. 1996). However, genetic gain per year is also known as genetic trend that measures varietal performance over year by comparing base or dominant varieties (Peng et al. 1999).

**Table 4.2** Recently identified QTLs for salt tolerance with controlling/responsible traits and phenotypic variation using different mapping populations at seedling and reproductive stages in rice

| Rice growth stage | QTL/gene name | Mapping population and parentage | Population size | Salinity tolerant trait/ index | Flanking marker | Positive allele and position (cM or Mb) | PVE (% $R^2$) | References |
|---|---|---|---|---|---|---|---|---|
| Seedling | *qSaltol* | RIL (IR29/Pokkali) | 80 | $Na^+$ uptake; $Na^+/K^+$ | RM140 - C1733S | Pokkali; 51.6–65.9 cM / 13.87 Mb | 39.2 | Bonilla et al. 2002 Thomson et al. 2010 |
| | *OsHKT1;5 (qSKC-1)* | Advanced backcross (Koshihikari*3/Nona Bokra) | 192 | Shoot $K^+$ concentration | K159 - K061 | Nona Bokra; 11.46 Mb | – | Ren et al. 2005 |
| | *qST1* | RIL (Milyang 23/ Gihobyeo) | 164 | Young seedling stage | Est12 -RZ569A | Gihobyeo; 40 cM | 27.8 | Lee et al. 2006 |
| | *qST3* | | | | RG179 - RZ596 | Milyang 23; 138 cM | 9.2 | |
| | *qSES1* | F2 (NERICA-L-19/ Hasawi) | 153 | SES | RM8094-RM582 | NERICA-L-19 | 19.6 | Bimpong et al. 2014 |
| | *qSES6* | | | | RM586-RM253 | NERICA-L-19 | 39.7 | |
| | *qSES10* | | | | RM228-RM333 | Hasawi | 30.7 | |
| | *qSES11* | | | | RM536-RM287 | Hasawi | 37.2 | |
| | *qSES2.1* | RIL (IR29/ Hasawi) | 142 | SES | id2004774–id2007526 | Hasawi; 64.8 cM | 11.1 | Bizimana et al. 2017 |
| | *qSES1.3* | | | SES | id1024972-id1023892 | Hasawi; 170.0 cM | 39.9 | Rahman et al. 2017 |
| | *qSES3.1* | F2 (capsule/ BRRI dhan29) | 94 | SES | RM5626-R3M53 | Capsule; 111.0 cM | 23.0 | Rahman et al. 2019b |

| | | | | | | | | |
|---|---|---|---|---|---|---|---|---|
| | *qSES12.3* | | | | RM252B RM27615-RM27877 | Capsule; 31.0 cM | 17.0 | |
| | *qSL7* | RIL (93–11/PA64s) | 132 | Higher shoot length | SNP7-191–SNP7226 | PA64s; 86.31 cM | 9.9 | Jahan et al. 2020 |
| | *qSUR5.2* | F2 (BRRI dhan49/Akundi) | 91 | Survival rate | SNP2291156 | Akundi; 10 | 14.9 | Maniruzzaman et al. 2022 |
| | *qNa6* | | | $Na^+$ conc. | SSIIA | Akundi; 26 | 16.8 | Maniruzzaman et al. 2022 |
| | *qK8* | | | $K^+$ conc. | SNP 4853081 | Akundi; 18.8 | 21.2 | Maniruzzaman et al. 2022 |
| | *qNaKR11* | | | $Na^+/K^+$ | IRGSP1-C11_27391141 | Akundi; 107.6 | 12.9 | Maniruzzaman et al. 2022 |
| Reproductive | *qNa1.7* | F2 (Cheriviruppu/Pusa basmati 1) | 218 | Na (mmol $g^{-1}$ dwt) | RM1349-RM7250 | Cheriviruppu; 31.06 Mb | 13.5 | Hossain et al. 2015 |
| | *qNaKR1.8* | | | Na/K ratio | RM1349-RM7250 | Cheriviruppu; 31.06 Mb | 11.0 | |
| | *qNFS2.1* | F2 (NSIC Rc222/BRRI dhan47) | 92 | Filled spikelet number | id2004774 | BRRI dhan47; 40.1 cM | 15.3 | Mondal et al. 2019 |
| | *qPFS2.1* | | | Filled spikelet (%) | id2013434 | BRRI dhan47; 122.1 cM | 18.4 | |
| | *qTS11.1* | | | Total spikelet (no) | id11000858 | BRRI dhan47; 14.0 cM | 15.8 | |
| | *qYLD2.1* | | | Yield | id2004774 | BRRI dhan47; 40.1 cM | 14.6 | |

*SES* overall phenotypic performance, *SL* shoot length, *Na* $Na^{+}$: concentration, $K^+$ concentration, *Na K-R* Na-K ratio, *Sur* survival, *Chr* chromosome number, *PVE* phenotypic variation explained

### 4.2.5 Advent of Salinity Breeding in Bangladesh

In 1995, BRRI Scientist Dr. Saleha Khatun did work on effects of salinity on seed set in rice with Dr. Timothy J Flower (Khatun and Flowers 1995). Breeding for salinity tolerance was started by Dr. M A Salam during the late 1990s; consequently moderately salt-tolerant varieties such as BRRI dhan40 and BRRI dhan41 were released in 2003 which were developed through conventional breeding.

#### 4.2.5.1 Important Milestones of Salinity Breeding, which Have Significantly Improved Rice Productivity and Cultivation Areas

1. Laying foundation of salinity research in collaboration with the University of Sussex-IRRI-BRRI in 1995.
2. IRRI-BRRI collaboration in 1999–2004 through Poverty Elimination Through Rice Research Assistance (PETRRA).
3. Farmers participation in variety selection initiated through participatory variety selection (farmers' breeding program) approach.
4. Physiological characterization of genetically fixed lines for isolating tolerant genotypes in 2002.
5. Collection of landraces from coastal saline zone in 2003.
6. A total of 17 salt-tolerant varieties released in Bangladesh. BRRI dhan47 is the first salt-tolerant milestone variety released in 2007 selected through PVS.
7. Momentous collaborative efforts from STRASA since 2008 facilitated the development and release of a number of salt-tolerant varieties such as BRRI dhan53, 54, 55, 61, 67, 73, and 78 through germplasm from Salt Tolerance Breeding Network (STBN).
8. Marker-assisted breeding for salinity tolerance began in 2004–2005.
9. Physiological and molecular characterization and genomic diversity explored novel salt-tolerant donors (Rahman 2010; Rahman et al. 2016).
10. *Saltol* introgression was performed in mega varieties BR11, BRRI dhan28, BRRI dhan29, and BRRI dhan49 using the donors FL378 (at BRRI) and FL478 at IRRI (Rahman 2010).
11. Studies on dissection of genetic mechanisms of salt tolerance using different donors for seedling (Rahman 2010; Rahman et al. 2017; Rahman et al. 2019a) and reproductive stage (Hossain et al. 2015; Mondal et al. 2019) tolerance.
12. Genetic gain enhancement through breeding modernization since 2015 and use of modern biometrical tools (BLUP: estimated breeding value (Quddus et al. 2019); Genomic selection) and holistic breeding approaches (Rahman et al. 2021).

To maintain the food security under this COVID-19 situation, horizontal expansion (increase of rice growing areas in high saline zone) of rice cultivation is needed especially in the coastal areas. Thus, large areas have to be brought under cultivation in high saline prone areas. Newly released salt-tolerant varieties BRRI dhan97 (IR83484-3-B-7-1-1-1) and BBRI dhan99 are suitable for these high saline zones (Fig. 4.1) where other salt-tolerant varieties did not survive.

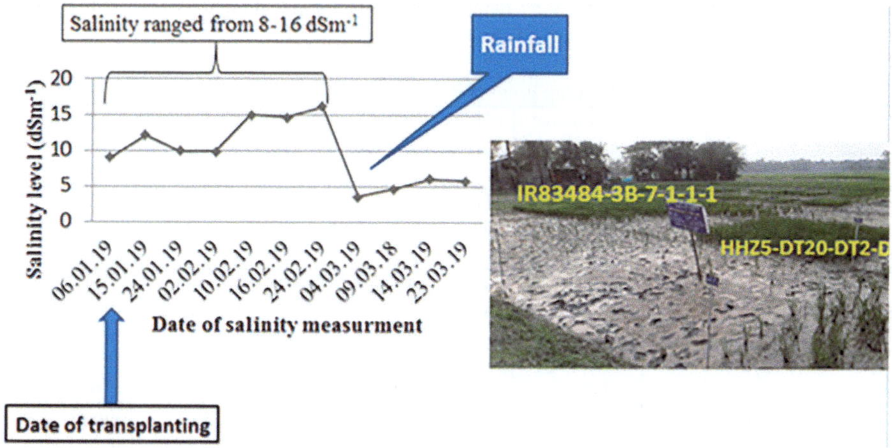


**Fig. 4.1** Salinity level of proposed variety trial varied at different locations of coastal zone. The degree of salinity (8–16 $dSm^{-1}$) was found highest at Kaliganj, Satkhira

Extent of both salinity level and areas is in the increasing trend ranging from 4.0 $dSm^{-1}$ with areas of 833,450 ha in 1973 and ~16.0 $dSm^{-1}$ within 10,156,260 ha areas during the last four decades.

However, promising breeding lines selected in OYT will be recycled to initiate next cycle of breeding for population improvement. Hopefully, following this transformed breeding approach, salinity tolerant varieties with enhanced grain yield and quality parameters will be developed.

### 4.2.6 Product Profile (PP) for Target Region

Product profile (PP) is a realistic roadmap for varietal development which addresses requirements and demands of plant breeders, farmers, millers/market, and consumers. Example of a product profile for salinity tolerant variety development for southern coastal zone in T. Aman season is shown in Table 4.3. Variety development based on formal PP will be able to replace market leading variety with the new one; thus, farmer could get early turnover from the new variety. However, the market leading variety may be or may not be a mega variety in target region. The way of developing PP is as follows:

- Designating the target region that uses common PP or trait requirements.
- Identifying the market leading variety and second important variety.
- Placing key traits into appropriate trait family.
- Estimating the amount of emphasis or research efforts needed to be given in the breeding program for selecting trait.

**Table 4.3** Product profile with market demand-led traits for developing salinity tolerant rice (STR) variety for target region under STR breeding program at BRRI

| **Salinity (T. Aman) Breeding Product Profiles (20% Breeding Program Focus)** | | | | | | |
|---|---|---|---|---|---|---|
| Country /Region: Salt affected coastal region and/Whole Bangladesh | | %Resource allocation for trait improvement | Market Leading Variety #1: BRRI dhan41 (Medium slender, high head rice yield, medium growth duration) | | | |
| Market Slot: T. Aman Rice (Medium slender to long slender grain) | | | Market Leading Variety #2: BR23 (Medium bold, high yield potential, photo period sensitive) | | | |
| **Trait Family** | **Key Economic Traits** | | **Trait value** | **Standard Variety Assessment** | **Trait Benchmarking Details** | **Trait availability in the Breeding Program** |
| Yield (Paddy) | Yield (>10% higher) | | 2-Must Have Trait | BRRI dhan41 | > BRRI dhan41 | Program has trait available |
| Maturity | Intermediate /long range | 10 | 2-Must Have Trait | BRRI dhan23 | <= BRRI dhan23 | Program is actively working with trait |
| Abiotic Stress Tolerance | Salinity tolerance | 25 | 2-Must Have Trait | BRRI dhan67 | > = BRRI dhan67 | Program is actively working with trait |
| Biotic Resistance (Fungal) | Blast | 20 | 2-Must Have Trait | BRRI dhan67 | Standard Evaluation Scale =< 3 | Trait Limited or NOT available |
| Biotic Resistance (Bacterial) | BLB | 10 | 3-Trait Values | BRRI dhan67 | Standard Evaluation Scale =< 3 | Program has trait available |
| Biotic Resistance (Insect) | BPH | 10 | 3-Trait Values | BRRI dhan67 | = Rathuheenati*(bph3, bph17)* | Trait Limited or NOT available |
| Consumer Traits | Amylose content | 15 | 2-Must Have Trait | BRRI dhan67 | => 24% | Program is actively working with trait |
| Yield (Economic)-Head Rice | Head rice recovery | 10 | 2-Must Have Trait | BRRI dhan67 | > 50% | Program is actively working with trait |
| Total | | 100 | | | | |

Legend: 1 = nice to have, 2 = must have traits, 3 = value added or game changer.

### 4.2.7 Variety Development

#### 4.2.7.1 Local Germplasm and Landrace Collection, Characterization for Salt Stress Tolerance, and Utilization

A total of 107 landraces was collected from southern coastal zone of Bangladesh. The genetic base of salt-tolerant donors needs to be broadened for developing climate-smart rice (Yadav et al. 2021) varieties for salt-affected regions with higher level of adaptation. All germplasm (107) were used for diversity analysis using a genome-wide set of 376 single nucleotide polymorphism (SNP) markers to identify and characterize novel sources of salt tolerance.

Three major clusters – the indica, aus, and aromatic subgroups – were identified. The largest group was indica, with the salt-tolerant Pokkali accessions in one subcluster, while Bangladeshi landraces, including Akundi, Ashfal, Capsule, Chikirampatnai, and Kutipatnai, were in a different subcluster. The salt-tolerant landrace Hasawi and Kalarata clustered into a distinct aus group near to indica. Allelic diversity study at the major QTL *Saltol* shows different alleles at the *Saltol* locus for Akundi, Ashfal, Capsule, Chikirampatnai, and Kutipatnai.

Sixty-nine landraces were further screened for physiological traits associated with salt stress at the seedling stage. Seven landraces such as Akundi, Ashfal, Capsule, Chikirampatnai, JataiBalam, Kalarata, and Kutipatnai uptake less Na and comparatively more K and maintain lower Na/K ratio in leaves. They efficiently restrict sodium transport root to the shoot. Newly identified salt-tolerant landraces are genetically and physiologically different from known donors (Pokkali and Nona Bokra). These landraces can be used to develop salt-tolerant varieties with higher tolerance (Rahman et al. 2016). These landraces may harbor novel sources of QTL/alleles for salt tolerance that will be useful in molecular breeding.

#### 4.2.7.2 Participatory Varietal Selection (PVS)

Participatory varietal selection (PVS) demonstrates the varietal/ line selection on the basis of farmers' need–/demand-based choice of varieties that generally differs from plant breeders' selection process. Plant breeders' evaluation of varietal performance – mostly following statistical designs and quantitative data based methods – is usually different from the farmers' choice (Ceccarelli et al. 2000; Mekbib 2006; Rahman et al. 2015). However, variety selection criteria may vary according to gender, environmental condition, market demand, and economic/social status (Defoer et al. 1997; Weltzien et al. 1998; Rahman et al. 2015). Until now, breeding objectives in different countries have not been properly focused on the opinions of farmers and their preferences and needs for the adverse growing conditions of their regions (Witcombe et al. 2006; Mekbib 2006; Islam et al. 2008a). To unravel this situation, participatory varietal selection is the important way of decentralized bottom-up breeding or farmer breeding approach that integrates farmers and their complex criteria for variety selection into varietal development programs (Weltzien et al. 2005; Ceccarelli and Grando 2007; Islam et al. 2008b; Weltzien et al. 2008; Thapa et al. 2009; Vom Brocke et al. 2010).

BRRI dhan47 (IR 63307-4B-4-3) was the first variety selected through PVS and released as salt-tolerant variety for *Boro* season in Bangladesh.

A total of 17 salt-tolerant rice varieties (15 from BRRI and 2 from BINA) were developed and released for transplanted Aman (T. Aman-RLR) and irrigated ecosystem (Boro-dry season) (Table 4.4).

#### 4.2.7.3 Marker-Assisted Selection

FL378 (IR66496-3R-78-1-1), a salinity tolerant recombinant inbred line derived from cross between IR29 and Pokkali, was used as donor for *Saltol*. FL378 had the Pokkali introgression on the chromosome 1 from RM1287 (10.9 Mb) to RM493 (12.2 Mb) for about 1.3 Mb, and its average tolerance score was around 4.7 (Thomson et al. 2010; Salam et al. 2007, Islam et al. 2012). The recurrent varieties were BR11, BRRI dhan28, and BRRI dhan29, and these three mega varieties of Bangladesh are widely grown in wet and dry season.

BRRI dhan28-*Saltol* seeds were developed at IRRI and FL478 as donor parents. Twelve introgression lines of BR11-*Saltol* and BRRI dhan28-*Saltol* were evaluated in salt-affected coastal district of Satkhira during dry season to investigate the salinity reaction. Among the 12 introgression lines of BR11-*Saltol*, BR8371-73-194-1-7 (7.2 $tha^{-1}$), BR8371-73-194-1-97 (7.2 t $ha^{-1}$), BR8371-18- 20-52-55 (7.2 $tha^{-1}$), BR8371-18-20-52-67 (7.0 $tha^{-1}$), and BR8371-18-20-52-124 (7.9 $tha^{-1}$) performed better than other lines. The level of salinity in the trial plot was 2–5 dS $m^{-1}$. Due to very low salinity stress, it performed better than the salt-tolerant check.

Moreover, *Saltol* QTL was introgressed into the genetic background of BRRI dhan49 (Hoque et al. 2015) and Pusa 44 and Sarjoo 52 (Krishnamurthy et al. 2020) through marker-assisted backcrossing. FL478 was used as a donor for *Saltol* QTL. A number of *Saltol* introgression lines (NILs: BRRI dhan49-*Saltol* lines) were developed (Hoque et al. 2015). Krishnamurthy et al. (2020) reported that the NILs PU99, PU176, PU200, PU215, PU229, PU240, PU241, PU244, PU252, and PU263 of Pusa44 and SAR17, SAR23, SAR35, SAR39, SAR77, SAR87, SAR123, and SAR136 NILs of Sarjoo52 exhibited salt tolerance with low salt injury score (SIS) of 3 or 5.

### 4.2.8 Seed Dissemination and Probable Impact

Bangladesh Rice Research Institute (BRRI) produces approximately 100 ton "breeder seed" annually of which around 7–10 tons of "breeder seed" of salt-tolerant varieties are produced.

Around 650,000 tons of "certified seed" can be produced from 100 tons of "breeder seed." This amount (650,000 tons) is almost double of the quality seed demand of Bangladesh. However, the government sector supplied about 35% of the quality seed, and the rest of the seed sources are the farmers, NGO's seed, private sectors (seed companies), traders, or local market (Kabir et al. 2015). Bangladesh has a huge potential market for quality rice seed; the estimated value of rice seed is around BDT 1650 crore ($290 million) (Hossain et al. 2002; Tulachan et al. 2012). Seed dissemination happens through the seed network like public sector (BRRI,

**Table 4.4** List of released salt-tolerant rice varieties with the information on pedigree, parentage and ecosystem/growing season that are cultivated in Bangladesh during T. Aman (RLR-wet) and Boro (dry) season

| Technologies/ products (variety name) | Parentage | Type of technology | Year of release |
|---|---|---|---|
| BR23 (BR716-7-2-1-1) | DA29/BR4 | Moderately salt-tolerant photosensitive *T. Aman* rice variety for rainfed lowland ecosystem | 1988 |
| BRRI dhan40 (BR5331-93-2-8-3) | IR4595-4-1-15/ BR10 | Moderately salt-tolerant *T. Aman* rice variety for rainfed lowland ecosystem | 2003 |
| BRRI dhan41 (BR5828-11-1-4) | BR23/BR1185-2B-16-1 | Moderately salt-tolerant *T. Aman* rice variety for rainfed lowland ecosystem | 2003 |
| BRRI dhan47[b] (IR 63307-4B-4-3) | IR51511-B-B-34-B/TCCP266–2-49-B-B-3 | Salt-tolerant *Boro* rice variety for irrigated ecosystem | 2007 |
| BRRI dhan53 (BR5778-156-1-3-HR14) | BR10/ BR23// BR847-76-1-1 | Salt-tolerant *T. Aman* (RLR) rice variety | 2010 |
| BRRI dhan54 (BR5999-82-3-2-HR1) | BR1185-2B-16-1/ BR548–128-1-3 | Salt-tolerant *T. Aman* (RLR) rice variety | 2010 |
| BRRI dhan55 (IR 73678-6-9-B: AS996) | IR64/*Oryza rufipogon* | Moderately salt, cold, and drought tolerant rice variety | 2011 |
| BRRI dhan28-*Saltol*[a] (IR 89573-84) | BRRI dhan28[b]3/ FL478 | Salt-tolerant *Boro* rice line for irrigated ecosystem | MABC product |
| BR11-*Saltol*[a] (IR 89574–7) | BR11[b]3/FL478 | Salt-tolerant *T. Aman* (RLR) rice line | MABC product |
| BRRI dhan61 (BR7105-4R-2) | IR64419-3B-4-3/ BRRI dhan29 | Salt-tolerant *Boro* rice variety/ irrigated rice | 2013 |
| BRRI dhan67 (BR7100-2R-6-6) | IR61247-3B-8-2-1/ BRRI dhan36 | Salt-tolerant *Boro*/irrigated rice variety | 2014 |
| BRRI dhan73 (IR78767-B-SATB1-28-3-24) | BRRI dhan40/ NSIC Rc106 (IR61920-3B-22-1-1) | Salt-tolerant *T. Aman* (RLR) rice variety | 2015 |
| BRRI dhan78 (IR77092-B-2R-B-10) | IR84645/IR84649 | Dual tolerant (salinity+*SUB1T. Aman* rice variety) | 2016 |
| BRRI dhan97 (IR83484-3-B-7-1-1-1) | IRRI 113/BRRI dhan40 | Salt-tolerant irrigated (Boro) rice | 2020 |
| BRRI dhan99 HHZ5-DT20-DT2-DT1 (GSR IR1-5-D20-D2-D1) | Huang-Hua-Zhan/ OM1723 | Salt-tolerant irrigated (Boro) rice | 2020 |
| Binadhan-8 (IR66946-3R-149-1-1) | IR29/Pokkali | Salt-tolerant irrigated (Boro) rice | 2010 |

(continued)

**Table 4.4** (continued)

| Technologies/ products (variety name) | Parentage | Type of technology | Year of release |
|---|---|---|---|
| Binadhan-10 (IR64197-3B-14-2) | IR42598-B-B-B-B-12/Nona Bokra | Salt-tolerant irrigated (Boro) rice | 2012 |

[a]First introgression of *Saltol* locus into the mega varieties (BRRI dhan28 and BR11) and developed two salt-tolerant rice genotypes, IR89573–84 (BRRI dhan28-*Salto*l) and IR89574–7 (BR11-*Saltol*), through marker-assisted breeding at IRRI that are under advanced stage of testing for release in Bangladesh and Myanmar
[b]A large number of salt-tolerant improved genotypes were grown in mother and baby trials of participatory varietal selection (PVS) in coastal areas of Bangladesh. Consequently, BRRI dhan47 (IR 63307-4B-4-3) was finally selected and released as salt-tolerant variety for *Boro* season in Bangladesh

BADC)-private sector (seed traders)/NGOs to farmers-market (seed dealer) to farmers and also farmer to farmer channel.

## 4.3 Conclusion

Out of the 17 salt-tolerant varieties, 12 (10 from BRRI and 2 from BINA) were developed and released through STBN collaboration under the STRASA platform. Novel donors for salinity tolerance were identified, and genetic mechanism of salt tolerance using these donors was reported. The newly identified donors and QTL will be used to develop high yielding salt-tolerant rice varieties through conventional and genome-assisted breeding approaches, respectively. Several major effect and consistent QTLs for salt tolerance have been identified, and there is huge potential to use those QTLs in MAB. Recently, two highly salt-tolerant varieties such as BRRI dhan97 (IR83484-3-B-7-1-1-1) and BRRI dhan99 (GSR IR1-5-D20-D2-D1) have been released in Bangladesh through STBN-STRSA and Green Super Rice collaboration, respectively. Several promising highly salt-tolerant products which can meet multi-stakeholder demand are in the evaluation stage which will be nominated for release as variety in Bangladesh.

## 4.4 Future Outlook

Stacking of new QTLs with previously reported genomic loci will help develop highly salt-tolerant varieties to expand rice cultivation in ~100,000 ha areas affected by the salinity class S4 (strong saline zone with EC:12–16 $dSm^{-1}$) where salt stress is a key problem for rice production during both dry and wet seasons in Bangladesh.

**Acknowledgments** This work was partly funded by the CGIAR Challenge Program on Water and Food (Competitive Project number 7) and the German Federal Ministry for Economic Cooperation and Development (BMZ), DFID, Bill and Melinda Gates Foundation, and government of Bangladesh.

## References

Akbar M, Khush GS, HillerIslambers D (1986) Genetics of salt tolerance in rice. In Rice Genetics I:399–409

Bernardo R (2010) Genomewide selection with minimal crossing in self-pollinated crops. Crop Sci 50(2):624–627

Bimpong IK, Manneh B, Diop B, Ghislain K, Sow A, Amoah NKA, Wopereis M (2014) New quantitative trait loci for enhancing adaptation to salinity in rice from Hasawi, a Saudi landrace into three African cultivars at the reproductive stage. Euphytica 200(1):45–60

Bizimana JB, Luzi-Kihupi A, Murori RW, Singh RK (2017) Identification of quantitative trait loci for salinity tolerance in rice (*Oryza sativa* L.) using IR29/Hasawi mapping population. J Genet 96(4):571–582

Bonilla P, Mackell D, Deal K, Gregorio G (2002) RFLP and SSLP mapping of salinity tolerance genes in chromosome 1 of rice (*Oryza sativa* L.) using recombinant inbred lines. Philipp Agric Sci 85(1):64–74

Ceccarelli S, Grando S (2007) Decentralized-participatory plant breeding: an example of demand driven research. Euphytica 155(3):349–360

Ceccarelli S, Grando S, Tutwiler R, Baha J, Martini AM, Salahieh H, Michael M (2000) A methodological study on participatory barley breeding I. Selection phase. Euphytica 111(2): 91–104

DasGupta S, Hossain M, Huq M, Wheeler D (2015) Climate change and soil salinity: the case of coastal Bangladesh. Ambio 44(8):815–826

Defoer T, Kamara A, De Groote H (1997) Gender and variety selection: farmers' assessment of local maize varieties in southern Mali. Afr Crop Sci J 5(1):65–76

Falconer, D. S., Mackay, T. F. & Frankham, R. (1996). Introduction to quantitative genetics (4th edn). Trends Genet, 12(7), 280

Flowers TJ, Koyama ML, Flowers SA, Sudhakar C, Singh KP, Yeo AR (2000) QTL: their place in engineering tolerance of rice to salinity. J Exp Bot 51(342):99–106

Grattan S, Zeng L, Shannon M, Roberts S (2002) Rice is more sensitive to salinity than previously thought. Calif Agric 56(6):189–198

Gregorio GB, Senadhira D (1993) Genetic analysis of salinity tolerance in rice (*Oryza sativa* L.). Theor Appl Genet 86(2):333–338

Gupta B, Huang B (2014) Mechanism of salinity tolerance in plants: physiological, biochemical, and molecular characterization. Int J Genomics 701596. https://doi.org/10.1155/2014/701596

Haque, T., Elias, S. M., Razzaque, S., Biswas, S., Khan, S. F., Jewel, G. N. A. & Seraj, Z. I. (2020). Natural variation in growth and physiology under salt stress in rice: QTL mapping in a Horkuch× IR29 mapping population at seedling and reproductive stages. bioRxiv. https://doi.org/10.1101/2020.03.01.971895

Hasanuzzaman M, Fujita M, Islam MN, Ahamed KU, Nahar K (2009) Performance of four irrigated rice varieties under different levels of salinity stress. Int J Integr Biol 6(2):85–90

Hoque ABMZ, Haque MA, Sarker MRA, Rahman MA (2015) Marker-assisted introgression of *Saltol* locus into genetic background of BRRI Dhan-49. Int J Biosci 6:71–80

Hossain M, Janaiah A, Husain AMM, Naher F (2002) Rice seed delivery system and seed policy in Bangladesh. CPD-IRRI Policy Brief 1

Hossain H, Rahman MA, Alam MS, Singh RK (2015) Mapping of quantitative trait loci associated with reproductive stage salt tolerance in rice. J Agron Crop Sci 201:17–31

Huda MM, Ferdausi A, Hassan L, Hossain MA (2017) Evaluation of the genetic variability and genetic advance for salt tolerance in rice genotypes. Progress Agric 28(4):262–270

Islam MR, Salam MA, Bhuiyan MAR, Rahman MA, Gregorio GB (2008a) Participatory variety selection for salt tolerant rice. Intl J BioRes 4(3):21–25

Islam MA, Rahman MA, Bhuiyan MAR, Uddin K, Sarker MRA, Yasmeen R, Rahman MS (2008b) BRRI dhan47: a salt tolerant variety for Boro season isolated through participatory variety selection for salt tolerant rice. Intl J Bio Res 5(1):1–6

Islam MR, Gregorio GB, Salam MA, Collard BC, Singh RK, Hassan L (2012) Validation of SalTol linked markersand haplotype diversity on chromosome 1 of rice. Molecular. Plant Breed 29;3(1)

Ismail AM, Horie T (2017) Genomics, physiology, and molecular breeding approaches for improving salt tolerance. Annu Rev Plant Biol 68:405–434

Jahan N, Zhang YU, Lv Y, Song M, Zhao C, Hu H et al (2020) QTL analysis for rice salinity tolerance and fine mapping of a candidate locus qSL7 for shoot length under salt stress. Plant Growth Regul 90(2):307–319

James RA, Blake C, Byrt CS, Munns R (2011) Major genes for $Na^+$ exclusion, *Nax1* and *Nax2* (wheat *HKT1; 4* and *HKT1; 5*), decrease $Na^+$ accumulation in bread wheat leaves under saline and waterlogged conditions. J Exp Bot 62(8):2939–2947

Kabir MS, Salam MU, Chowdhury A, Rahman NMF, Iftekharuddaula KM, Rahman MS et al (2015) Rice vision for Bangladesh: 2050 and beyond. Bangladesh Rice J 19(2):1–18

Khatun S, Flowers TJ (1995) Effects of salinity on seed set in rice. Plant Cell Environ 18(1):61–67

Koyama ML, Levesley A, Koebner RM, Flowers TJ, Yeo AR (2001) Quantitative trait loci for component physiological traits determining salt tolerance in rice. Plant Physiol 125(1):406–422

Krishnamurthy SL, Pundir P, Warraich AS, Rathor S, Lokeshkumar BM, Singh NK, Sharma PC (2020) Introgressed *Saltol* QTL lines improves the salinity tolerance in rice at seedling stage. Front Plant Sci 11:833

Lee SY, Ahn JH, Cha YS, Yun DW, Lee MC, Ko JC, Eun MY (2006) Mapping of quantitative trait loci for salt tolerance at the seedling stage in rice. Mol Cells 21(2):269

Lin HX, Zhu MZ, Yano M, Gao JP, Liang ZW, Su WA, Chao DY (2004) QTLs for $Na^+$ and $K^+$ uptake of the shoots and roots controlling rice salt tolerance. Theor Appl Genet 108(2):253–260

Maniruzzaman S, Rahman MA, Hasan M, Rasul MG, Molla AH, Khatun H, Akter S (2022) Genetic mapping to detect stringent QTLs using 1k-RiCA SNP genotyping platform from the new landrace associated with salt tolerance at the seedling stage in Rice. Plan Theory 11(11):1409

Mekbib F (2006) Farmer and formal breeding of sorghum (*Sorghum bicolor* (L.) Moench) and the implications for integrated plant breeding. Euphytica 152(2):163–176

Mirdar Mansuri R, Shobbar ZS, Babaeian Jelodar N, Ghaffari M, Mohammadi SM, Daryani P (2020) Salt tolerance involved candidate genes in rice: an integrative meta-analysis approach. BMC Plant Biol 20(1):1–14

Mohammadi R, Mendioro MS, Diaz GQ, Gregorio GB, Singh RK (2014) Genetic analysis of salt tolerance at seedling and reproductive stages in rice (*Oryza sativa*). Plant Breed 133(5):548–559

Mondal S, Borromeo TH, Diaz MGQ, Amas J, Rahman MA, Thomson MJ, Gregorio GB (2019) Dissecting *qtl*'s for reproductive stage salinity tolerance in rice from BRRI dhan 47. Plant Breeding Biotechnol 7(4):302–312

Munns R (2005) Genes and salt tolerance: bringing them together. New Phytol 167:645–663

Munns R, Gilliham M (2015) Salinity tolerance of crops–what is the cost? New Phytol 208(3): 668–673

Peng S, Cassman KG, Virmani SS, Sheehy J, Khush GS (1999) Yield potential tRends of tropical rice since the release of IR8 and the challenge of increasing rice yield potential. Crop Sci 39(6): 1552–1559

Quddus MR, Rahman MA, Jahan N, Debsharma SK, Disha RF, Hasan MM et al (2019) Estimating pedigree-based breeding values and stability parameters of elite rice breeding lines for yield under salt stress during the Boro season in Bangladesh. Plant Breeding and Biotechnol 7(3): 257–271

Rahman MA, Bimpong IK, Bizimana JB, Pascual ED, Arceta M, Swamy BP, Singh RK (2017) Mapping QTLs using a novel source of salinity tolerance from Hasawi and their interaction with environments in rice. Rice 10(1):1–17

Rahman MA, Thomson MJ, Shah-E-Alam M, de Ocampo M, Egdane J, Ismail AM (2016) Exploring novel genetic sources of salinity tolerance in rice through molecular and physiological characterization. Ann Bot 117(6):1083–1097

Rahman MA, Khatun H, Sarker MRA, Hossain H, Quddus MR, Iftekharuddaula KM, Kabir MS (2021) Enhancing abiotic stress tolerance to develop climate-smart rice using holistic breeding approach. Cereal Grains 2:91

Rahman MA (2010) Marker-assisted backcrossing of *Saltol*, discovery of additional quantitative trait loci and assessment of allelic variability in *Saltol* in rice germplasm. Ph.D. dissertation. Bangladesh Agricultural University, p 326

Rahman MA, Thant AA, Win M, Tun MS, Moet P, Thu AM, Singh RK (2015) Participatory varietal selection (PVS): a "bottom-up" breeding approach helps rice farmers in the Ayeyarwady Delta, Myanmar. SABRAO J Breed Genet 47(3)

Rahman MA, Quddus MR, Jahan N, Rahman MA, Sarker MRA, Hossain H, Iftekharuddaula KM (2019a) Field rapid generation advance: an effective technique for industrial scale rice breeding program. The Experiment 47(2):2659–2670

Rahman MA, Thomson MJ, De Ocampo M, Egdane JA, Salam MA, Ismail AM (2019b) Assessing trait contribution and mapping novel QTL for salinity tolerance using the Bangladeshi rice landrace capsule. Rice 12(1):1–18

Rahnama A, James RA, Poustini K, Munns R (2010) Stomatal conductance as a screen for osmotic stress tolerance in durum wheat growing in saline soil. FPB 37(3):255–263

Ray PS, Islam MA (2008) Genetic analysis of salinity tolerance in rice. Bangladesh J Agric Res 33(4):519–529

Ren ZH, Gao JP, Li LG, Cai XL, Huang W, Chao DY, Lin HX (2005) A rice quantitative trait locus for salt tolerance encodes a sodium transporter. Nat Genet 37(10):1141–1146

Ribaut JM, Hoisington D (1998) Marker-assisted selection: new tools and strategies. Trends Plant Sci 3(6):236–239

Rozema J, Flowers T (2008) Crops for a salinized world. Science 322(5907):1478–1480

Salam MA, Rahman MAR, Bhuiyan K, Uddin MRA, Sarker R, Yasmeen MS (2007) BRRI dhan 47: a salt-tolerant variety for the boro season. International Rice Research Notes (IRRN) 32.1/ 2007; page 42

Singh RK, Flowers TJ (2010) 36 the physiology and molecular biology of the effects of salinity on Rice. In: Pessarakli M (ed) Third edition of "handbook of plant and crop stress". Taylor and Francis, Florida, pp 899–939

Singh RK, Gregorio GB, Jain RK (2007) QTL mapping for salinity tolerance. Physiol Mol Biol Plants 13:87–99

Souleymane O, Salifou M, Hamidou M, Manneh B, Danquah E, Ofori K (2017) Genes action in salinity tolerance and the implication in rice breeding. J Plant Breeding Genetics 5(3):115–120

Sun BR, Fu CY, Fan ZL, Chen Y, Chen WF, Zhang J, Li C (2019) Genomic and transcriptomic analysis reveal molecular basis of salinity tolerance in a novel strong salt-tolerant rice landrace Changmaogu. Rice 12(1):1–15

Tanksley SD (1993) Mapping polygenes. Annu Rev Genet 27(1):205–233

Thapa DB, Sharma RC, Mudwari A, Ortiz-Ferrara G, Sharma S, Basnet RK, Joshi KD (2009) Identifying superior wheat cultivars in participatory research on resource poor farms. Field Crop Res 112(2):124–130

Thomson MJ, de Ocampo M, Egdane J, Rahman MA, Sajise AG, Adorada DL, Ismail AM (2010) Characterizing the *Saltol* quantitative trait locus for salinity tolerance in rice. Rice 3(2):148–160

Tuberosa R, Salvi S (2006) Genomics-based approaches to improve drought tolerance of crops. Trends Plant Sci 11(8):405–412

Tulachan PM, Bashar MK, Islam MR, Chanda D, Portal C (2012) Value chain analysis of rice seeds in Bangladesh Los Baños. International Rice Research Institute, Philippines, p 46

Vom Brocke K, Trouche G, Weltzien E, Barro-Kondombo CP, Gozé E, Chantereau J (2010) Participatory variety development for sorghum in Burkina Faso: farmers' selection and farmers' criteria. Field Crop Res 119(1):183–194

Weltzien E, Whitaker ML, Rattunde HFW, Dhamotharan M, Anders MM (1998) Participatory approaches in pearl millet breeding. In: Witcombe JR, Virk DS, Farrington J (eds) Seeds of choice: making the most of new varieties for small farmers. Oxford and IBH Publishing Co. Pvt. Ltd., Delhi, pp 143–170

Weltzien E, Vom Brocke K, Rattunde F (2005) Planning plant breeding activities with farmers. In: Christinck A, Weltzien E, Hoffmann V (eds) Setting breeding objectives and developing seed systems with farmers, A handbook for practical use in participatory plant breeding projects. Margraf Publishers GmBH Scientific Books, Weikersheim, Germany, pp 123–152

Weltzien E, Kanouté M, Toure A, Rattunde F, Diallo B, Sissoko I, Siart S (2008) Participatory identification of superior sorghum varieties using multi-locational trials in two zones in Mali. Cah Agric 17(1):134–139

Witcombe JR, Joshi KD, Gyawali S, Musa AM, Johansen C, Virk DS, Sthapit BR (2006) Participatory plant breeding is better described as highly client-oriented plant breeding. I. Four indicators of client-orientation in plant breeding. Exp Agric 41(3):299–319

Yadav S, Sandhu N, Dixit S, Singh VK, Catolos M, Mazumder RR, Kumar A (2021) Genomics-assisted breeding for successful development of multiple-stress-tolerant, climate-smart rice for southern and southeastern Asia. Plant Genome 14(1):e20074

Ye G, Collard BCY, Zhao XQ, Nissila E (2013) Enhancing rice breeding efficiency: the role of breeding. Sabrao J Breed Genet 45(1):143–158

Zou J, Liu A, Chen X, Zhou X, Gao G, Wang W, Zhang X (2009) Expression analysis of nine rice heat shock protein genes under abiotic stresses and ABA treatment. J Plant Physiol 166(8): 851–861

Zou J, Liu C, Liu A, Zou D, Chen X (2012) Overexpression of OsHsp17. 0 and OsHsp23. 7 enhances drought and salt tolerance in rice. J Plant Physiol 169(6):628–635

# Genetic Improvement of Rice for Inland Saline and Alkaline Areas of North-Western India

# 5

S. L. Krishnamurthy, Suman Rathor, B. M. Lokeshkumar, A. S. Warriach, N. M. Vinaykumar, Parbodh C. Sharma, R. K. Gautam, and R. K. Singh

**Abstract**

Rice is one of the leading crops in salt affected region and being affected by several stresses such as salinity and sodicity during crop growth stages is reducing economic yield. Rice sensitive to salt stress at two stages namely seedling and reproductive stage. However, genetic variation is more in rice for salt stress hence, we can identify the new donor and lines to enhance the salt tolerance at both the stages. We identified 241 tolerant lines among 7502 rice lines tested at seedling stage. We developed many mapping populations for salt tolerance at reproductive stage are being used for identification of QTLs. We pyramided the Saltol (Seedling stage) and qSSIFSH8 (spikelet fertility) QTLs into high yielding varieties namely, Pusa44, PR114 and Sarjoo52. We developed 13 salt tolerant rice varieties and distributed to private companies, farmers and etc. We produced 1056 quintals and 3126 quintals of breeder and truthfully labelled seeds of salt tolerant rice varieties during last 22 years (2000–2021). These salt tolerant rice varieties covered area of app. 1.02 m ha covered every year. Another major notable impact of using salt tolerant rice varieties is bringing back the

S. L. Krishnamurthy (✉) · S. Rathor · B. M. Lokeshkumar · A. S. Warriach · P. C. Sharma
ICAR-Central Soil Salinity Research Institute, Karnal, India

N. M. Vinaykumar
Kuvempu University, Shankaraghatta, Shimogga, India

R. K. Gautam
ICAR-Central Soil Salinity Research Institute, Karnal, India

ICAR- National Bureau of Plant Genetic Resources, New Delhi, India

R. K. Singh
ICAR-Central Soil Salinity Research Institute, Karnal, India

International Center for Biosaline Agriculture, Dubai, United Arab Emirates


R. K. Singh et al. (eds.), *Genetic Improvement of Rice for Salt Tolerance*,
https://doi.org/10.1007/978-981-99-3830-8_5

unproductive barren or salt affected lands into the fold of productive normal lands. Therefore, faster conversion of degraded lands into normal ones with the use of resilient varieties can ensure more area and thus increased food productivity from these reclaimed lands.

**Keywords**

Rice · Salinity · Sodicity · QTLs · Genetic improvement

## 5.1 Introduction

The presence of salts in soil and water poses a serious constraint to crop productivity especially in those areas where irrigation is essentially required. Agricultural salinity affects about 900 million hectares of land globally and therefore poses a formidable task of taking up agriculture and enhancing productivity in these areas. About 400 million ha of land in South and Southeast Asia are covered by problematic soils where rice is the staple crop (FAO 2013). In India, studies on mapping of salt affected areas indicate that about 6.73 m ha land area is salt affected out of which 3.77 and 2.96 m ha are affected, respectively, by sodic and saline soils, respectively. In Haryana, a total of 0.23 million ha (mha) of land is affected by salts, out of which 0.049 mha of land is saline and 0.18 mha of land is sodic. Moreover, groundwater in almost all the districts of Haryana is facing the problem of salinity and residual sodium carbonates (RSC). Harnessing the potential of such salt affected areas can play a significant role in increasing and sustaining our food security. Moreover, these disadvantageous agroecosystems are often inhabited by resource-poor inhabitants, and any improvement in the agricultural productivity is directly linked with poverty alleviation.

## 5.2 Historical Perspective

The Central Soil Salinity Research Institute (CSSRI) is a premier research institute dedicated to pursue interdisciplinary research on salinity/alkalinity management and use of poor quality irrigation waters in different agroecological zones of the country. The government of India constituted an Indo-American team to assist the Indian Council of Agricultural Research to develop a comprehensive water management program for the country. As a follow-up of these recommendations, it was decided to establish the Central Soil Salinity Research Institute under Fourth Plan period. The institute started functioning at Hisar (Haryana) on March 1, 1969. Later on, it was decided to shift this Institute to Karnal during October 1969.

The Central Soil Salinity Research Institute is situated in Karnal, Haryana, and is 125 km north of Delhi. The institute campus is located in northwest part of Karnal City on Kachhwa Road on the right bank of the Western Yamuna Canal. The head office is situated at coordinates of 29° 43′ N latitude, 76° 58′ E longitude, and

245 meter elevation above the mean sea level. The climate of Karnal is semiarid with hot summer and cold winter. Nearly 80% of total rainfall occurs during the monsoon (only up to September) with a few showers in the winter.

## 5.3 Salient Research Work Done

### 5.3.1 Collection of Exotic and Indigenous Germplasm in Rice

Right from its inception, CSSRI Karnal collected and evaluated app. 7500 germplasm lines consisting of introductions, landraces, and improved type material of rice developed in India and at International Rice Research Institute, Philippines. The material was screened for salt tolerance through diverse sodicity and salinity conditions in Karnal, Gudha, Mundalana, Sarswati, Kaithal, and Nain farms as well as state of the art precisely controlled sodic and saline microplot facility at CSSRI Karnal. This massive work was also facilitated by different national and international programs from time to time. This resulted in the development and release of 13 salt tolerant rice varieties mostly suited to sodic and inland saline soils of the country. We have identified the 90 elite rice genotypes which are tolerant to salinity and sodicity including 13 salt tolerant rice varieties and 4 genetic stocks.

### 5.3.2 Creation of Variability in Rice

We developed more cross combinations by involving diverse parents' viz. exotic lines, landraces, genetic stock and elite lines with respect to salt tolerance, yield, quality, and pest and disease resistance in rice. We also developed thousands of cross combinations and evaluated different segregating populations in salt stress conditions. The best among the segregating populations in each cross combinations were selected and carried forward for next generations.

### 5.3.3 Efficient Screening and Selection Strategies in Precisely Controlled Field Conditions and Microplot Setup

#### 5.3.3.1 Microplots

Experiments on rice in the precisely controlled microplots filled with soils of desired levels of salinity and sodicity were conducted. Use of this system enables precision, control, and monitoring of stress levels enabling more accurate and efficient genotypic evaluation besides minimizing problems caused by soil heterogeneity. We have a unique and extensive setup of microplots and lysimeters which is well monitored and characterized for natural saline and sodic fields, providing the most reliable and actual field conditions for the selection of breeding materials. We have made extensive use of these facilities for efficient screening of large germplasm and populations besides conducting basic and fundamental studies.

A total of 225 recombinant inbred lines (RILs) derived from CSR 27/MI 48 was carried out in 3 environments {normal salinity, moderate salinity (EC ~ 6), and high salinity (EC ~ 10) conditions} for 3 years for agro-morphological and biochemical traits. This mapping population was used to map QTL for salt ion concentrations in different tissues and salt stress susceptibility index (SSI) for spikelet fertility, grain weight, and grain yield. Eight significant QTL intervals were mapped on chromosomes 1, 8, and 12 for the salt ion concentrations, and a QTL controlling SSI for spikelet fertility was colocated in one of these intervals on chromosome 8 (Pandit et al. 2010).

A systematic phenotyping of 225 recombinant inbred lines (RILs) derived from CSR 11/MI 48 was carried out in 3 environments {normal sodic, moderate sodic ($pH_2$ ~ 9.5), and high sodic ($pH_2$ ~ 9.9) conditions} for 3 years for agro-morphological traits. Three season data were pooled for further analysis and grouped the RILs into different groups based on the grain yield SSI in moderate and high sodic stress conditions. A total of 25, 103, and 97 RILs were clustered into low, medium, and high SSI, respectively, in moderate sodic stress condition. Similarly, 17, 103, and 105 RILs were clustered into low, medium, and high SSI, respectively, in high sodic stress condition (Tiwari et al. 2016). DNA was extracted from each of the extreme RILs and combined in equal quantities to prepare extreme tolerant and extreme sensitive bulks and analyzed along with the two parents using all the polymorphic markers. A total of five regions on chromosomes 1, 2, 8, 9, and 10 showing tight linkage with the trait were identified. Two of these locations match with QTLs previously mapped on chromosomes 1 and 8. Genetic map of the three novel QTLs on chromosomes 2, 9, and 10 has been prepared to identify their precise location and effect (Tiwari et al. 2016).

A total of 180 diverse rice accessions were evaluated for salinity tolerance at reproductive stage for morphological and physiological parameters. We recognized 28 significant marker-trait associations with 19 associations for $Na^+$, $K^+$, and $Na^+/K^+$ uptake in stem and leaves. The phenotypic variance for these associations ranged from 5.12 to 13.37%.

We evaluated 140 $F_2$ populations of PS5/CSR10 for salinity tolerance at reproductive stage in rice for 18 different morphological, biochemical, and physiological traits. A total of 39 significant QTLs are identified along with QTL*qGY-2* for grain yield on chromosome 2 (Ammar et al. 2009). Several QTLs responsible for salinity tolerance were also reported (Mohammadi et al. 2013; Bimpong et al. 2014; Hossain et al. 2015; Kumar et al. 2015; Pundir et al. 2021; Warraich et al. 2020; Warraich et al. 2021; Mazumder et al. 2020).

#### 5.3.3.2 Glasshouse and Pots

We screened 7502 rice germplasms for seedling stage salinity tolerance in hydroponic solution and identified 241 high tolerant (salt injury score 3) rice accessions (Table 5.1). These are being used in breeding program (Krishnamurthy et al. 2014; Krishnamurthy et al., 2015; Krishnamurthy et al. 2016a, d, e; Chowdhury et al. 2016; Babu et al. 2017a, 2017b; Ravikiran et al. 2018).

**Table 5.1** List of rice germplasm evaluated for salinity tolerance at seedling stage

| Years | Evaluated rice germplasm | Tolerant to salinity |
|---|---|---|
| 2012–2013 | 724 | 38 |
| 2013–2014 | 400 | 0 |
| 2014–2015 | 658 | 6 |
| 2015–2016 | 1000 | 6 |
| 2016–2017 | 1202 | 23 |
| 2017–2018 | 1000 | 27 |
| 2018–2019 | 1000 | 83 |
| 2019–2020 | 1022 | 10 |
| 2020–2021 | 496 | 48 |
| **Total** | **7502** | **241** |

#### 5.3.3.3 Transgenic Glass House

We screened 1000 transgenic progenies of IR 64 for salinity and sodicity tolerance in the controlled transgenic glass house facility and selected the best two transgenic progenies for salinity and sodicity tolerance (Joshi et al. 2020). The serine hydroxymethyltransferase 3 (SHMT3) gene was identified in CSR 27 and validated for salinity tolerance (Mishra et al. 2019).

### 5.3.4 Categorizing the Accessions and Varieties Using Threshold Stress Level, Slope, and Tolerance Indices

Salt tolerance curves for salinity and sodicity stresses for rice were developed. This information was quiet useful in planning screening of germplasm for salt stress in different crops. Stress tolerance index (STI) was found to be a better selection criterion than stress susceptibility index (SSI) for grain yield (Ali et al. 2013; Krishnamurthy et al. 2016b). SSI is based on the yield minimization under stress compared to nonstress environment and reflects the stress intensity (SI) as well (Fischer and Maurer 1978):

$$\mathrm{SSI} = 1 - (Ys/Yp)/SI$$

where *Xs* is the mean stress yield over genotypes and *Xp* is the mean nonstress yield over genotypes.

STI is for identification of the genotypes with higher yield and stress tolerance potentials. It considers the potential yield under nonstress, yield under stress condition, and stress intensity (Fernandez 1993).

$$\mathrm{STI} = Yp \times Ys/(\mathrm{Xp})^2$$

where *Yp* and *Ys* are the respective grain yield of a genotype under nonstress and stress conditions, respectively, and *Xp* is the mean yield of all genotypes under nonstress conditions.

### 5.3.5 Selection under Different Stress Levels and Situations

Selection of the genetic materials at different levels of stress under different situations is important so as to select genotypes with wider adaptation to salt stress. Accordingly, selection of genotypes was made at pH from 8.0 to pH 10.0, and different levels of salinity were created and maintained under microplot, and field plots were earmarked for evaluation.

### 5.3.6 Involvement of Physiological Parameters and Element/Microelement Toxicities

A number of physiological parameters are required to select for salt tolerance in rice. We identified different rice genotypes with different tolerance mechanisms like $Na^+$ excluder, tissue tolerant to $Na^+$, $K^+$ miner and low Na/K uptake. Low Na/K uptake is one of the most important parameters for salt tolerance. Physiological mechanisms influencing tolerance include various attributes like Na exclusion, tissue tolerance, low Na/K ratio, and efficient salt partitioning ability within plant to retain harmful salts in functionally less active organs like vacuoles and older plant parts. Negative effects of salts are also offset by higher uptake of K, Zn, and P. Since earlier results showed that no single mechanism could confer the absolute tolerance, pyramiding genes for diverse physiological mechanisms into one genetic background has been initiated, and good breeding populations have been generated. A wide spectrum of germplasm in rice has been evaluated and categorized based on tissue tolerance, $Na^+$ exclusion, $K^+$ uptake, $Na^+$/ $K^+$ ratio, and reproductive stage tolerance for using these as donors for augmenting breeding program (Table 5.2).

### 5.3.7 Selection Based upon Multiple Salt Tolerance Indices

We identified different traits and stress indices which are strongly associated with yield in salinity and sodicity environments. Breeder can use those traits for indirect selection to improve the yield along with other traits in stress environments. We estimated correlations of SSI and STI with morphological traits, it was found that STI could develop significant correlation with many morphological traits in all the three environments than SSI, thus revealing the higher utility of STI in indirect selection. Therefore, STI can be used as an effective selection criterion to improve grain yield and other plant traits (Krishnamurthy et al. 2014; Krishnamurthy et al. 2016b).

### 5.3.8 Knowledge of Yield Components Correlated with Tolerance and Yield

Systematic work was carried out to find out yield components correlated with tolerance in rice with different salt stress situations. Differential expression of traits

**Table 5.2** Salt tolerance mechanism based grouping of rice genotypes

| Group1 | Group 2 | Group 3 | Group 4 | Group 5 | Group 6 |
|---|---|---|---|---|---|
| Tissue tolerance to $Na^+$ | $Na^+$ excluder | $K^+$ miners | Low cl Uptake | Low $Na^+/K^+$ ratio | Early vigor |
| SR26B, CSR21, CSR26, CSR27, Pokkali, IR4630–22–2-1-3, TCCP266–2–49-2B-3, GR11 | CSR1, PAC-831, CSR8, CSR10, CSR13, CSR19, CSR20, CSR22, CSR23, CSR24, CSR25 | Hathwan, Swarndhan, Achhi, CSR23, IR36, IR4630-20-22-2–5–1-3, CSR21, GR11, SR26B, CSR1, CSR10, CSR11, CSR20, Panvel23 | HKR128, Jaya, CSR10, PR108, CSR19 | PR108, Achhi, CSR1, CSR19, ADT36, IR4630–22–2-5-1-3, CSR18, SLR1214, CSR10, IR42 | CSR-92–5, 89-H1–1-3, 91-H2–6-B-2, 92-H5–4, KR1–24, CSR11, CSR1, CSR27 |

under normal and salt stress would be based on the salt concentration in soil. Grain yield per plant showed positive significant association with plant height, total tillers, productive tillers, panicle length, and biological yield per plant and harvest index under normal environment, whereas grain yield showed positive significant association with biological yield and harvest index under salinity stress (Krishnamurthy et al. 2016b). Hence, selection would be entirely different under normal and saline environments. Efforts were made to determine genetic variance for desired characteristics (Chaudhary et al. 2021). Information on heritability, variability, and nature of gene action of different yield attributes were also taken into account for planning breeding program and selection strategies.

### 5.3.9 Use of Stability Parameters

We assessed the stability of rice genotypes across five coastal saline and seven alkaline locations during 2011–2013. Genotypes CSR 2 K 262, CSR27, NDRK11–4, NDRK11–3, NDRK11–2, CSR 2 K 255, and PNL 1–1–1-6-7-1 were identified as specifically adapted to favorable locations.

The genotype CSR 2 K 262 has outyielded in both coastal saline and alkaline locations over national checks (Krishnamurthy, et al., 2016c; Krishnamurthy et al. 2017; Krishnamurthy et al. 2021a; Krishnamurthy et al., 2021c). Similarly, many rice lines received from different NARES partners and IRRI Philippines were evaluated in across nine coastal saline and nine alkaline locations during 2014–2015 and identified IR 87938-1-1-2-1-3-B and IR 87938-1-2-2-1-3-B genotypes with 60–80% yield advantage across salt stress locations. These could be used for commercial cultivation in specific environments of salt affected soils in India.

### 5.3.10 Development of Salt Tolerant Varieties, Genetic Stocks, and Elite Breeding Lines

Many elite rice genotypes were identified at CSSRI after thorough systematic phenotyping at microplot and field conditions (Table 5.3). Presently, ICAR CSSRI Karnal registered four salt tolerant genetic stocks, namely, CSR47 (Krishnamurthy et al. 2019a), CSR51 (Krishnamurthy et al. 2020a), CSR53 (Krishnamurthy et al. 2019a), and CSR59 (Krishnamurthy et al. 2021a).

Many projects on salt tolerance in rice, functional genomics in rice, establishment of data in rice, and transfer of major *Saltol* QTL (Gregorio 1997; Krishnamurthy et al. 2015; Krishnamurthy et al. 2020a) to high yielding rice varieties were carried out at CSSRI with external fund (Table 5.4).

## 5.4 Infrastructure and Facilities Developed at CSSRI (Pot Houses, Green Houses Including Transgenic Glasshouse, Microplots, and Fields)

### 5.4.1 Field Evaluation

This is the real site for validation of crop germplasm or varieties for salt tolerance for making practical impact. However, it suffers from inherent problem of uneven distribution of salts in natural field conditions resulting in spatial variability to adversely affect the screening efficiency and reliability. The original screening work was started in the naturally occurring barren fields, which remained uncultivated for an unknown period. That is why not much success could be achieved in this endeavor in the past. Considering this problem, modifications in the layout plans, designs of experiments, and replicated experiments along with other screening aids enhanced the screening efficiency. For field experiments, it is desirable to have two to five long rows of 10–30 m with space planting for each genotype to minimize the effect of spatial variability. For lesser number of genotypes (<30), randomized block design with at least three replications is ideal. For more number of genotypes, we may resort to incomplete block design or augmented design in which a set of check varieties is repeated many times without repeating test entries. The initial soil sampling for stress monitoring within a field allows the "local control" which is a prerequisite for any experiment. Accordingly within replication, stress variation should be minimum compared to between replications.

**Table 5.3** List of elite breeding lines developed for salt tolerance at CSSRI, Karnal

| SL no. | Variety/ elite lines | Parentage | $pH_2$ | EC (dS/m) |
|---|---|---|---|---|
| 1 | CSR 1 | Local selection | 9.8–10.2 | 6–11 |
| 2 | CSR 2 | Local selection | 9.8–10.2 | 6–11 |
| 3 | CSR 3 | Local selection | 9.8–10.2 | 6–11 |
| 4 | CSR 5 | TKM 6/IR8 | 9.0–9.5 | < 6.0 |
| 5 | CSR 8 | CSR 1 mutant | 9.0–9.6 | < 7.0 |
| 6 | CSR 9 | CSR 1/bas. 370//CSR 5 | 9.7–9.9 | < 9.0 |
| 7 | CSR 10 | M40-431-24-114/Jaya | 9.8–10.2 | 6–11 |
| 8 | CSR 11 | M40-431-24-114/Bas.370 | 9.8–10.2 | 6–11 |
| 9 | CSR 12 | CSR 1/Bas.370//CSR 5 | 9.2–9.8 | < 7.0 |
| 10 | CSR 13 | CSR 1/Bas.370//CSR 5 | 9.2–9.9 | < 9.0 |
| 11 | CSR 14 | Milyang 23/Milyang 30 | 9.2–9.8 | < 8.0 |
| 12 | CSR 15 | Sipi 661,044/Sipi651020 | 9.2–9.8 | < 8.0 |
| 13 | CSR 16 | CSR 1 Mutant | 9.2–9.8 | < 8.0 |
| 14 | CSR 17 | IR19661-131-1-2/IR9129-209-2-2-2-1 | 9.2–9.8 | < 8.0 |
| 15 | CSR 18 | RPA 5829/CSR 5 | 9.2–9.8 | < 8.0 |
| 16 | CSR 19 | CSR 1/Bas. 370 | 9.2–9.8 | < 8.0 |
| 17 | CSR 20 | CSR 5/Palaman 579 | 9.2–9.8 | < 8.0 |
| 18 | CSR 21 | IR5657-33-2/IR4630-22-2-5-1-3 | 9.8–10.0 | < 9 |
| 19 | CSR 22 | IR64//IR4630-22-2-5-1-3/ IR9764-45-2-2 | 9.6–9.9 | < 10 |
| 20 | CSR 23 | IR64//IR4630-22-2-5-1-3/IR9764-45-2-2 | 9.8–10.0 | < 10 |
| 21 | CSR 24 | IR58/CHETTIVIRUPPU | 9.6–9.9 | < 10 |
| 22 | CSR 25 | IR17494-32-3-1-1-3/IR4432-52-6-4 | 9.8–10.0 | < 10 |
| 23 | CSR 26 | NONA BOKRA/IR5657-33-2 | 9.8–10.0 | < 9 |
| 24 | CSR 27 | NONA BOKRA/IR5657-33-2 | 9.6–9.9 | < 10 |
| 25 | CSR 28 | IR42/IR4630–22–2-5-1-3 | 9.8–10.0 | < 10 |
| 26 | CSR 29 | IR14632-22-3/IR19799-17-03-01-2001 | < 10 | < 10 |
| 27 | Basmati CSR 30 | BR4-10/Basmati 386 | < 9.5 | < 6 |
| 28 | CSR 31 | Somaclonal variant of Pokkali (TCCP266-2-49-2B-3) | 9.6–9.8 | <7 |
| 29 | CSR 32 | KDM105/IR4630–22–2-5-1-3// IR20925-33-3-1-1-28 | 9.6–9.8 | <7 |
| 30 | CSR33 | IR64//IR4630-22-2-5-1-3/IR9764-45-2-2 | 9.6–9.8 | <7 |
| 31 | CSR34 | Mahsuri//IR11418-19-2-3 | 9.6–9.8 | <7 |
| 32 | CSR35 | CSR18/Panvel 2 | 9.6–9.8 | <7 |
| 33 | CSR36 | CSR13/Panvel 2//IR36 | 9.8–10.0 | < 10 |
| 34 | CSR37 | CSR11/Bas. 370 | 9.6–9.8 | <7 |
| 35 | CSR38 | Pusa 33/CSR13//CSR12 | 9.6–9.8 | <7 |
| 36 | CSR39 | IR10206-29-2-1/IR29337-36-3 | 9.6–9.8 | <7 |
| 37 | CSR40 | TCCP6533-8-4-2-AC204-2/IR10206-29-2-1 | 9.6–9.8 | <7 |
| 38 | CSR41 | IR9884-54-3-1E-P1/ KuatikPutih | 9.6–9.8 | <7 |
| 39 | CSR42 | IR37255-21-3-3-2/TCCP266-2-49-2B-3 | 9.6–9.8 | <7 |

(continued)

**Table 5.3** (continued)

| SL no. | Variety/ elite lines | Parentage | $pH_2$ | EC (dS/m) |
|---|---|---|---|---|
| 40 | CSR43 | KDML 105 / IR 4630-22-2-5-1-3// IR 20925- 33-3-1-28 | 9.6–9.8 | <7 |
| 41 | CSR44 | KDML 105 / IR 4630-22-2-5-1-3// IR 20925- 33-3-1-28 | 9.6–9.8 | <7 |
| 42 | CSR45 | IR2070-245-3-15-6/IR944-102-2-3-2 | 9.6–9.8 | <7 |
| 43 | CSR 46 | IR72 / CSR 23 | 9.6–9.9 | <10 |
| 44 | CSR 47 | Jaya / CSR 13 | 9.6–9.9 | <10 |
| 45 | CSR 48 | Jaya / CSR 23 | 9.6–9.8 | <9 |
| 46 | CSR 49 | IR5711-3B-7-B-9-1-8/NDRK 5032 | 9.6–9.8 | <7 |
| 47 | CSR 50 | *O perrenis*/ CSR21 | 9.6–9.8 | <7 |
| 48 | CSR 51 | CSR23/CSR27 | 9.8–10.0 | < 10 |
| 49 | CSR52 | CSR23/CSR27 | 9.8–10.0 | < 10 |
| 50 | CSR53 | CSR23/CSR27 | 9.8–10.0 | < 10 |
| 51 | CSR54 | CSR10/CSR27 | 9.8–10.0 | < 10 |
| 52 | CSR 55 | CSR23/CSR27 | 9.8–10.0 | < 10 |
| 53 | CSR 56 | CSR21/CSR10 | 9.8–10.0 | < 10 |
| 54 | CSR 57 | CSR21/CSR10 | 9.8–10.0 | < 10 |
| 55 | CSR 58 | CSR27/CSR11 | 9.8–10.0 | < 10 |
| 56 | CSR 59 | CSR27/CSR11 | 9.8–10.0 | < 10 |
| 57 | CSR 60 | IR 4630–22–2-5-1-3/IR05N204 | 9.8–10.0 | < 10 |
| 58 | CSR 61 | CSR11/MI48 | 9.8–10.0 | < 10 |
| 59 | CSR62 | CSR 2013-IR42-25 | 9.6–9.8 | <7 |
| 60 | CSR63 | CSR 2013-IR 42-20 | 9.6–9.8 | <7 |
| 61 | CSR64 | TRY1/PB1 | 9.6–9.8 | <7 |
| 62 | CSR65 | TRY1/PB1 | 9.6–9.8 | <7 |
| 63 | CSR66 | CSR27/CSR30 | 9.6–9.8 | <7 |
| 64 | CSR67 | CSR23/CSR27 | 9.6–9.8 | <7 |
| 65 | CSR68 | CSR23/CSR27 | 9.6–9.8 | <7 |
| 66 | CSR69 | Pusa44/FL478//Pusa44*3 | 9.5–9.6 | <7 |
| 67 | CSR70 | Pusa44/FL478//Pusa44*3 | 9.5–9.6 | <7 |
| 68 | CSR71 | Pusa44/FL478//Pusa44*3 | 9.5–9.6 | <7 |
| 69 | CSR72 | Sajoo52/ FL478// Sajoo52*3 | 9.5–9.6 | <7 |
| 70 | CSR73 | Sajoo52/ FL478// Sajoo52*3 | 9.5–9.6 | <7 |
| 71 | CSR74 | CSR27/BPT5204 | 9.5–9.6 | <7 |
| 72 | CSR75 | CSR27/BPT5204 | 9.5–9.6 | <7 |
| 73 | CSR76 | CSR27/MI48 | 9.7–9.8 | <7 |
| 74 | CSR77 | CSR27/MI48 | 9.7–9.8 | <7 |
| 75 | CSR78 | IR4630/CSR11 | 9.7–9.8 | <7 |
| 76 | CSR79 | IR4630/CSR11 | 9.7–9.8 | <7 |
| 77 | CSR80 | IR10F548/IR 55179-3B-11–3 | 9.7–9.8 | <7 |

(continued)

**Table 5.3** (continued)

| SL no. | Variety/ elite lines | Parentage | $pH_2$ | EC (dS/m) |
|---|---|---|---|---|
| 78 | CSR81 | Selection in basmati CSR 30 | 9.7–9.8 | <7 |
| 79 | CSR82 | CHERIVIRUPPU/IR 10205–37–1-3 | 9.7–9.8 | <7 |
| 80 | CSR83 | CSR30/CSR36 | 9.5–9.6 | <6 |
| 81 | CSR84 | CSR30/CSR36 | 9.5–9.6 | <6 |
| 82 | CSR85 | CSR11/CSR27 | 9.7–9.8 | <7 |
| 83 | CSR86 | CSR30/CSR36 | 9.5–9.6 | <6 |
| 84 | CSR87 | IR 5 / IR 52713-2B-8-2B-1-2 | 9.7–9.8 | <7 |
| 85 | CSR88 | CSR30/CSR36 | 9.5–9.6 | <6 |
| 86 | CSR89 | Local selection of basmati CSR30 | 9.7–9.8 | <7 |
| 87 | CSR90 | CSR30/CSR23 | 9.5–9.6 | <6 |
| 88 | CSR91 | CSR30/CSR23 | 9.5–9.6 | <6 |
| 89 | CSR92 | Fedearroz 50 / SHZ-2 // PSBRc 82 / PSBRc 158 //// IR 77298-14-1-2-10 / IR 4630-22-2-5-1-3 // IR45427-2B-2-2B-1-1 / Samba Mahsuri + Sub 1 | 9.5–9.6 | <6 |
| 90 | CSR93 | IR 71730-51-2/NSIC Rc106 | 9.5–9.6 | <6 |
| 91 | CSR94 | Fedearroz 50 / SHZ-2 // PSBRc 82 / PSBRc 158 //// IR 77298-14-1-2-10 / IR 4630-22-2-5-1-3 // IR45427-2B-2-2B-1-1 / Samba Mahsuri + Sub 1 | 9.5–9.6 | <6 |

### 5.4.2 Screening in Microplots/Lysimeters

To obviate the problems of soil heterogeneity and spatial variability, a state of the art facility of mini-salt fields called as microplots or lysimeters was created at CSSRI, Karnal (Fig. 5.1). These are having the desired and varying levels of controlled salinity and sodicity environments. It is possible to create and maintain desired levels of sodicity and salinity in these microplots simulating field conditions minus the soil heterogeneity. The plot size in these plots is kept small because of a good "local control" over microenvironment to accommodate more number of genotypes. This constitutes the landmark facility for salinity research and crop improvement. Currently, the breeding programs at CSSRI, Karnal, are actively using this facility in rice, wheat, and mustard improvement for salt tolerance. However, large populations particularly segregating or large number of lines cannot be screened due to space limitation, though the most promising material should be preferred under this facility. These facilities also form excellent plate form for phenotyping the stabilized mapping populations for molecular tagging of genes controlling sodicity/salinity tolerance.

**Table 5.4** List of projects carried out at CSSRI Karnal

| Sl. no. | Name of project | Funded by | Project duration |
|---|---|---|---|
| *Institute funded projects* | | | |
| 1 | NRMACSSRISIL202001801009. Genetic improvement of rice genotypes for salt tolerance using conventional and molecular breeding methods | Institute | July 2020 June 2025 |
| 2 | Development of better rice genotypes with tolerance to coastal salinity | Institute CSSRI | June 2021–May 2025 |
| 3 | NRMACSSRISIL201501100887. Development of rice genotypes for salt tolerance in rice: Conventional and molecular approaches | Institute | July 2015–June 2020 |
| 4 | Development of salt tolerant rice (*Oryza sativa* L.) genotypes for coastal salinity | Institute | July 2018–June 2021 |
| 5 | P1–2009/CIS4.6-ISR-30/0150. Genetic improvement of salt tolerance in rice | Institute | July 2010–June 2015 |
| 6 | NRMACSSRISIL201300600851. Physiological and biochemical basis of salinity and drought stresses in rice and wheat | Institute | June 2013–May 2018 |
| *External funded projects* | | | |
| 1 | Introgression of blight and blast resistance genes in salt tolerant basmati CSR30 rice variety through marker assisted backcross breeding (MABB). **SERB, DST** | SERBDST, New Delhi | March 2021–March 2024 |
| 2 | NRMACSSRICOL202000400995. Trait discovery-rainfed ecosystem | IRRI, Philippines | October 2019-March 2022 |
| 3 | NRMACSSRICOL202000500996. Rice grain quality and nutritional quality: Low arsenic rice, bioactives in rice and low GI rice | IRRI, Philippines | September 2019-March 2022 |
| 4 | NRMACSSRICOL202000600997. Climate smart management practices. Sub project 2.1 and 2.4 | IRRI, Philippines | September 2019-March 2022 |
| 5 | NRMACSSRICOL202000700998. Climate smart management practices. Sub project 2.4 | IRRI, Philippines | September 2019-March 2022 |
| 6 | From QTL to variety: Genomics assisted introgression and field evaluation of rice varieties with genes/QTLs for yield under drought, flood, and salt stress | DBT, new Delhi | March 2018–March 2022 |
| 7 | Development of high zinc rice varieties | IRRI, Philippines | April 2017–March 2022 |
| 8 | Molecular genetic analysis of resistance/tolerance in rice wheat chickpea and mustard including sheath blight complex genomics: Rice component | ICAR | April 2015–March 2026 |
| 9 | Network project on functional genomics and genetic modification in crops (NPFGGM): Salt tolerance in rice | ICAR | April 2017–March 2026 |
| 10 | Identification of traits, genes, and physiological mechanisms to the development of climate smart varieties for unfavorable environment | IRRI, Philippines | July 2018–March 2020 |

(continued)

**Table 5.4** (continued)

| Sl. no. | Name of project | Funded by | Project duration |
|---|---|---|---|
| 11 | Consortium research platform (CRP) on agro-biodiversity – Component 2 – Evaluation of rice germplasm accessions against biotic/abiotic stresses | ICAR | April 2015–March 2021 |
| 12 | Stress tolerant rice for poor farmers of Africa and South Asia (phase III) | BMGF-IRRI | April 2014 - March 2019 |
| 13 | **Contract research:** Evaluation of BAYER rice hybrids under salinity stress phase II | Ms. Bayer bio science Ltd. | July 2015–June 2019 |
| 14 | Novel genetic stocks: Multi-parent advanced generation intercrosses (MAGIC) among diverse genotypes to facilitate gene discovery for various traits in rice | BMGF-IRRI | February 2015–December 2016 |
| 15 | National Project on Transgenics in crops – Salt tolerance in rice: Functional genomics | ICAR | October 2010–March 2017 |
| 16 | Establishment of National Rice Resource Database | DBT | August 2010 - March 2017 |
| 17 | From QTL to variety: Marker assisted breeding of abiotic stress tolerant rice varieties with major QTLs for drought, submergence, and salt tolerance | DBT | April 2010–March 2015 |
| 18 | Development and evaluation of salt tolerant transgenic rice | DBT | October 2010–May 2013 |
| 19 | Stress tolerant rice for poor farmers of Africa and South Asia (phase II) | BMGF-IRRI | April 2011–March 2014 |
| 20 | Enhancing and stabilizing the productivity of salt affected areas by incorporating genes for tolerance of abiotic stresses in rice | BMZ-Germany | 2008 January 2010 |
| 21 | Evaluation and breeding of rice varieties for salinity, alkalinity, and submergence stress tolerance | Institute CSSRI | 1999 2010 |
| 22 | Development of technologies to harness the productivity potential of salt affected areas of indo-Gangetic, Mekong, and Nile River basins | IRRI-CPWF | 2004 2008 |
| 23 | Salt affected soils and breeding for salt resistance crops (rice, Indian mustard, and gram) | Indo-Sussex | 1993 1999 |

### 5.4.3 Screening in Pots

For more controlled study of the individual plant response under a constant stress, round porcelain pots of 20 or 30 cm diameter, with a capacity of 8 or 16 kg soil with a provision to allow or plug off leaching from bottom, are used. Pot studies also allow for periodic uprooting of plants at a given growth stage for specific studies.

### 5.4.4 Screening in Saline Solution Culture

Solution culture technique is being used in two ways, one for mainly screening up to seedling stage and second up to maturity. In the first category, 7-day-old seedlings

**Fig. 5.1** Sodic (**a** and **b**) and saline (**c** and **d**) microplot facility at CSSRI, Karnal

grown on nonstress modified Yoshida culture solution (Table 5.5) are transferred to the desired level of stress either in the bread box with perforated lid or perforated Styrofoam with mesh. This modified culture solution has $KH_2PO_4$ and $K_2HPO_4$ in place of $NaH_2PO_4$ as sodium salt within culture solution which may increase the Na ion concentration. Initial salinization of culture solution with 50 mM NaCI (equivalent to about 4.9 $dSm^{-1}$) for 3 days followed by increase in stress level up to 70 to 100 mM (approx. 6.8–9.5 $dSm^{-1}$) is good enough for screening of genotypes for salinity at seedling stage. After 10 to 15 days, treated seedlings are used for the morphological, physiological, or biochemical analyses (Yoshida et al. 1971).

The adult plants uprooted from nursery can also be screened for reproductive stage tolerance under this saline hydroponics but with large spacings between plants inserted in the saline medium through holes on the thermocol sheets. Alternatively, for the adult stage screening for tolerance, plants are grown in 0.5 kg capacity soil filled pots, and the desired salinity level of rhizosphere is maintained through time automated circulatory irrigation system. However, the screening under solution culture requires frequent changing of nutrient solution with maintenance of pH around 4.5 for better nutrient availability. Recently, growing seeds of large number of segregating progenies of rice on the nylon mesh tops floating on nutrient solution under glass house conditions was started. This is followed by transferring about 25-day-old seedlings on the highly salinized (12–14 dS/m) nutrient medium to select

**Table 5.5** Chemical composition of rice culture solution (modified Yoshida solution)

| Stock | Regent | $g\ l^{-1}$ |
|---|---|---|
| 1 | $NH_4NO_3$ | 91.4 |
| 2 | $K_2SO_4$ | 71.4 |
| 3 (a) | $KH_2PO_4$ | 23.1 |
| (b) | $K_2HPO_4$ | 4.3 |
| 4 | $CaC1_26H_2O$ | 175 |
| 5 | $MgSO_47H_2O$ | 324 |
| 6 | Minor nutrients | |
| (a) | $MnCl_24H_2O$ | 1.5 |
| (b) | $(NH_4)_6\ Mo_7O_{24}4H_2O$ | 0.074 |
| (c) | $H_3BO_3$ | 0.93 |
| (d) | $ZnSO_47H_2O$ | 0.035 |
| (e) | $CuSO_45H_2O$ | 0.03 |
| 7 or | FeNaEDTA[a] | 10.5 |
| 8 | FeSO4 (made fresh) | 2.5 |

[a]replace regularly
1.25 ml each stock solution per liter culture solution pH 4.5 to be adjusted with nitric acid

few but most robust tolerant progenies for seedling stage tolerance. The few surviving seedling stage tolerant plants are then planted in pots having normal soil for advancing their generations during offseason under environmentally controlled glass house conditions. This method could be extremely useful particularly for those locations where only one crop of rice can be taken due to winter season.

## 5.4.5 Screening in Trays

This method is particularly useful for rapid screening of large number of lines but is restricted only up to seedling stage. For this, salt affected soil is filled in shallow depth in the wooden or metal germination trays provided with a polythene sheet lining on the inner side. These are quite handy for screening lines with appropriate control on salinity, sodicity, and moisture. The trays can also be conveniently shifted from one growth chamber to another.

### 5.4.5.1 Rice Molecular Laboratory

Molecular markers are becoming important tools for accelerating breeding progress. In view of this, equipment pertaining to molecular markers work have been added to the Rice Biotechnology Lab at CSSRI, Karnal, under different externally funded projects (ICAR, DBT, DST, and BMGF).

### 5.4.5.2 Varietal Development

CSSRI has developed seven salt tolerant rice varieties for salt affected soils (Table 5.6). CSSRI, Karnal, also led the country wide program on rice improvement

**Table 5.6** CVRC released salt tolerant rice varieties developed by CSSRI, Karnal

| Name of variety | CSR10 | CSR13 | CSR23 | CSR27 | Basmati CSR30 | CSR36 | CSR 43 |
|---|---|---|---|---|---|---|---|
| Parentage | M40-431-24-114/ Jaya | CSR1 / Bas.370 // CSR5 | IR64 // IR4630-22-2-5-1-3/ IR 964-45-2-2 | NONA BOKRA / IR565-33-2 | BR4-10 / Pak. Basmati | CSR13/ Panvel 2// IR36 | KDML105 / IR 4630-22-2-5-1-3//IR 20925-33-3-1-28 |
| IET No. | 10349 | 10348 | 13769 | 13765 | 14720 | 17340 | 18259 |
| Year of release | 1989 | 1998 | 2004 | 1998 | 2001 | 2005 | 2011 |
| Plant height (cm) | 85 | 115 | 115 | 115 | 155 | 110 | 100 |
| Maturity days | 120 | 145 | 130 | 125 | 155 | 140 | 110 |
| Tolerance limits | | | | | | | |
| • Salinity | | | | | | | |
| • (ECe:dS/m) | <11.0 | < 9.0 | <10.0 | <10.0 | <7.0 | <11.0 <10.0 | <9.0 |
| Sodicity ($pH_2$) | <10.2 | <10.0 | <9.9 | < 9.9 | <9.5 | | <10.0 |
| Yield (t/ha) | | | | | | | |
| Nonstress | >6.0 | >6.0 | >6.5 | >6.5 | >3.0 | >6.5 | >6.5 |
| Salt stress | >3.0 | >3.0 | >4.0 | >4.0 | >2.0 | >4.0 | >3.5 |
| Grain type | Short bold | Long slender | Long slender | Long slender | Basmati type | Long slender | Long slender |
| Recommended ecology | Acts as biological amendment for highly deteriorated sodic and inland saline soils | Sodic and inland saline soils of UP, Haryana, Gujarat, and Maharashtra | Sodic soils of Haryana, UP, and coastal saline areas of Maharashtra, Gujarat, Tamil Nadu, Kerala, and West Bengal | Sodic and coastal saline soils of India | First salt tolerant basmati variety developed for sodic areas of UP, Haryana, and Punjab. | Sodic soils of Haryana, UP, and Pondicherry | Sodic soils of UP |

| Name of variety | CSR46 | CSR49 | CSR52 | CSR56 | CSR60 | CSR76 |
|---|---|---|---|---|---|---|
| Parentage | IR72/CSR23 | IR 5711-3B-7-B-9-1-8/NDRK 5012 | CSR 23/CSR 27 | CSR 21 / CSR 10 | IR4630-22–2-5-1-3/ IR05N204 | CSR 27/MI48 |
| IET no. | 18710 | 20329 | 23206 | 24537 | 25378 | 27070 |
| Year of release and notification | 2018 | 2019 | 2019 | 2018 | 2018 | 2022 |
| Plant height (cm) | 110 | 115 | 90 | 100 | 110 | 95 |
| Maturity days | 110 | 130 | 120 | 125 | 125 | 125 |
| Tolerance limits | | | | | | |
| • Salinity. | | | | | | |
| • (ECe:dS/m) | <9.0 | <8.0 | <8.0 | <8.0 | <8.0 | <7.0 |
| Sodicity ($pH_2$) | <9.9 | <9.9 | <9.9 | <9.9 | <9.9 | <9.9 |
| Yield (t/ha) | | | | | | |
| Nonstress | >7.0 | >7.5 | >7.5 | >7.5 | >7.5 | >7.5 |
| Salt stress | >4.0 | >3.5 | >3.5 | >4.1 | >4.3 | >4.5 |
| Grain type | Long slender | Medium Slender | Long bold | Long bold | Long slender | Long slender |
| Recommended ecology | Sodic soils of Uttar Pradesh | Sodic soils of Uttar Pradesh | Sodic soils of Uttar Pradesh | Sodic soils of Uttar Pradesh and Haryana | Sodic soils of Uttar Pradesh and Pondicherry | Sodic soils of Uttar Pradesh |

for salt affected soils encompassing different salinity centers which resulted in the development of other genetic stock with salt tolerant rice varieties.

Recently, ICAR CSSRI Karnal has released six salt tolerant rice varieties, namely, CSR46 (Krishnamurthy et al. 2019c), CSR49 (Krishnamurthy et al. 2021b), CSR52 (Krishnamurthy et al. 2021a), CSR56 (Krishnamurthy et al. 2019a), CSR60 (Krishnamurthy et al. 2019b), and CSR76, to enhance the productivity of salt affected soils in India.

## 5.5 Genetic Stock Development

(a) **Alkali tolerant rice genetic stock CSR47:** The rice (*Oryza sativa* L.) line CSR 47 (CSR 2 K 232) is an alkaline tolerant high yielding, developed using Jaya and CSR 13 as parent through pedigree breeding method at the Division of Crop Improvement, Central Soil Salinity Research Institute, Karnal, Haryana. The line CSR 47 (IET 20328) was developed for alkaline areas where pH is up to 9.9. Of the two parents, CSR 13 is tolerant to alkalinity (pH ~9.9) and another parent Jaya has high yield potential. Hence, this culture combines the high yield from Jaya and tolerance from CSR 13 for alkalinity. It possessed long bold grains and 95 days' flowering duration. This is a semidwarf culture, with green foliage, erect flag leaf, compact panicle and complete panicle exertion.

(b) **Saline tolerant rice genetic stock CSR53:** The line CSR 53 (Bulk 18) was developed for saline areas where ECe is up to 10.0 dS/m from the cross CSR23/CSR27 through bulk breeding method at the Division of Crop Improvement, Central Soil Salinity Research Institute, Karnal, Haryana. This is a medium duration, semidwarf culture, with green foliage, erect flag leaf, compact panicle, medium slender grains, and complete panicle exertion. The CSR53 has medium slender grain with tolerance to salinity; therefore, plant breeder can use this line for transfer of medium slender grain with salinity tolerance in BPT 5204 background without much breeding effort (Krishnamurthy et al. 2019b).

(c) **Saline and alkaline tolerant rice genetic stock CSR51:** Developed and registered rice genetic stock CSR51, which is tolerant to both salinity and alkalinity. This line could be used as a donor for development of high yielding rice varieties with long slender grains with lesser use of resources. This germplasm can be used as a genetic stock for future breeding programs aiming at development of high yielding salt tolerant rice varieties for saline and alkaline soils (Krishnamurthy et al. 2020a).

(d) **Saline and alkaline tolerant rice genetic stock CSR59:** Developed and registered rice genetic stock CSR59 which is tolerant to salinity and alkalinity. This line could be used as a donor for development of high yielding rice varieties with long slender grains with lesser use of resources. This germplasm can be used as a genetic stock for future breeding programs aiming at development of high yielding salt tolerant rice varieties for saline and alkaline soils (Krishnamurthy et al. 2021a).

## 5.6 Current Research Priorities and Progress

### 5.6.1 Transfer of Salt Tolerance to High Yielding Varieties Using Conventional Breeding Programs

Different segregating populations were developed using high yielding varieties, salt tolerant varieties, salt tolerant genetic stock, drought tolerant varieties, landraces, and rice varieties with good grain quality (Table 5.3 and Table 5.6). These segregating populations were screened and selected under high salinity (EC ~ 7.0 dS/m) in saline microplot, Nain farm (EC ~ 16 ds/m), and sodic stress environments. The selection was carried out from each segregating population by considering the yield, quality, tolerance, and other traits to further screening/evaluation and advance the generation (Krishnamurthy et al. 2016b).

### 5.6.2 Transfer of Salt Tolerance to Popular Varieties through Marker Assisted Backcross Method in Rice

Near isogenic lines (NILs) of PUSA 44, PR 114, and Sarjoo 52 for saltol have been developed for seedling stage salinity tolerance in rice using marker assisted back-cross method. We identified ten NILs in the background of Pusa 44 and eight NILs in the background of Sarjoo52 (Krishnamurthy et al. 2020b). We also introgressed Saltol QTL into PB1509 (Yadav et al. 2020) and PB1121 (Babu et al. 2017a, 2017b).

### 5.6.3 Identification QTLs for Salinity/Sodicity Tolerance Stresses at Reproductive Stages

We identified QTL for SSI spikelet fertility in RIL of CSR27/MI 48 at chromosome 8 (Pandit et al. 2010). The fine mapping of this QTL is in progress. We carried out extensive and systematic phenotyping for salt tolerance in rice in microplots (salinity and alkalinity) and salt stress field conditions for reproductive stage tolerance. We performed the SNP genotyping of mapping population with good coverage of 12 linkage groups in rice. We also identified QTLs for salinity tolerance at seedling stage in Kalajoha landrace (Mazumder et al. 2020). We identified the QTLs for salinity tolerance at reproductive stage using $F_2$s of PS5/CSR10 (Pundir et al. 2021). We also identified marker-trait association for salt tolerance using GWAS panel in rice (Kumar et al. 2015; Warraich et al. 2020; Warraich et al. 2021).

### 5.6.4 Development of National Rice Minicore set

We established national rice data base for 9000 rice germplasm for 30 traits and developed a core set of 1500. This core set is used for evaluation in salinity stresses and other stresses.

### 5.6.5 Coordination, Planning, and Execution of Breeding for Salt Tolerance in Rice across Saline and Alkaline Locations in India

The Salinity Tolerance Breeding Network (STBN) under Stress Tolerance Rice for Poor Farmers of Africa and South Asia (STRASA) project funded by the Bill and Melinda Gates Foundation (BMGF) helped many plant breeders to meet at a single platform. They collected advanced breeding lines of rice for salinity/alkalinity from different research station and evaluated across saline and alkaline locations in India. Around 30–40 entries were conducted at 8 saline and 7 alkaline locations, respectively. The experiment was carried out in different salt stress environments across India with the main objective of identifying the best tolerant and high yielding line in salinity, alkalinity, and across the salt (saline + alkaline) stress environments. Accordingly, rice genotypes, namely, CSR2K262, CSR27, NDRK11–4, NDRK11–3, NDRK11–2, CSR 2 K 255, and PNL 1–1–1-6-7-1, were identified as specifically adapted to favorable locations. The genotype CSR2K262 outyielded for both coastal saline and alkaline locations and CSR 36 for alkaline locations over national checks.

## 5.7 Seed Dissemination and its Impact

CSSRI has developed 13 salt tolerant rice varieties distributed to private companies, farmers, etc. We produced 1056 quintals and 3126 quintals of breeder and truthfully labeled seeds of salt tolerant rice varieties during last 22 years (2000–2021). These salt tolerant rice varieties covered an area of app. 1.02 m ha every year.

These salt tolerant rice varieties covered about 1.6 m ha of salt affected land and estimated to earn USD 4732 million annually to national exchequer. These salt tolerant varieties have been adopted by land reclamation corporations, benefiting 430,000 famers. With the introduction of these salt tolerant varieties, about 0.6 million tons of gypsum was saved, which would have cost of USD 250,000. Salt tolerant rice varieties facilitated in manage the salt affected soils and sustainable agriculture under salt affected soils.

## 5.8 Future Outlook

- Breeding rice varieties possessing higher salt tolerance, disease resistance, and higher yield potential.
- Mapping of QTLs conferring tolerance to sodicity and salinity at reproductive stage and their introgression and pyramiding through molecular marker assisted breeding.
- Multiple abiotic stress tolerance in rice thorough marker assisted breeding.
- Harnessing transgenes and gene editing technology for higher salt tolerance.
- Harnessing heterosis for salt tolerance in rice hybrids.

- Germplasm collection, conservation, and characterization of the material from the salinity "hotspots" for knowing the salinity adaptive traits.
- Popularization of salt tolerant varieties in the target salt stress areas through large scale seed production and distribution.

## 5.9 Conclusion

CSSRI has developed 13 salt tolerant rice varieties distributed to many private companies, farmers, etc. and also developed many genetic stocks, mapping populations, and elite breeding lines which have been used for identification of different mechanisms involved in salinity tolerance, identification of novel QTLs for salt tolerance, and biochemical mechanisms for salt tolerance. We identified many salt tolerant donors and used for further breeding programme to enhance the salinity tolerance level. CSSRI, Karnal, also led the country wide program on rice improvement for salt affected soils encompassing different salinity centers which resulted in the development of other salt tolerant rice varieties. Another major notable impact of using salt tolerant rice varieties is bringing back the unproductive barren or salt affected lands into the fold of productive normal lands. Therefore, faster conversion of degraded lands into normal ones with the use of resilient varieties can ensure more area and thus increased food productivity from these reclaimed lands.

**Acknowledgments** Authors thank the directors of CSSRI, Karnal, for the support and facilities. Authors also thank the different funding agencies for funds. The authors sincerely thank the Bill and Melinda Gates Foundation for funding support under the STRASA project (IRRI-ICAR collaborative project), ICAR-Incentivizing Research in Agriculture, "From QTL to Variety" (grant no. BT/PR/14544/AGR/02/745/2010), through the Department of Biotechnology, government of India, DST, for funding for the project "Introgression of Blight and Blast Resistance Genes in Salt Tolerant Basmati CSR30 Rice Variety Through Marker Assisted Backcross Breeding (MABB)" code – CRG/2020/003078, ICAR – Network Project on Transgenic in Crops (NPTC)-Functional Genomics for the funding, and first author sincerely thanks ICAR for funding under Lal Bahadur Shastri Outstanding Young Scientist Award.

## References

Ali S, Gautam RK, Mahajan R, Krishnamurthy SL, Sharma SK, Singh RK, Ismail AM (2013) Stress indices and selectable traits in SALTOL QTL introgressed rice genotypes for reproductive stage tolerance to Sodicity and salinity stresses. Field Crop Res 154:65–73

Ammar MHM, Pandit A, Singh RK, Sameena S, Chauhan MS, Singh AK, Singh NK (2009) Mapping of QTLs controlling Na+, K+ and CI− ion concentrations in salt tolerant indica rice variety CSR27. J Plant Biochem Biotechnol 18(2):139–150

Babu NN, Krishnan SG, Vinod KK, Krishnamurthy SL, Singh VK, Singh MP, Singh AK (2017a) Marker aided incorporation of Saltol, a major QTL associated with seedling stage salt tolerance, into *Oryza sativa* 'Pusa basmati 1121'. Front Plant Sci 8:41

Babu NN, Vinod KK, Krishnamurthy SL, Krishnan SG, Yadav A, Bhowmick PK, Singh AK (2017b) Microsatellite based linkage disequilibrium analyses reveal Saltol haplotype fragmentation and identify novel QTLs for seedling stage salinity tolerance in rice (*Oryza sativa* L.). J Plant Biochem Biotechnol 26(3):310–320

Bimpong IK, Manneh B, Diop B, Ghislain K, Sow A, Amoah NKA, Wopereis M (2014) New quantitative trait loci for enhancing adaptation to salinity in rice from Hasawi, a Saudi landrace into three African cultivars at the reproductive stage. Euphytica 200(1):45–60

Chaudhary K, Krishnamurthy SL, Singh N, Sharma PC (2021) Salinity induced impact on morphological traits of rice at reproductive stage. Oryza 58(2):287–294

Chowdhury AD, Haritha G, Sunitha T, Krishnamurthy SL, Divya B, Padmavathi G, Sarla N (2016) Haplotyping of rice genotypes using simple sequence repeat markers associated with salt tolerance. Rice Sci 23(6):317–325

FAO, IFAD, & WFP (2013) The state of food insecurity in the world 2013. In: The multiple dimensions of food security. FAO, Rome

Fernandez GC (1993) Effective selection criteria for assessing plant stress tolerance. In: Proceeding of the International Symposium on Adaptation of Vegetables and other Food Crops in Temperature and Water Stress, Aug. 13-16, Shanhua, Taiwan, pp 257–270

Fischer RA, Maurer R (1978) Drought resistance in spring wheat cultivars. I. Grain yield responses. Aust J Agric Res 29(5):897–912

Gregorio GB (1997) Tagging salinity tolerance genes in rice using amplified fragment length polymorphism (AFLP). Ph.D. Thesis,. University of the Philippines, Los Baños, Laguna

Hossain H, Rahman MA, Alam MS, Singh RK (2015) Mapping of quantitative trait loci associated with reproductive-stage salt tolerance in rice. J Agron Crop Sci 201(1):17–31

Joshi R, Sahoo KK, Singh AK, Anwar K, Pundir P, Gautam RK, Singla-Pareek SL (2020) Enhancing trehalose biosynthesis improves yield potential in marker-free transgenic rice under drought, saline, and sodic conditions. J Exp Bot 71(2):653–668

Krishnamurthy S, Gautam RK, Sharma PC, Sharma DK, Singh YP (2019b) CSR 53 (bulk 18) (IC0619320; INGR17029), a rice (*Oryza sativa*) germplasm with tolerance to salinity stresses up to ECe 10.0 dS/m. IJPGR 32(2):240–241

Krishnamurthy SL, Pundir P, Singh YP, Sharma SK, Sharma PC, Sharma DK (2015) Yield stability of rice lines for salt tolerance using additive main effects and multiplicative interaction analysis - AMMI. J Soil Salinity and Water Quality 7(2):98–106

Krishnamurthy SL, Gautam RK, Sharma PC, Sharma DK (2016a) Effect of different salt stresses on agro-morphological traits and utilisation of salt stress indices for reproductive stage salt tolerance in rice. Field Crop Res 190:26–33

Krishnamurthy SL, Gautam RK, Sharma PC, Sharma DK, Singh YP, Singh RK (2019a) CSR 47 (CSR 2K 232) (IC0639318; INGR17068), a Rice (*Oryza sativa*) germplasm with tolerance to alkalinity stresses up to pH 9.9. IJPGR 32(2):275–275

Krishnamurthy SL, Pundir P, Warraich AS, Rathor S, Lokeshkumar BM, Singh NK, Sharma PC (2020b) Introgressed saltol QTL lines improves the salinity tolerance in rice at seedling stage. Front Plant Sci 11:833

Krishnamurthy SL, Sharma PC, Gautam RK, Sharma DK, Singh YP (2020a) CSR51 (bulk 212) (IC0619319; INGR19001), a Rice (*Oryza sativa*) germplasm for tolerant to alkalinity stresses up to ph 9.9 and salinity stresses up to ece 10.0 ds/m with long slender grain. Indian J Plant Genetic Resour 33(2):240–241

Krishnamurthy SL, Sharma PC, Gautam RK, Singh YP (2021a) CSR 59 (IC627947; INGR20075), a Rice (*Oryza sativa*) germplasm tolerant to salinity stresses upto ECe 10.0 dS/m with long bold grain

Krishnamurthy SL, Sharma PC, Gautam RK, Singh VK, Singh YP, Sharma DK, Lokeshkumar BM, Suman Rathor A, Warraich S, Vinaykumar NM (2021b) Notification of crop varieties and registration of germplasm: variety CSR52. Indian J Genet 81(3):486

Krishnamurthy SL, Sharma PC, Sharma DK, Singh YP, Mishra VK, Burman D, Singh RK (2021c) Additive main effects and multiplicative interaction analyses of yield performance in rice genotypes for general and specific adaptation to salt stress in locations in India. Euphytica 217(2):1–15

Krishnamurthy SL, Sharma PC, Sharma DK, Ravikiran KT, Singh YP, Mishra VK, Burman D, Buddheswar M, Mandal S, Sarangi SK, Gautam RK, Singh PK, Manohara KK, Marandi BC,

Padmavathi G, Vanve PB, Patil KD, Thirumeni S, Verma OP, Khan AH, Tiwari S, Geetha S, Shakila M, Gill R, Yadav VK, Roy SKB, Prakash M, Bonifacio J, Abdelbagi MI, Gregorio GB, Rakesh KS (2017) Identification of mega-environments and rice genotypes for general and specific adaptation to saline and alkaline stresses in India. Sci Rep 7:1–14. https://doi.org/10.1038/s41598-017-08532-7

Krishnamurthy SL, Sharma PC, Ravikiran KT, Basak N, Vineeth TV, Singh YP, Sarangi SK (2016c) G×E Interaction and Stability Analysis for Salinity and Sodicity Tolerance in Rice at Reproductive Stage. J Soil Salinity and Water Quality 8(2):54–64

Krishnamurthy SL, Sharma PC, Sharma SK, Batra V, Kumar V, Rao LVS (2016d) Effect of salinity and use of stress indices of morphological and physiological traits at the seedling stage in rice. Indian J Exp Biol 54(12):843–850

Krishnamurthy SL, Sharma PC, Singh YP, Gautam RK, Singh RK, Sharma DK, Sharma SK, Mishra VK, Warraich AS (2019c) Notification of crop varieties and registration of germplasm: variety CSR46 (IET 18710). Indian J Genet. 79(2):511–512

Krishnamurthy SL, Sharma SK, Kumar V, Tiwari S, Batra V, Singh NK (2014) Assessment of genetic diversity in rice genotypes for salinity tolerance using Saltol markers of chromosome 1. Indian J Genet Plant Breeding 74(2):243–247

Krishnamurthy SL, Sharma SK, Kumar V, Tiwari S, Singh NK (2016e) Analysis of genomic region spanning Saltol using SSR markers in rice genotypes showing differential seedlings stage salt tolerance. J Plant Biochem Biotechnol 25(3):331–336

Krishnamurthy SL, Sharma SK, Sharma DK, Sharma PC, Singh YP, Mishra VK, Singh RK (2016b) Analysis of stability and G× E interaction of rice genotypes across saline and alkaline environments in India. Cereal Res Commun 44(2):349–360

Kumar V, Singh A, Mithra SA, Krishnamurthy SL, Parida SK, Jain S, Mohapatra T (2015) Genome-wide association mapping of salinity tolerance in rice (*Oryza sativa*). DNA Res 22(2):133–145

Mazumder A, Rohilla M, Bisht DS, Krishnamurthy SL, Barman M, Sarma RN, Mondal TK (2020) Identification and mapping of quantitative trait loci (QTL) and epistatic QTL for salinity tolerance at seedling stage in traditional aromatic short grain rice landrace Kolajoha (*Oryza sativa* L.) of Assam, India. Euphytica 216(5):1–18

Mishra P, Jain A, Takabe T, Tanaka Y, Negi M, Singh N, Rai V (2019) Heterologous expression of serine hydroxymethyltransferase-3 from rice confers tolerance to salinity stress in *E. coli* and Arabidopsis. Front Plant Sci 10:217

Mohammadi R, Mendioro MS, Diaz GQ, Gregorio GB, Singh RK (2013) Mapping quantitative trait loci associated with yield and yield components under reproductive stage salinity stress in rice (*Oryza sativa* L.). J Genet 92(3):433–443

Pandit A, Rai V, Bal S, Sinha S, Kumar V, Chauhan M et al (2010) Combining QTL mapping and transcriptome profiling of bulked RILs for identification of functional polymorphism for salt tolerance genes in rice (*Oryza sativa* L.). Mol Genet Genomics 284(2):121–136

Pundir P, Devi A, Krishnamurthy SL, Sharma PC, Vinaykumar NM (2021) QTLs in salt rice variety CSR10 reveals salinity tolerance at reproductive stage. Acta Physiol Plant 43(2):1–15

Ravikiran KT, Krishnamurthy SL, Warraich AS, Sharma PC (2018) Diversity and haplotypes of rice genotypes for seedling stage salinity tolerance analyzed through morpho-physiological and SSR markers. Field Crop Res 220:10–18

Tiwari S, Sl K, Kumar V, Singh B, Rao AR, Mithra SVA et al (2016) Mapping QTLs for salt tolerance in rice (*Oryza sativa* L.) by bulked segregant analysis of recombinant inbred lines using 50K SNP chip. PLoS One 11(4):e0153610

Warraich AS, Krishnamurthy SL, Sooch BS, Lokeshkumar BM, Vinaykumar NM, Dushyanthkumar BM, Sharma PC (2021) Population structure and association studies for reproductive stage salinity tolerance in rice (*Oryza sativa* L.). Indian J Exp Biol (IJEB) 59(05):328–338

Warraich AS, Krishnamurthy SL, Sooch BS, Vinaykumar NM, Dushyanthkumar BM, Bose J, Sharma PC (2020) Rice GWAS reveals key genomic regions essential for salinity tolerance at reproductive stage. Acta Physiol Plant 42(8):1–15

Yadav AK, Kumar A, Grover N, Ellur RK, Krishnan SG, Bollinedi H et al (2020) Marker aided introgression of 'Saltol', a major QTL for seedling stage salinity tolerance into an elite basmati rice variety 'Pusa basmati 1509'. Sci Rep 10(1):1–15

Yoshida S, Forno DA, Cock JH (1971) Laboratory manual for physiological studies of rice. In: Laboratory manual for physiological studies of rice. IRRI, Las Banos, Laguna, p 83

# Genetic Improvement of Rice for Saline Tolerance Under Coastal Agroecosystem of Tamil Nadu

6

R. M. Kathiresan, M. Prakash, and K. R. Saravanan

**Abstract**

The responses of rice to salt stress are complex, and it depends on the duration and type of salt stress, crop development stage, and other factors. It is relatively tolerant to both salinity and alkalinity during the germination stage. Salinity work on rice is carried out at different states of India, including Tamil Nadu. The present research work was carried out to evaluate 66 entries during 2016–2017, 88 entries during 2017–2018, and 82 rice entries during *Kharif* 2018–2019 for saline tolerance using augmented block design at Faculty of Agriculture, Annamalai University, Annamalai Nagar, Cuddalore, Tamil Nadu, India. The yield and yield attributing traits recorded significant variability among all the genotypes studied. The grain yield per hectare recorded maximum with CSSR 2016-IR18–17, CARI DHAN7, CSSR 2016-IR18–12, CSSR 2016-IR18–7, CSR2016-IR18-9, CST 7-1, CARI Dhan 10, CR3881-M-3-1-5-3-1-1, CSR2016-IR18-10, CSR-17817, CSR11-143, CSR-C27SM- 117, JK-95, CSR-2748-4441-66 and CSR-2748-4441-133. High heritability and genetic advance as percent of mean were observed for all the characters except productive tillers per plant. This indicated the predominance of additive gene action for all the traits. These genotypes were found to be promising for majority of the traits studied and recorded yield of more than 5 tons/ha under salinity stress condition at Annamalai Nagar location. Hence, these genotypes can be utilized for further crop improvement work for salinity tolerance. Apart from the above, Annamalai

R. M. Kathiresan
Department of Agronomy, Faculty of Agriculture, Annamalai University, Chidambaram, Tamil Nadu, India

M. Prakash (✉) · K. R. Saravanan
Department of Genetics and Plant Breeding, Faculty of Agriculture, Annamalai University, Chidambaram, Tamil Nadu, India



R. K. Singh et al. (eds.), *Genetic Improvement of Rice for Salt Tolerance*,
https://doi.org/10.1007/978-981-99-3830-8_6

Nagar Centre serves as a voluntary center for All India Coordinated Rice Improvement Project (AICRIP) and contributed for "Stress Tolerant Rice for Africa and South Asia" (STRASA). This chapter discusses about the saline tolerance work done in rice in Tamil Nadu, in particular at Annamalai Nagar.

**Keywords**

Salinity screening · Salt tolerance · Rice · *Oryza sativa* L · Coastal salinity

## 6.1 Introduction

Rice is a monocot and cereal grain, which provides 21% of global human per capita energy and 15% of per capita protein. Rice is one of the staple foods and the most important cereal crop which contributes more to our food requirements annually. It accounts for about 43% of total food grain production in the country. It is the primary source of energy for more than half of the world as complex carbohydrates. Depending on the strain of rice, it contains considerable amounts of fiber, protein, vitamin B, iron, and manganese.

Salt stress is one of the major abiotic stresses, which adversely affect the crop productivity. Soil salinity has been a key abiotic constraint affecting the crop production by reducing growth, development, and yield of the plant. The effect of salinity was also related to the stage of plant development at which salinity was imposed, concentration and nature of the salt, and the duration of salinization. Breeding of saline tolerant varieties necessitates the understanding of salinity and the characters affected (Hoang et al. 2016; Singh et al. 2021). One of the major prominent issues with soil toxicity is the presence of excessive amounts of salt in many rice-growing regions (Khan et al. 2016). The availability of genetic divergence in the crop determines the effectiveness of any breeding program.

According to Yeo and Flowers (1982), rice genome has a well-known source of salt-tolerant genes that can be used to enhance salinity tolerance in high-yielding varieties. Even though the production and consumption of rice is higher in Asia, the adverse environmental conditions critically threaten the rice production. Rice is highly sensitive to salinity specifically at the early vegetative and late reproductive stages. Therefore, studying the responses of crop at the morphological, physiological, biochemical, and molecular level is an effective strategy. Hence, understanding the mechanisms behind the salinity such as osmotic stress and osmolytes; ion exclusion, inclusion, and compartmentation; antioxidant response; and hormonal regulation becomes vital. Using the conventional and molecular breeding approaches is a prerequisite for effective management and to develop salt-tolerant cultivars in rice.

In India, salt affected area accounts for 8.5 million hectares, with a 30–50% yield loss (Babu et al. 2005). In Tamil Nādu, salt affected area accounts for 0.037 million hectares, and among them, 0.029 million hectares are moderately affected and 0.008 million hectares are highly affected by salinity. Salinity resistant varieties, viz.,

TRY1, TRY2, TRY3, TRY4, and CO43, are usually cultivated in coastal areas in Tamil Nādu, and among these, CO43 yielded an average of 6.3 t/ha, 21.5% higher than IR20. On the other hand, TRY 4 had an average grain yield of 5730 kg per hectare (ha) with 7% increased yield over TRY 3.

The objective of any plant breeder includes selection from a generated or natural population for one or several characters. To improve the grain yield, evaluation of germplasm is the most important aspect because yield as such is controlled by many genes. Increasing the yield may ultimately lead to the extinction of landraces and wild rice varieties (Solis et al. 2020; Jayabalan et al. 2022). Hence, we have to characterize the genetic diversity of existing landraces of rice which helps in identifying the salt-tolerant germplasm (Quan et al. 2018; Haque et al. 2021). The present chapter briefly discusses about saline tolerance work undertaken on rice in Tamil Nadu and, in particular, in the coastal ecosystem of Tamil Nadu including AICRIP and "Stress Tolerance Rice for Africa and South Asia" (STRASA) work undertaken.

## 6.2 Background of Salinity Research in Tamil Nādu

In Tamil Nadu, the problem of salinity exists both in coastal areas and inlands. The soils of coastal areas offer many unique problems like salinization, water logging, clay pan formation, and seawater inundation. These soils occur all along the east coast from Chengalput, South Arcot, Tanjore, and Ramnad to Tirunelveli. It has been estimated that about one lakh hectares in the coastal belt pose salinity problems. In the coastal area bordering Thanjavur district, salinity problem exists in parts of the taluks of Sirkali, Mayuram, Nagappatinam, Pattukottai, and Thiruthuraipoondi. The entire area is cultivated mainly with long duration rice crop, except in Sirkali taluk where double crop is followed. In all other taluks, only a single crop of rice is raised.

Salinity problem exists in the taluks of Tiruvandari, Ramanathapuram, and Mudukulathur in Ramanathapuram district. Most of the area in this belt is mono cropped with rice both under irrigated and rainfed conditions. In the coastal belt from Ramanathapuram to Mandapam, a fragipan has developed about 30 cm below the soil surface, and rice is grown under semidry conditions in such soils. It has been reported that many traditional varieties have been slowly replaced by HYVs, narrowing their number to hundreds. This has led to the loss of the genetic pool of many landraces that can adapt to the ongoing climate change in Tamil Nadu. Tamil Nadu has now been left with nearly a hundred traditional landraces of rice that have the capability to cope with ongoing climate change. The local rice varieties, namely, Kallundai Samba, Suran Kuruvai, and Arupatham Kuruvai, were cultivated for a period of 1 year, covering one season, i.e., purely under rainfed conditions with organic methods of crop cultivation.

Breeding of saline tolerant lines has emerged as an important aspect to overcome the problem of salt stress. So far, breeders make use of several salt-tolerant genotypes like Nona Bokra, Pokkali, and FL 478. However, the indirect detrimental characters present in these genotypes lead to the difficulty in cultivation of these landraces. Almost all the conventional breeding methods have been followed for the development of the salt-tolerant materials, i.e., introduction, selection, hybridization, mutation, and shuttle breeding approach. But, the conventional breeding approaches have a lot of challenges as it takes more time to select the desirable lines. Most widespread use of markers has helped in the marker assisted backcrossing (MAB) of major genes into already established varieties, mega-varieties (which occupy a large area within the country and across the countries), or elite cultivars. These markers could reduce the linkage drag around the target gene and also recover the recurrent parent background within less number of generations in comparison with conventional breeding. IRRI has already used marker assisted selection for *saltol* QTL, which is responsible for seedling stage salinity tolerance in rice, to facilitate the breeding for salt-tolerant cultivars. So, in recent days, the researchers are employing in introgressing QTL for salinity tolerance from the landraces to the elite varieties which can be a solution to this problem.

With the advancement in the field of marker assisted selection (MAS), it is possible to introgress the QTL in a lesser period of time. Using the strategy, several improved versions of rice varieties have been developed. Recently, TRY 5, a mutant of TRY 2, was released which is suited for saline soil conditions and can be harvested 10–15 days earlier than the existing crops that were grown in similar conditions. It is suitable for kuruvai (June–July), late samba (September–October), and navarai (December–January) cultivation. It has an average yield of more than 5 tons per hectare. It has 64% milling and 54% head rice recovery.

List of salt-tolerant rice varieties released in Tamil Nadu

| Sl. no. | Name of the variety | Special feature | Released year |
|---|---|---|---|
| 1. | CO 43 | Fine grain, tolerant to alkalinity and salinity | 1982 |
| 2. | TRY 1 | Suitable for both saline and alkaline soils. Tolerant to water inundation. Moderately resistant to blast. Medium bold grain. Performs well even under zinc deficient soils | 1995 |
| 3. | TRY (R) 2 | Suitable for both saline and alkaline soils. Short duration. Moderately resistant to blast. Long slender grain. Also suitable for semidry system of cultivation | 2002 |
| 4. | TRY 3 | High grain yield over ADT (R) 46 (10.8%). Highly suitable for "Idli" making. Moderately tolerant to sodicity. Medium bold grain. High milling (71.3%) and head rice recovery (66.0%). Resistant to leaf folder, stem borer, and BPH. Resistant to blast, brown spot, sheath rot, and sheath blight | 2010 |
| 5. | TRY 4 | Average grain yield of 5730 kg per hectare (ha) with 22% increased yield. High milling (68.1%) with high head rice recovery (57.2%). Resistant to major pests such as leaf folder, stem borer, and gall midge and major diseases such as blast and brown spot | 2021 |

## 6.3 Rice Salinity Breeding at Annamalai Nagar, Tamil Nādu

The Annamalai University popularized Annamalai Uvar Nel (AU 1) and AU 2 which comes up well under salinity. AU 2, a salt-tolerant variety, was developed from the Faculty of Agriculture, Annamalai University, and notified in 1989. It was a selection from IR 1561-228-1-2 / IR 1737, having 102 days for 50% flowering. Salinity tolerance is a special characteristic of these varieties. These varieties, AU 1 and AU 2, are semidwarf in nature (about 95–115 cm in height) with erect, non-lodging habit and moderately photo insensitive with a duration of 105 days in Kuruvai and 115 days in Navarai. These varieties can yield 5.5–6.2 t $ha^1$ in normal soil with good quality water. However, when irrigated entirely with water of salinity level of 2 $dSm^1$ during Kuruvai season, they can yield around 3.0 t $ha^1$. As a part of AICRP program, AICRIP trials IVT-1, IVT-2, AVT-1, and AVT-2 were conducted with several entries from 2009 to 2022. The best entries were selected and utilized in further breeding programs.

In another study, ten promising lines, viz., IRLON GSR-9, IRLON GSR-5, CSR-2016-IR-18-18, CR-3878-245-2-4-1, CSR-11–143, CARIDhan-7, CSR-2748-441-195, CR-3437-1*200-83, JK-58, and NDRK-11-20, were shortlisted and subjected to salinity screening under seedling stage. Different saline concentrations were prepared by using four salts, viz., NaCl, $Na_2SO4^{2-}$' $CaCl_2$, and $MgSO_4^{2-}$ (by mixing NaCl 5 meq, $Na_2O_4^{2-}$1.75 meq, $CaCl_2$–2.5 meq, and $MgSO_4^{2-}$ 0.75 meq), as per the protocol outlined by Sehrawat et al. 2014. Screening was done for these genotypes to check their performance (Roshni et al. 2022).

Many unique genes were identified in CARIDhan-7 which were categorized under genes responsive to stress, responsive to abiotic stimulus. Hence, CARIDhan-7 was taken for comparative transcriptomics under control and saline induced conditions at EC 6 $dSm^{-1}$, belonging to the pathways like "MAPK signaling pathway-plant", "Phenylpropanoid biosynthesis", "Plant hormone signal transduction", "Diterpenoid biosynthesis" and "Starch and sucrose metabolism", "Diterpenoid biosynthesis", "MAPK signaling pathway-plant" and "alpha-Linolenic acid metabolism". Some novel high affinity potassium transporters, cation hydrogen exchangers, ABC transporters, glutathione S-transferases, UDP-glucosyl transferase, 1,3-beta-glucosidase, IAA homologs, heat shock proteins, aquaporins, LEA proteins, F-box proteins, sugar transporters, and stress responsive proteins were found to be significantly expressed in the stressed plant. These pathways were reported to be playing a major role in inducing stress tolerance in plants under salinity conditions. All these results added a note to the saline tolerance mechanism which is prevalent in crops: ion exclusion, tissue tolerance, and somatic tolerance (Roshni et al. 2022).

The major mechanism of saline tolerance exhibited by the genotype under the study was through ion exclusion through activation of various ion transportation related genes. The data mobilized in the present study can support future studies in the understanding of perspectives of gene ontology and transcriptional regulation and networking pathways for salinity tolerance (Roshni et al. 2022).

Stress Tolerant Rice for Africa and South Asia (STRASA) work was carried out with salinity tolerance breeding during 2014–2019 at Annamalai University, Annamalai Nagar. From 2016 to 2019, 66 entries with 6 check varieties (Pusa-44, CSR-10, CSR-27, CSR-36, CST-7-1, and TRY -2 (LC)), 88 entries with 6 checks (CSR-27, CSR-36, CST-7-1, Pusa-44, SambaMashuriSub-1, and TRY-1 (LC)), and 82 entries with 7 checks (CSR 27, NSICRC222, CSR 36, CST 7–1, CSR 10, Pusa 44, TRY-3 (LC)) were evaluated during *Kharif* 2016–2017, 2017–2018, and 2018–2019, respectively, at Annamalai Nagar, Cuddalore district, Tamil Nadu, India (11° 24' N latitude, 79° 44′ E longitude, +5.79 m ASL) under salinity stress with the pH of 7.8 to 8.3 (soil ECe, 4 dS/m; water ECe, 6.4 to 8.2 dS/m) in clayey loam to sandy loam soil near the seashore area. This is in agreement with the earlier findings of Kashyap et al. (2020b) and Snehi et al. (2022). The southwest monsoon, with an onset in June and lasting up to August, brings scanty rainfall. Bulk of the rainfall is received during the northeast monsoon in the months of October, November, and December.

The experiments were conducted in augmented block design with six blocks, with the main objective of identifying the best tolerant and high-yielding line under saline stress condition. The data on various traits (days to 50% flowering, plant height (cm), panicle length, number of productive tillers/plants, grain yield) recorded was analyzed by applying Mahalanobis $D^2$ to determine genetic divergence among genotypes (Mahalanobis 1936).

During trial period of *Kharif* 2016, 2017, and 2018, total rainfall received was 499.8 mm, 1857.1 mm, and 1124.8 with 40, 61, and 48 rainy days, respectively. The received rainfall during 2016 was 52% lower when compared to average rainfall of Annamalai Nagar location, whereas the received rainfall during 2017 and 2018 were higher when compared with average rainfall of our station. There were heavy rains coupled with high tide during November, resulting in submergence and water stagnation for 6 to 7 days. The salinity stress was experienced from flowering stage onward to most of the cultures (Roshni et al. 2019). The average maximum and minimum temperatures were 32.71 °C to 33.78 °C and 24.10 °C to 24.8 °C, respectively (Table 6.1).

Analysis of variance for 2016, 2017, and 2018 revealed significant differences among the entries for all the characters studied (Table 6.2). During 2016–2017, plant height ranged from 74 cm (KR 15003) to 157 cm (CR3899-134-2-6-4) with a mean height of 106 cm, days to 50% flowering ranged from 73 days (CSR-10) to 145 days (TRY 2) with a mean of 98.7 days, productive tillers ranged from 7 (CR3878-245-1-9-2) to 22 (CR3881-4-1-6-3-4-1) with a mean of 9.4, panicle length ranged from 15 cm (KR 15014) to 27 cm (TRY-2) with a mean length of 22.2 cm, and grain yield ranged from 1380 kg/ha (CR 2850- S-2B-12-1-2-1) to 5470 kg/ha (CR3881-M-3-1-5-3-1-1), with a mean yield of 3760 kg/ha (Table 6.2) .

During 2017–2018, days to 50% flowering ranged from 78 days to 131 days among the 88 rice genotypes. The minimum days to 50% flowering was recorded in PAU 3835-36-6-3-3-4 (78 days), followed by CR 3878-245-2-4-1 (81 days), CSR 2016-IR18-7 (82 days), CSR2016-IR18-6 (83 days), CSR 2016-IR18-1 (84 days), and CSR-2748-441-193 (85 days). The plant height at maturity revealed significant

**Table 6.1** Weather data during 2016–2018 at Annamalai Nagar, Annamalai University, Tamil Nadu, India

| Weather parameters | June | July | August | September | October | November | December | Total |
|---|---|---|---|---|---|---|---|---|
| | 2016 | | | | | | | |
| No. of rainy days | 5 | 5 | 8 | 6 | 4 | 7 | 5 | 40 |
| Total rainfall (mm) | 55.0 | 41.8 | 167.2 | 86.6 | 52.5 | 51.8 | 44.9 | 499.8 |
| Avg. maximum temperature (°C) | 35.1 | 35.3 | 35.8 | 34.3 | 34.2 | 30.6 | 30.7 | – |
| Avg. minimum temperature (°C) | 25.8 | 25.4 | 24.8 | 24.9 | 25.2 | 22.8 | 19.8 | – |
| Sunshine hours | 5.2 | 5.6 | 6.7 | 4.8 | 6.6 | 5.9 | 6.3 | |
| Wind velocity | 5.1 | 4.5 | 5.1 | 5.8 | 2.5 | 3.6 | 3.7 | |
| *Year* | *2017* | | | | | | | |
| No. of rainy days | 7 | 6 | 10 | 6 | 8 | 17 | 7 | 61 |
| Total rainfall (mm) | 80.8 | 120.2 | 240.4 | 48.8 | 372.7 | 762.6 | 231.6 | 1857.1 |
| Avg. maximum temperature (°C) | 36.8 | 36.1 | 33.4 | 33.1 | 32.4 | 29.4 | 28.6 | – |
| Avg. minimum temperature (°C) | 26.2 | 25.6 | 25.9 | 25.0 | 24.7 | 23.9 | 22.3 | – |
| Sunshine hours | 6.1 | 3.6 | 5.6 | 4.6 | 4.7 | 5.4 | 3.5 | |
| Wind velocity | 6.2 | 5.0 | 5.8 | 3.7 | 2.6 | 2.5 | 2.4 | |
| *Year* | *2018* | | | | | | | |
| No. of rainy days | 3 | 5 | 7 | 3 | 10 | 15 | 5 | 48 |
| Total rainfall (mm) | 73.3 | 68.0 | 195.0 | 26.9 | 302.4 | 375.0 | 84.2 | 1124.8 |
| Avg. maximum temperature (°C) | 35.6 | 36.1 | 35.3 | 34.2 | 31.1 | 29.5 | 29.1 | – |
| Avg. minimum temperature (°C) | 26.1 | 26.1 | 25.0 | 24.9 | 24.8 | 23.7 | 22.3 | – |
| Sunshine hours | 5.9 | 7.7 | 4.8 | 6.1 | 4.4 | 4.6 | 5.0 | |
| Wind velocity | 6.5 | 6.1 | 5.3 | 3.2 | 2.2 | 2.0 | 2.6 | |

**Table 6.2** Evaluation of rice entries under coastal salinity during 2016, 2017, and 2018

| Year (entries + checks) | Mean | Range | CD | CV |
|---|---|---|---|---|
| *2016–2017* | *$EC_{lw}$ ~ dS/m = 3.4–4.2* | | *pH: 7.9–8.3* | |
| Plant height (cm) | 106.5 | 74–157 | 1.93 | 0.58 |
| Days to 50% flowering | 98.7 | 73–145 | 13.56 | 10.59 |
| Productive tillers/plant | 9.4 | 7–22 | 2.14 | 7.6 |
| Panicle length (cm) | 22.2 | 15–27 | 1.24 | 2.56 |
| Grain yield (t/ha) | 3.76 | 1.38–5.47 | 1.12 | 8.78 |
| *2017–2018* | *$EC_{lw}$ ~ dS/m = 3.8–4.5* | | *pH: 7.8–8.5* | |
| Plant height (cm) | 107.61 | 69–137 | 4.49 | 2.61 |
| Days to 50% flowering | 94.89 | 78–131 | 7.10 | 4.67 |
| Productive tillers/plant | 13.12 | 8–23 | 3.61 | 17.17 |
| Panicle length (cm) | 22.58 | 18–27 | 3.18 | 8.79 |
| Grain yield (t/ha) | 3.70 | 1.45–5.98 | 1.57 | 11.56 |
| *2018–2019* | *$EC_{lw}$ ~ dS/m = 6.4–8.2* | | *pH: 7.8–8.3* | |
| Plant height (cm) | 93.47 | 36–152 | 16.92 | 2.89 |
| Days to 50% flowering | 88.76 | 65–118 | 10.18 | 5.04 |
| Productive tillers/plant | 11.78 | 6–21 | 4.71 | 15.93 |
| Panicle length (cm) | 19.51 | 9–27 | 2.71 | 7.56 |
| Grain yield (t/ha) | 3.90 | 1.95–5.89 | 2.01 | 13.45 |

variation among the 88 rice genotypes. The minimum plant height was recorded in CSR 2016-IR18–8 (69 cm), followed with increase in CSR 36 (80.3 cm), PUSA 44 (89 cm), Sambha Sub - 1 (90 cm), and CR 2859-S-B-2-1-B-7 (91 cm). Similar findings were made by Sara et al. (2019) and Kashyap et al. (2020b) (Table 6.2).

The number of productive tillers per plant ranged from 8 to 23 among the 88 rice genotypes. The maximum number of productive tillers per plant was recorded in CR 3912-221-3-1 (23) followed by CR 3882-7-1-6-2-2-1 (19), Sambha Sub-1 (19), CSR 2016-IR 18-17 (17), CARI Dhan 6 (16), and CSR 27 (15) (Table 6.2). Panicle length (cm) ranged from 18 cm to 27 cm among the 88 rice genotypes. The maximum panicle length was recorded in RP-5683-101-85-30-2 3-1 (27 cm), followed by CSR 2016-1R18-11 (26 cm), CSAR 1620 (25 cm), TR 05043 (24 cm), Sambha Sub-1 (23 cm), and RP 5690-20-6-3-2-1 (22 cm), respectively, while minimum panicle length was recorded in RAU720 (18 cm). Kashyap et al. (2020a) and Krishnamurthy et al. (2022) also made similar observations.

The grain yield per plant ranged from 1450 kg to 5980 kg among the rice genotypes. The maximum grain yield per plant was recorded in CSSR 2016-IR18-17 (5980 kg) followed by CSSR 2016-IR18–12 (5890 kg), CSSR 2016-IR18–7 (5870 kg), CST 7-1 (5850 kg), CARI Dhan 7 (5780 kg), and CARI Dhan 10 (5670 kg). The minimum grain yield per plant was recorded in IR84649-31-4-1-3B-CR3397-S-B-4B-1 (1450 kg) (Table 6.2). A similar trend of observations was recorded for all the parameters studied during 2018–2019 also. Gangaiah (2019) also reported a similar trend in rice (Table 6.3).

**Table 6.3** Promising entries for grain yield during 2016, 2017, and 2018 at Annamalainagar location

| 2016–2017 | 2017–2018 | 2018–2019 |
|---|---|---|
| CR3881-M- 3-1-5-3-1-1 | CSSR 2016-IR18–17 | CARI Dhan 7 |
| CSR-C27SM – 117 | CSSR 2016-IR18–12 | CSSR 2016-IR18–9 |
| JK – 95 | CSSR 2016-IR18–7 | CSSR 2016-IR18–10 |
| CSR-2748-4441-66 | CST 7-1 | CST 7-1 |
| CSR-2748-4441-133 | CARI Dhan 7 | CSAR17817 |
| – | CARI Dhan 10 | CSR 11–143 |

## 6.4 Conclusion

From the studies conducted over a period of years, higher grain yield per hectare was recorded with CSSR 2016-IR18-17, CARI DHAN7, CSSR 2016-IR18-12, CSSR 2016-IR18-7, CSR2016-IR18-9, CST 7-1, CARI Dhan 10, CR3881-M-3-1-5-3-1-1, CSR2016-IR18-10, CSAR-17817, CSR11-143, CSR-C27SM-117, JK-95, CSR-2748-4441–66, and CSR-2748-4441-133. These genotypes were found to be promising for majority of the traits studied and attained yield more than 5 tons/ha under salinity stress condition of Annamalai Nagar location. These genotypes can be utilized for identifying saltol resistant varieties through selection or hybridization for salinity tolerance. The STRASA project has helped in uplifting farmers from poverty, and the project gets completed during March 2019. The coverage of these rice varieties is progressively increasing every year because of its salinity tolerance, early maturity and suitability to local cropping systems, good taste, and good market price.

## References

Babu S, Anbumalarmathi J, Sheeba A, Yogameenakshi P, Rangasamy P (2005) Stability in performance of salt tolerant rice hybrids. Oryza 42:222–224

Gangaiah B (2019) Rice hybrids and high yielding varieties for augmenting public distribution system demand in Andaman & Nicobar Islands. India Int J Curr Microbiol App Sci 8(6): 3204–3211

Haque MA, Rafii MY, Yusoff MM, Ali NS, Yusuff O, Datta DR, Ikbal MF (2021) Advanced breeding strategies and future perspectives of salinity tolerance in rice. Agronomy 11(8):1631

Hoang TML, Tran TN, Nguyen TKT, Williams B, Wurm P, Bellairs S, Mundree S (2016) Improvement of salinity stress tolerance in rice: challenges and opportunities. Agronomy 6(4):54

Jayabalan S, Rajakani R, Kumari K, Pulipati S, Hariharan RVG, Venkatesan SD, Venkataraman G (2022) Morpho-physiological, biochemical and molecular characterization of coastal rice landraces to identify novel genetic sources of salinity tolerance. Plant Physiol Biochem 187: 50–66

Kashyap A, Yadav VK, Singh P, Singh PK, Shweta K (2020a) Mean and Euclidian cluster analysis for salt tolerance rice genotypes under alkaline soil condition. Int J Curr Microbiol App Sci 9(11):359–367

Kashyap A, Prasanna HC, Kumari N, Mishra P (2020b) Understanding salt tolerance mechanism using transcriptome profiling and de novo assembly of wild tomato solanum chilense. Sci Rep 10:15835

Khan S, Javed MA, Jahan N, Manan FA (2016) A short review on the development of salt tolerant cultivars in rice. Int J Public Health Sci (IJPHS) 5(2):201–212

Krishnamurthy SL, Lokeshkumar BM, Rathor S, Warraich AS, Yadav S, Gautam R, Sharma PC (2022) Development of salt-tolerant rice varieties to enhancing productivity in salt-affected environments. Environmental Sciences Proceedings 16(1):30

Mahalanobis PC (1936) On the generalized distance in statistics. Proc National Institute of Science (India) 2:49–55

Quan R, Wang J, Hui J, Bai H, Lyu X, Zhu Y, Huang R (2018) Improvement of salt tolerance using wild rice genes. Front Plant Sci 8:2269

Roshni AM, Saravanan KR, Harikrishnan M (2019) Screening for seedling stage salinity tolerance and comparative transcriptome analysis in Rice, Oryza sativa L M Sc (Ag.) Thesis, Annamalai Univerity,168

Roshni AM, Saravanan KR, Harikrishnan M, Anandan R, Prakash M (2022) Screening for seedling stage salinity tolerance and comparative transcriptome analysis in Rice, *Oryza sativa* L. IJEB 60:471–480

Sara GS, Saravanan KR, Abinaya S (2019) Studies on screening genotypes for salinity stress tolerance in rice genotypes (*Oryza sativa* L.). Plant Arch 19(2):2415–2419

Sehrawat N, Yadav M, Bhat K, Sairam R, Jaiwal P (2014) Evaluation of mungbean genotypes for salt tolerance at early seedling growth stage. Biocatal Agri Biotechnol 3:108–113

Singh RK, Kota S, Flowers TJ (2021) Salt tolerance in rice: seedling and reproductive stage QTL mapping come of age. Theor Appl Genet 134:3495–3533

Snehi S, Bhutia RN, Prakash NR (2022) Breeding Rice cultivars suitable for coastal regions of India. Food and Scientific Reports 3(4):54–58

Solis CA, Yong MT, Vinarao R, Jena K, Holford P, Shabala L, Chen ZH (2020) Back to the wild: on a quest for donors toward salinity tolerant rice. Front Plant Sci 11:323

Yeo AR, Flowers TJ (1982) Accumulation and localization of sodium ions within the shoots of rice (*Oryza sativa* L.) varieties differing in salinity resistance. Physiol Plant 56:343–348

Yadav S, Mondal MK, Shew A, Jagadish SK, Khan ZH, Sutradhar A, Bhandari H, Humphreys E, Bhattacharya J, Parvin R, Rahman M (2020) Community water management to intensify agricultural productivity in the polders of the coastal zone of Bangladesh. Paddy Water Environ 18:331–343

# Genetic and Agronomic Approaches for Higher Rice Productivity in Sundarbans Delta Region

7

Sukanta K. Sarangi, D. Burman, S. L. Krishnamurthy, and R. K. Singh

**Abstract**

Rice is the lifeline of the farming community in the Sundarbans region of the Ganges Delta. Monsoon rain-dependent rice cultivation during the wet season (*Kharif*; June–November) and groundwater-irrigated rice during the dry season (*Boro*; December–May) form the major cropping activity. Rice cultivation is challenged by several abiotic (soil and water salinity, soil acidity, drought, flood, storms, cyclones) and biotic (insects, diseases, weeds, rodents) stresses. An increase in population density, environmental stresses, and climatic hazards in the Sundarabans delta region necessitates higher productivity by deploying tolerant high-yield genotypes and adaptive management strategies. Development of high-yielding stress-tolerant genotypes requires rigorous multilocation and multiyear testing to confirm suitability. In this chapter, evaluation results of several breeding lines over locations and years are reported. Apart from the best genotypes, commensurate agronomic management strategies involving producing robust seedlings, balanced nutrition of crop and soil, and adaptive cropping systems are also required to cope with the increasing challenges as well as to achieve higher income.

S. K. Sarangi (✉) · D. Burman
ICAR-Central Soil Salinity Research Institute, Regional Research Station, Canning Town, West Bengal, India
e-mail: S.Sarangi@icar.gov.in

S. L. Krishnamurthy
ICAR-Central Soil Salinity Research Institute, Karnal, Haryana, India

R. K. Singh
International Center for Biosaline Agriculture, Dubai, United Arab Emirates





R. K. Singh et al. (eds.), *Genetic Improvement of Rice for Salt Tolerance*,
https://doi.org/10.1007/978-981-99-3830-8_7

**Keywords**

Salinity · Stress score · Sundarbans · Rice · Genotypes · Nutrient management

## 7.1 Introduction

In the coming decades, the farm families of India need to produce more food grains and other agricultural commodities under conditions of shrinking per capita land and water resources and expanding biotic as well as abiotic stresses. India needs to increase its rice production at the rate of 2.0 million tons per year until 2050 to meet its food security. By 2025, 15 out of 75 million hectares of Asia's flood-irrigated rice crop will experience water shortage (Tuong and Bouman 2003). To meet the burgeoning food demand due to the increasing population pressure, more foods have to be produced per unit of cultivated area by increasing the land and water resource use efficiency in manifolds.

Rice is grown in many regions across India. India has about 44 million hectares of area, with a production of 110 million metric tons of rice (average of 2014–2015 to 2018–2019) and an average productivity of 2.5 t $ha^{-1}$ (Government of India 2020). Rice cultivation could be found in almost all regions of India having the necessary warmth and abundant moisture favorable to its growth, mainly subtropical rather than hot or cold. In India, rice is grown in different regions; however, rice is more a choice crop in coastal regions, due to high rainfall, food habits, and topographic congeniality. The Sundarbans delta region situated in West Bengal, India, and Bangladesh is a typical rice-growing area both during wet (monsoon) and dry seasons (pre- and post-monsoon). The soils are mostly saline, rainfall is very high (about 1800 mm per annum), and topography is low-lying saucer shaped. The low-land rice ecosystem of this region suffers from stagnant flooding (SF) during the wet season, which persists for about 4 months (July–October) (Photo 7.1a). During the dry season as the soil dries, soil salinity and cracking gradually develop on the soil layer (Photo 7.1b). Alluvial soils are rich in potash but are deficient in organic matter, nitrogen, and phosphorus. In the Gangetic delta, rice is grown in saline soils of Sundarbans with several challenges. Recent developments in research

**Photo 7.1** (**a**) Stagnant flooding of rice fields during the wet season. (**b**) Soil salinity and cracking during the dry season in the Sundarbans delta region

and innovations brought considerable opportunities for productive agriculture by introducing abiotic stress-tolerant rice varieties (STRVs), tailored management practices, integrated farming systems, crop diversification, and, more recently, the use of information technology for monitoring and access to knowledge (Ismail et al. 2022). Broad aspects to tackle the challenges in rice production in this agroecosystem are by genetic improvement and agronomic management options.

## 7.2 Genetic Approaches to Achieve Higher Rice Productivity

### 7.2.1 Multilocation Selection of Better Genotypes

In multilocation trials in the Sundarbans delta region during wet and dry seasons of 2008–2014, the best identified genotypes for the wet season, yielding >4 t $ha^{-1}$ grain yield, were CSRC(D)-12-8-12, Geetanjali, Amal-Mana, and Swarna-Sub 1, whereas for the dry season, the best entries were Gontra Bidhan-2, BINA dhan8, Boby, Annada, N Sankar, BRRI dhan47, IR 76393-2B-7-1-1-3-1, Lal Minikit (WGL 20471), Parijat, Super Sankar, and Rasi (Burman et al. 2018). New salt-tolerant rice varieties will contribute to the productivity of more than 20 mha of farmland suited to rice production in Asia, which are affected by excess salt and other salt-related problems (Gregorio et al. 2014). For the wet season, farmers preferred genotypes that were tall (140–170 cm), of long duration (maturity 160–170 days), lodging resistant, and high yielding as these traits are required for the lowland situation where SF stresses are frequent. During the dry season, farmers preferred genotypes that are high yielding, salt tolerant, and early maturing (115–130 days) with long, slender, and good quality grains for better market price (Burman et al. 2018).

Under SF situation, rice variety Amal-Mana combined with a fertilizer dose of 50–20-10 kg N-$P_2O_5$-$K_2O$ and 5 tons of farmyard manure (FYM) per hectare, transplanting two seedlings per hill at a spacing of 15 cm × 15 cm resulted in higher grain yield (4.51 t $ha^{-1}$) than that of the farmers' management and varieties (2.55 t $ha^{-1}$) (Sarangi et al. 2016). For areas affected by flash floods, CR 1009-Sub 1, BR 11-Sub 1, and Swarna-Sub 1 could be recommended for cultivation (Sarangi et al. 2020a).

### 7.2.2 Salinity-Tolerant Breeding Network for the Identification of Superior Genotypes

For increasing rice productivity in saline soils, a multilocation and multiyear evaluation is indispensable for identifying genotypes with consistency in adaptation to a particular agroclimatic condition (Krishnamurthy et al. 2017). Salinity Tolerant Breeding Network (STBN) trial was conducted at the experimental farm of the Indian Council of Agricultural Research (ICAR)-Central Soil Salinity Research Institute (CSSRI), Regional Research Station (RRS), Canning Town during *Kharif*

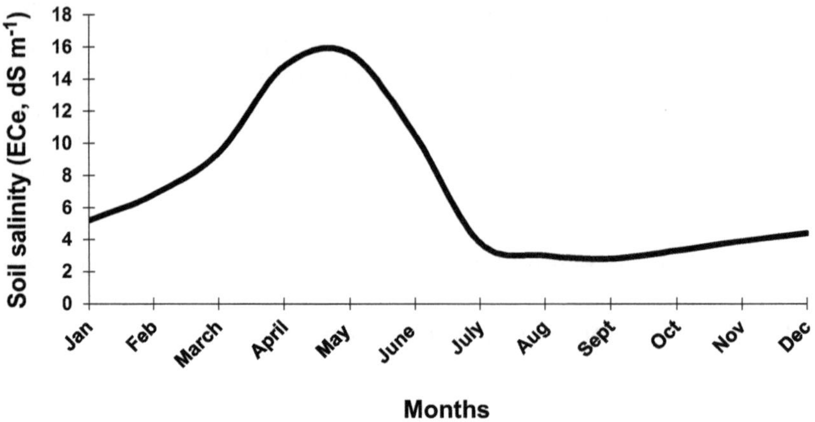


**Fig. 7.1** Seasonal variability of soil salinity in the Sundarbans delta region

2016–2018. ICAR-CSSRI, RRS at Canning Town is located in the coastal region of Ganges Delta (Sundarbans delta), about 46 km Southeast of Kolkata. The cropping pattern has been observed to be primarily mono-cropped with rice (80% of agricultural land) in *Kharif*. Most of the lands remain fallow (about 80% of agricultural land) during *Rabi* season due to severe stress of salinity and scarcity of good quality irrigation water. The soils of the experimental site have hyperthermic temperature and aquic moisture regimes. The experimental soil was heavy in texture (silty clay) and normal in soil reaction. Soil salinity during the dry months was very high (ECe: 9–12 $dSm^{-1}$). However, it varied from 3.65 to 6.87 $dSm^{-1}$ during *Kharif* season.

Soil salinity is one of the major constraints for agricultural production in the study areas of the coastal salt-affected region of Sundarbans (coastal areas of Ganges Delta, West Bengal), especially during dry season (pre- and post-monsoon period). Salinity builds up due to gradual drying up of lands after monsoon season and attains maximum in the month of April–May (Fig. 7.1b). It builds up in the soil due to upward capillary movement of saline groundwater present at shallow depth and occasional inundation with saline water from the river. Farmers in this region usually identify the salinity stress by visual observation of salt deposition on the soil surface. July, August, and September are the rainiest months of the year (Table 7.1), during which the rice crop often suffers from submergence.

The experiment during *Kharif* 2016 consisted of 66 rice genotypes including five check varieties (PUSA 44, CSR 10, CSR 27, CSR 36, and CSR 7–1) and one local check variety (Amal-Mana). At the reproductive stage, stress score varied from 1 (CR3903-161-1-3-2, CR3881-4-1-6-3-4-1, CR3909-192-1-7-2, CR2838-1-S-2B-9-1, CR2860-S-B-189-1-1-1, CR3884-244-8-5-11-1-4, TR 05043, NDRK 11-24, CSR-2748-4441-193, CSR-2748-4441-104, KR 15014, CSR 7-1 (check), and Amal-Mana (local check)) to 5 (CR3878-245-1-9-2, CSR-2748-4441-133, PAU 3835-12-1-1-1, RP-320-4-3-2-1, RAU 1397-14, and CARI Dhan 6) with a mean value of 3 (Fig. 7.2). The grain yield of the entries varies from 956.4 to

**Table 7.1** Weather data during 2016–2018 at Canning Town, West Bengal, India

| Weather parameters | June | July | August | September | October | November | December |
|---|---|---|---|---|---|---|---|
| | 2016 | | | | | | |
| No. of rainy days | 13 | 22 | 20 | 20 | 8 | 3 | 0 |
| Total rainfall (mm) | 217.9 | 502.4 | 619.0 | 244.0 | 27.5 | 48.2 | 0 |
| Avg. maximum temperature (°C) | 33.9 | 29.3 | 31.8 | 32.19 | 31.86 | 29.6 | 26.6 |
| Avg. minimum temperature (°C) | 26.9 | 24.7 | 26.6 | 26.56 | 24.75 | 18.7 | 15.0 |
| Sunshine hours | 6.7 | 2.4 | 4.5 | 5.15 | 5.8 | 7.4 | 6.3 |
| Wind velocity | 4.1 | 2.7 | 2.6 | 0.6 | 0.3 | 1.1 | 1.9 |
| | 2017 | | | | | | |
| No. of rainy days | 17 | 24 | 23 | 10 | 15 | 5 | 3 |
| Total rainfall (mm) | 170.8 | 567.4 | 420.3 | 173.6 | 420.8 | 34.6 | 28.8 |
| Avg. maximum temperature (°C) | 35.0 | 31.7 | 32.8 | 33.7 | 31.4 | 29.3 | 25.7 |
| Avg. minimum temperature (°C) | 27.4 | 26.7 | 27 | 27.2 | 24.9 | 19.1 | 15.6 |
| Sunshine hours | 5.4 | 2.7 | 3.8 | 4.6 | 5.5 | 6.9 | 5.9 |
| Wind velocity | 7.3 | 5.8 | 5.1 | 3.7 | 4.1 | 2.0 | 2.0 |
| | 2018 | | | | | | |
| No. of rainy days | 16 | 22 | 20 | 11 | 06 | 01 | 02 |
| Total rainfall (mm) | 416.0 | 446.4 | 229.8 | 164.6 | 110.4 | 4.6 | 15.4 |
| Avg. maximum temperature (°C) | 34.9 | 32.5 | 32.8 | 33.4 | 32.6 | 30.3 | 25.7 |
| Avg. minimum temperature (°C) | 26.5 | 26.6 | 26.9 | 26.7 | 23.6 | 19.1 | 13.4 |

(continued)

**Table 7.1** (continued)

| Weather parameters | June 2016 | July | August | September | October | November | December |
|---|---|---|---|---|---|---|---|
| Sunshine hours | 4.7 | 3.4 | 4.6 | 4.7 | 5.7 | 4.7 | 6.6 |
| Wind velocity | 5.1 | 6.2 | 5.4 | 4.2 | 2.6 | 1.5 | 2.1 |

5388.7 kg ha$^{-1}$ with a mean yield of 3523.9 kg ha$^{-1}$ (Fig. 7.3). Out of all the entries, a higher grain yield was recorded from CSR-2748-197 (4748.1 kg ha$^{-1}$), CSR 36 (check) (4830.5 kg ha$^{-1}$), CR3903-161-1-3-2 (5009.5 kg ha$^{-1}$), CR3881-M-3-1-5-1-1-1 (5320.1 kg ha$^{-1}$), CR3884-244-8-5-6-1-1 (5373.1 kg ha$^{-1}$), CR3881-4-1-6-3-4-1 (5380.7 kg ha$^{-1}$), and Amal-Mana (local check) (5388.7 kgha$^{-1}$).

During *Kharif* 2017, the trial consisted of 88 rice genotypes including six check varieties (CSR 10, CSR 27, CSR 36, CST7–1, PUSA44, and Sambha Sub 1) and one local check variety (Amal-Mana). At the reproductive stage, stress score varied from 1 (KR15006, CR3884-244-8-5-6-1-1, CR 2839-1-S-10-B2-B-43-2B-1, CR 3882-7-1-6-2-2-1, RAU 1397-14, TR 09030, IR 83421-6-B-3-1-1, CR 3364-S-2B-14-2B-1, RP 5690-20-6-3-2-1, CSR RIL-01-IR165, CSR 2016-IR18-3, CSR 2016-IR 18-16, CST7-1 (check), and Amal-Mana (local Check)) to 4 (CSAR 1604, RP-5683-101-85-30-2-3-1, CR3881-M-3-1-5-1-1-1, TR 05043, PAU 4254-14-1-2-2-2-4-1, RP 5440-302-100-7-6-3-2, CSR- 2748-4441-193, CSR 2016-IR18-6, CSR 2016-IR18-11, CSR 2016-IR18-14, RAU 720, PUSA44 (check)) with a mean value of 4. The grain yield of the entries varies from 919.1 to 5401.7 kg ha$^{-1}$ with a mean yield of 3576.8 kg ha$^{-1}$ (Fig. 7.4). Out of all the entries, a higher grain yield was recorded with RP 5694-36-9-5-1-1 (5011.8 kg ha$^{-1}$), CR 3903-161-1-3-2 (5015.8 kg ha$^{-1}$), RP-320-4-3-2-1 (5090.7 kg ha$^{-1}$), CR 2859-S-B-2-1-B-7 (5098.6 kg ha$^{-1}$), CR 2851-S-B-1-B-B-1 (5209.1 kg ha$^{-1}$), CR 2839-1-S-10-B2-B-43-2B-1 (5359.0 kg ha$^{-1}$), and Amal-Mana (local check) (5401.7 kg ha$^{-1}$).

During *Kharif* 2018, the experiment consisted of 82 rice genotypes including six check varieties (CSR 10, CSR 27, CSR 36, CST7-1, PUSA44, and NSICRC222) and one local check variety (Amal-Mana). At the reproductive stage, stress score varies from 1 (CSR2016-IR18-15, CSR2016-IR18-17, CSR2016-IR18-1, CSR2016-IR18-2, CSR2016-IR18-3, CSRRIL-01-IR75, CSRRIL-01-IR165, CSR2016-IR18-11, CSR2016-IR18-14, CR3878-245-2-4-1, IR83421-6-B-3-1-1/CR3364-S-2B-14-2B, PAU3835-36-6-3-3-4, CSAR1620, CARI Dhan 6, JK-58, JK-238, NDRK 11-27, CR2851-S-1-6-2B-4-1, RAU1525-1-2, PAU5564-18-1-2, IRLON-GSR5, CSR10 (check), CST7–1 (check), Pusa 44 (check), and Amal-Mana (local check)) to 5 (CR3881-4-1-3-7-2-3) with a mean value of 2. The grain yield of the entries varies from 1217.0 to 4771.9 kg ha$^{-1}$ with a mean yield of 3644.9 kg ha$^{-1}$(Fig. 7.5). Out of all the entries, a higher grain yield was recorded with CARI Dhan 6 (4607.5 kg ha$^{-1}$), IRLON-GSR5 (4613.4 kg ha$^{-1}$), KR15075 (4648.9 kg ha$^{-1}$), CSR2016-IR18–15 (4670.6 kg ha$^{-1}$), and Amal-Mana (local check) (4771.9 kg ha$^{-1}$).

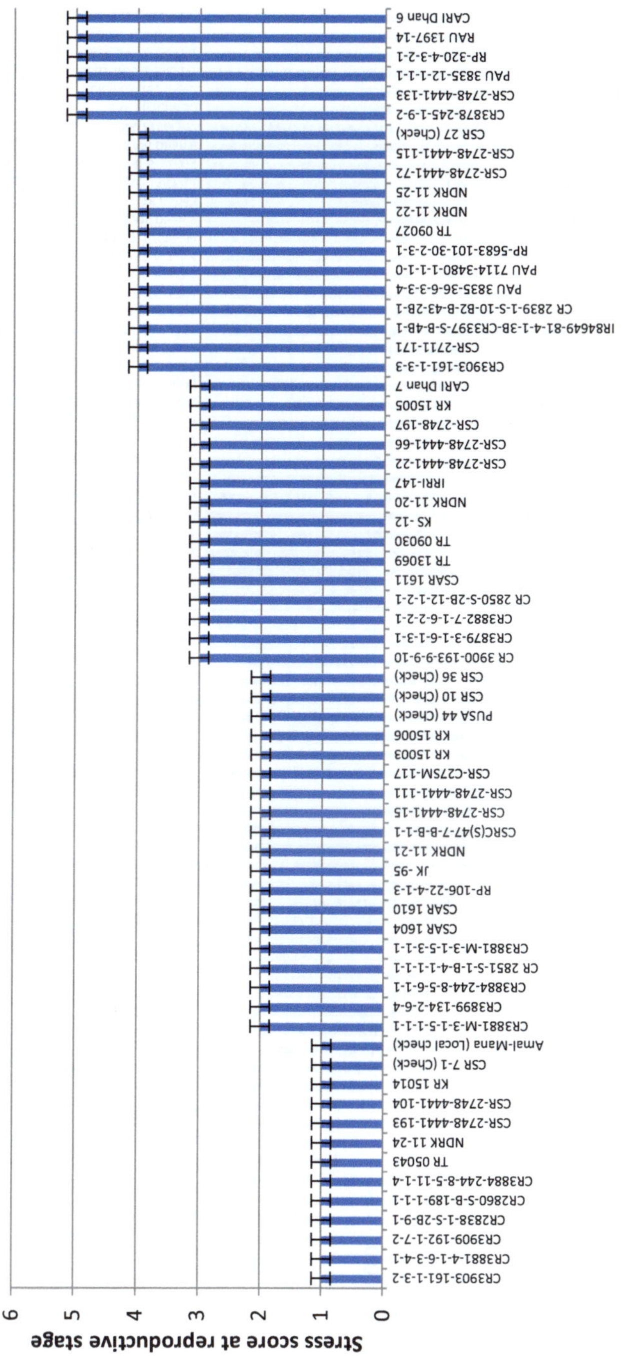


**Fig. 7.2** Stress score of rice genotypes at the reproductive stage during *Kharif* 2016

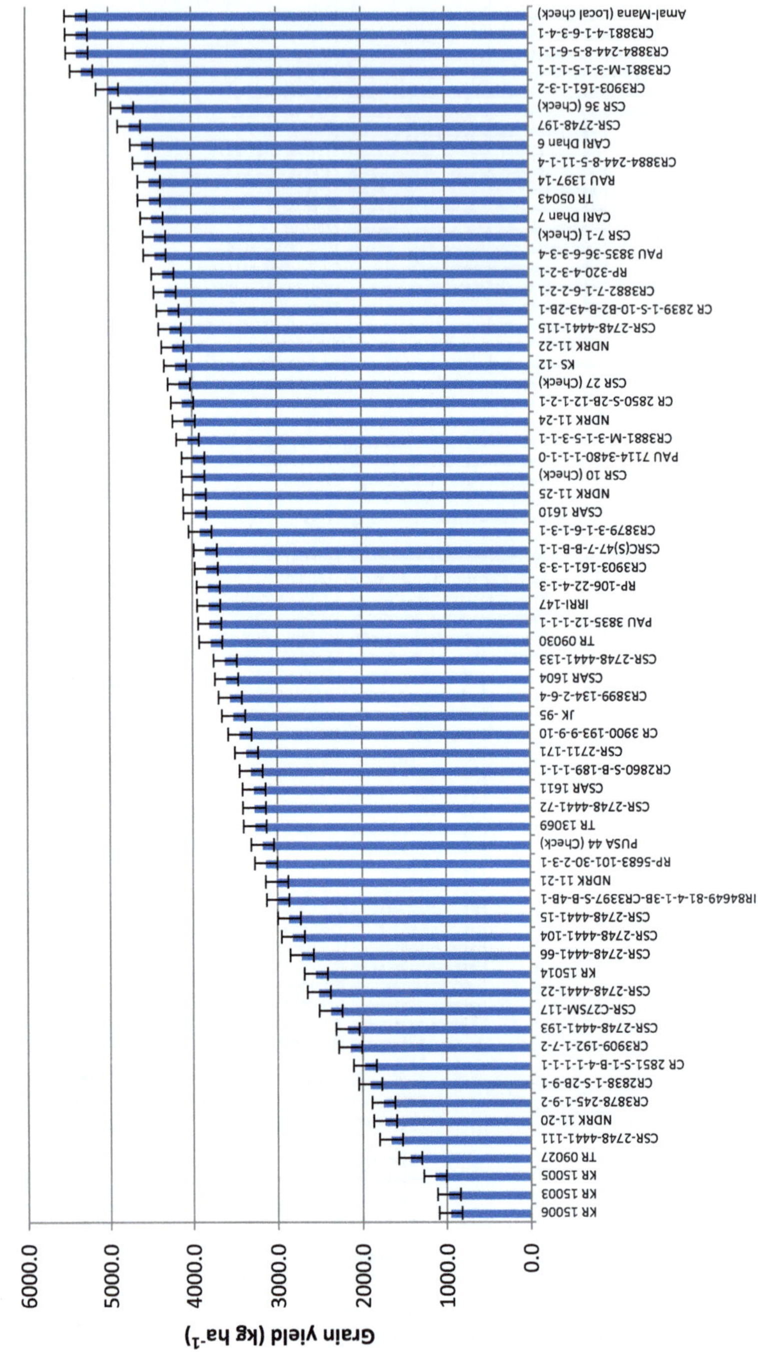


**Fig. 7.3** Grain yield of 66 entries under STBN trial during *Kharif* 2016

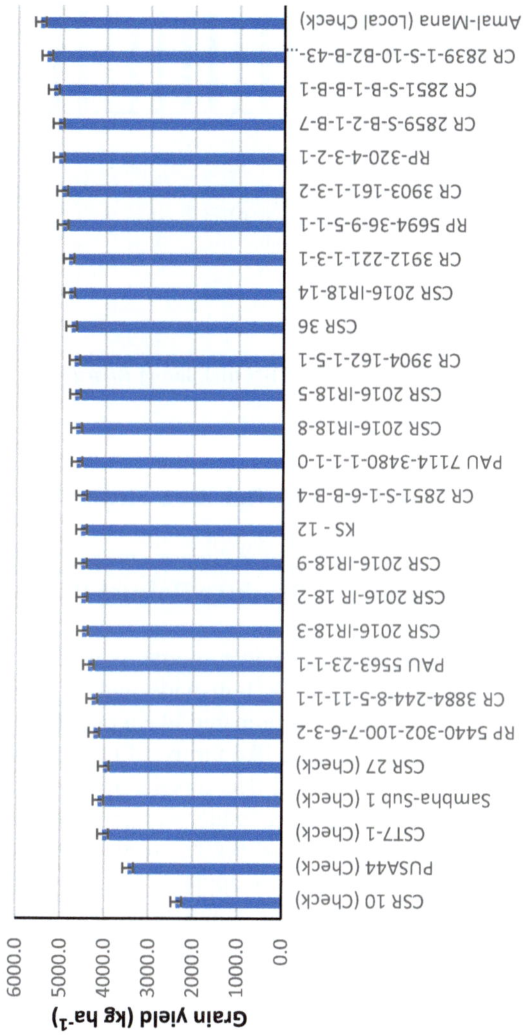


**Fig. 7.4** Grain yield of selected entries under STBN trial during *Kharif* 2017

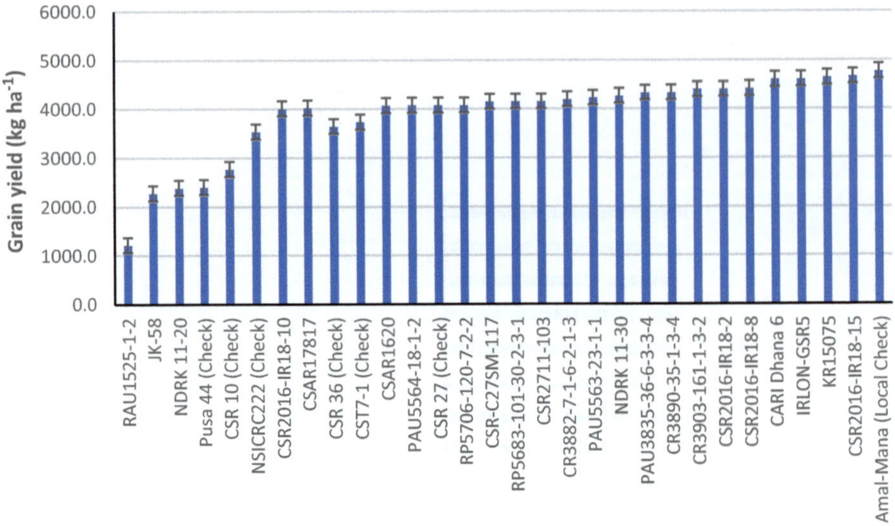


**Fig. 7.5** Grain yield of selected entries under STBN trial during *Kharif* 2018

## 7.3 Agronomic Approaches to Achieve Higher Rice Productivity

### 7.3.1 Nursery Management to Produce Robust Rice Seedlings

Nursery area should be chosen close to a water source, with good drainage facility and free from shade. Seed bed should be 1 m wide and of a convenient length (maximum 10 m). The wet nursery bed is prepared like a puddled rice field but with extra care. Application of organic material like farmyard manure (FYM) is essential for a healthy nursery bed. FYM is the well-decomposed mixture of dung with litter and leftover material from roughages and fodder fed to cattle. The N, P, and K concentrations of FYM are 0.50, 0.25, and 0.40%, respectively. At least, 500 kg FYM for a nursery area of 1000 $m^2$ should be applied. The use of FYM significantly increased soil organic matter, soil N and P status, hydraulic conductivity, water-holding capacity, and decreased soil bulk density, thus creating a good environment for growth and development of rice. After puddling and leveling, beds are made to a height of 25–30 cm, and the top is properly leveled to avoid submergence of seedlings due to heavy rain after emergence.

To produce vigorous seedlings, seeds of good quality with high germination percentage (90%) are sown uniformly using a seed rate of 25 g $m^{-2}$. A proper wider spacing for seedlings in the nursery enhances seedling growth because of better light interception and higher nutrient and moisture uptake. Seedlings thus produced establish faster after transplanting, survive better, and exhibit quick post-submergence recovery. During the nursery sowing, the pre-germinated seeds may be scattered uniformly with a mild force onto the raised bed to fix them in the puddled soil. This would help seeds to anchor quickly and not to be carried away by the irrigation/floodwater. The nursery bed should be well protected from damage by birds, squirrels, other animals, and pests.

It is always preferable to raise the nursery during the wet season in high lands with good water control. Irrigation should be well managed according to the length of seedlings, and excess water should be drained out. Excess water may lead to the formation of taller and weaker seedlings, and many times, prolonged flooding also causes seedling death. It is recommended to maintain a water level of 2–3 cm initially for 10 days. When there is a sudden heavy rain causing submergence of the seedbed, excess water should not be completely drained in one go, as the entire shoot along with the leaves will touch the muddy soil resulting in seedling mortality due to decomposition. In this case, the water should be gradually drained as the leaves float over the water, then the seedlings are allowed to put forth new leaves above the water level; subsequently, the seedlings get stronger, and then, the rest of the excess water should be drained. When the water level recedes and there is no further rain, progressively raise the water level until seedlings are ready to be pulled for transplanting.

The application of balanced nutrients in the seedbed has several advantages. In the relatively small area of the nursery, even small amounts of fertilizers can make a big difference. Healthier seedlings cope better with the transplanting shock and perform better when conditions are suboptimal. The application of a small amount of nutrients at 5–3-1.5 kg of N-$P_2O_5$-$K_2O$ + 500 kg FYM in a nursery area of 1000 $m^2$ is enough to produce healthy and robust seedlings with high seedling vigor for 1 hectare of main field for transplanting (Sarangi et al. 2015). The full dose of phosphorus and potassium and 50% of nitrogen are to be applied as basal at sowing. The rest of the nitrogen should be applied in two equal splits at 15 days of sowing and 7 days before uprooting. In the soils where zinc is deficient, two sprays of zinc sulfate (0.5% $ZnSO_4$ + 0.25% Ca $(OH)_2$ mixture), i.e., 50 g $ZnSO_4$ + 25 g Ca hydroxide mixed in 10 liters of water for 100 $m^2$ area after 10 days' interval from sowing may be applied. Adding organic sources such as FYM not only supplies macronutrients but also essential micronutrients such as Fe, Mn, Cu, and Zn.

### 7.3.2 Nitrogen Fertilization Management Options

Nitrogen (N) in the form of neem-coated urea is the preferred source over prilled urea, and it should to be avoided as basal dose just at the time of transplanting. Instead, the same should be applied 1 week after transplanting when there is clear weather and no possibility of heavy rain within a couple of days (Sarangi et al. 2020b). The N fertilizer may be placed into the soil (1.5–2.0″) or may be applied in the form of mud-ball for better efficiency. Combined sources (organic + inorganic) of N will produce better yields.

**Table 7.2** Nitrogen-containing inorganic fertilizers

| Name of the fertilizer | N % | Remarks |
|---|---|---|
| A) Urea | 44.0–46.0 | • Urea, ammonium sulfate nitrate, calcium ammonium nitrate, and calcium nitrate are generally used. Among which cost for 1 kg nitrogen is cheaper in urea.<br>• Ammonium sulfate should not be mixed with seed as germination of seed will be affected.<br>• Ammonium sulfate nitrate should not be mixed with lime and should not be applied along with the seed.<br>• Urea is an ideal fertilizer for foliar application, and its contact with seed should be avoided. |
| B) Ammonium sulfate | 19.9–20.0 | |
| C) Ammonium sulfate calcium nitrate | 26.0 | |
| D) Ammonium nitrate | 25.0 | |
| E) Ammonium chloride | 25.0 | |
| F) Calcium nitrate | 13.0–15.0 | |
| G) Sodium nitrate | 16.0 | |
| H) Calcium cyanamide | 21.0 | |
| I) Anhydrous ammonium | 82.0 | |
| J) Ammonium nitrate | 32.0–35.0 | |

The N fertilizer application is better onto a dry soil surface. But in the transplanted rice crop, as the condition of soil wetness is a prerequisite, N fertilizer is applied onto the saturated or muddy soil. The water stagnation is to be maintained immediately to curtail N losses in the form of $NH_3$ volatilization. The inundation should be completed over the applied area within 5 days. The larger N rate should not be applied into the floodwater because the rice plant does not have an extensive root system at this stage or the capacity to take up N quickly enough to compete with the N-loss processes. Fertilizer N applied at midseason, at the proper times and in the proper amounts, is taken up in 3 to 7 days with 65% to 80% efficiency. At the reproductive growth stage, the rice plant develops an extensive root system near the soil surface. Midseason N application should be well timed according to plant development and type of variety (Zhang et al. 2013); that is, applied at any time from panicle initiation (i.e., beginning of internode elongation or green ring) to differentiation (i.e., the top internode has a 1.5 cm spacing).

Heavy textured clayey soils can fix part of the preflood N fertilizer and make it momentarily unavailable for plant uptake (Smith and Dilday 2002). During the midseason, the N fertilizer is practically applied onto the field water as the water level reaches the peak depth. Some of the applied N is lost through leaching (Singh and Craswell 2021) before it's taken up by the rice plant. Therefore, it is recommended for split application rather than bulk supply. Recently, various N diagnostic techniques such as chlorophyll (SPAD) meter and GreenSeeker have been standardized for optimum N fertilizer application as per the plant need. The source of the fertilizer plays a significant role in increasing nutrient use efficiency. The commonly used N fertilizer sources are given in Table 7.2.

### 7.3.3 Phosphorus Fertilizer Management Options

Adequate P nutrition of rice is essential since it is required for energy storage and transfer within the plant (Tahir et al. 2019). In addition to the metabolic functions, P has been observed to increase root growth and promote early maturity, straw strength, crop quality, and disease resistance. Total P uptake increases in conjunction with total dry matter. Several P fertilizer sources (Table 7.3) are effectively utilized in rice production containing three plant nutrients, viz., phosphorus, sulfur, and calcium. Single superphosphate (SSP) is the most common one and typically contains 16% P, 11% S, and 21% Ca. Diammonium phosphate (DAP) contains 18% N and 20% P (46% $P_2O_5$) and is commonly used either as preplant or preflood application because of the additional N. P fertilizer can be combined with the N fertilizer and applied at the four- to five-leaf growth stage before the increase in the depth of stagnant flooding. DAP as a source of P is frequently used in this condition since it contains a substantial amount of $NH_4$-N (15.5%). In P-fixing soils, an adequate amount of P fertilizer or manure application is necessary in order to provide sufficient plant-available P (Vinod and Heuer 2012). On soils that have a history of P deficiency or are highly responsive to P fertilization, split applications have been utilized successfully. One-half of the P fertilizer is applied prior to seeding, followed by the remaining amount applied prior to flooding. Phosphorus fertilizer should be applied directly to the rice crop at preplant or preflood on responsive soils, since P fertilizer applied to previous crops may be unavailable to rice during critical growth stages. Phosphatic biofertilizers such as phosphate-solubilizing bacteria (PSB), which solubilize insoluble P to soluble forms, could augment P fertilizer input to rice crops (Stephen et al. 2015).

### 7.3.4 Potassic Fertilizer Management Options

Potassium deficiency has not been documented on clay soils used for rice production in this rice-growing region. These heavy textured alluvial soils are typically very

**Table 7.3** N, P, and K content of compound fertilizers

| Compound fertilizers | | | | |
|---|---|---|---|---|
| Form | N | $P_2O_5$ | $K_2O$ | Companies |
| A Diammonium phosphate (DAP). | 18 | 46 | 0 | GFCL, SPIC, PPL |
| B Urea ammonium phosphate (Gromor). | 28 | 28 | 0 | CFL, FACT, ZACL FACT |
| C Nitrophosphate. | 24 | 24 | 0 | Mahadhan |
| NPK | 10 | 26 | 26 | IFFCO |
| | 17 | 17 | 17 | MFL |
| | 15 | 15 | 15 | RCF |
| | 19 | 19 | 19 | ZACL |
| | 14 | 35 | 14 | CFL, ZACL |

Complex fertilizers are costlier than the straight fertilizers high in exchangeable $K^+$ and contain significant amounts of clay-fixed or nonexchangeable K. However, to maintain K balance in high-intensity cropping systems or under high-yield targets, K applications in maintenance doses are advised. In intensively cultivated soils, rice crop canopy suffering from K deficiency may not show greenness even with top dressing of N fertilizer. The commonly used K fertilizer in the Sundarbans region is muriate of potash (MOP), which is potassium chloride (KCl) containing 60% potash ($K_2O$). MOP is a mined product and an imported fertilizer for India.

### 7.3.5 Micronutrient and Agronomically Important Nutrient Fertilization

Cropping system intensification and high system yield targets require balanced crop nutrition including micronutrients. Often land leveling and modifications remove surface soil that contains the bulk of soil nutrients and expose subsoils that may have poor fertility or undesirable characteristics (e.g., low organic matter content, textural changes, saline/sodic/acidic subsoils). Maintenance of organic matter in the rhizosphere is very important to ensure micronutrient availability to crops. Zinc deficiency normally occurs early during vegetative growth, and very little yield loss may be experienced if Zn deficiency of seedling rice is quickly corrected and proper management is followed. Complete removal of the floodwater is required for plant recovery when Zn deficiency is severe. Recommendations for rescue situations of Zn-deficient rice are to (1) drain the flood, (2) dry the soil, (3) watch for evidence of new shoot (emerging green leaf) and root (white roots) growth, (4) apply a Zn fertilizer solution to rice foliage, (5) apply ammonium sulfate, and (6) reflood. Application of ammonium sulfate is for growth stimulation and to account for some N loss resulting from draining and drying the soil. Ammonium sulfate is used instead of urea because urea causes a brief increase in soil pH from hydrolysis of the urea fertilizer, which may further aggravate the Zn deficiency. Foliar application of Zn fertilizer without draining the flood may be practical only when Zn deficiency is mild and/or found quickly. The two most widely used Zn fertilizers are $ZnSO_4$ and Zn-EDTA (chelated Zn).

After oxygen, silicon (Si) is the second most abundant mineral in the earth's crust. Though Si is not considered an essential element for many plant species, it is considered a "beneficial" element for rice crops. Adequate Si nutrition improves straw strength, increases photosynthetic efficiency, and imparts disease and insect tolerance. The critical level of Si in rice plants tissue is less than 5% Si, rice straw at maturity should have an optimum amount of Si (8%–10%), and the critical soil concentration for the occurrence of Si deficiency is 40 mg Si $kg^{-1}$ (Dobermann and Fairhurst 2000).

### 7.3.6 Integrated Nutrient Management to Boost Rice Yield in Saline Soils

Soils of Sundarbans delta region are degraded due to the effect of salinity and acidity. Soil salinity (ECe) peaks to about 16 $dSm^{-1}$, and groundwater salinity (EC) reaches to 7 $dSm^{-1}$ during the driest months (April–May) severely affecting the dry season cropping (Sarangi and Islam 2019). Highly acidic (pH $< 4.0$) saline soils are occasionally found at a number of places in Sundarbans region of coastal West Bengal (Sarangi and Maji 2017). Therefore, an integrated nutrient management approach could sustain the rice yield and improve soil health. Green manuring with *Dhaincha* (*Sesbania* spp.) during the *Kharif* season significantly increases grain yield of rice. However, in acid saline soils of Sundarbans, application of soil amendments (lime/rock phosphate) along with green manuring boosts the rice yield to 5.66–6.17 t $ha^{-1}$ compared to 3.79–3.96 t $ha^{-1}$ under control (Table 7.4). Out of the three-year study in the Sundarbans, it was observed that in the first year, application of both lime and rock phosphate along with green manuring recorded significantly higher grain yield (6.17 t $ha^{-1}$) of rice. During the subsequent 2 years, either application of lime or rock phosphate along with green manuring sustains the yield of *Kharif* rice (ICAR-CSSRI, CSI4CZ 2020).

During the dry season, integrated nutrient management involving inorganic fertilizer and *Azolla* (a free-floating freshwater fern provides organic matter under decomposition and N-fixation by the blue-green algae *Anabaena,* present in the lobes of *Azolla* leaves) and farmyard manure (FYM) significantly increases grain yield of rice (Sarangi et al. 2014). In the second year of study, the grain yield was significantly higher (4.41–5.02 t $ha^{-1}$) under integrated application of a recommended dose of fertilizer and Azolla/FYM application compared to application of RDF alone (3.77 t $ha^{-1}$).

**Table 7.4** Grain yield (t $ha^{-1}$) of rice during the *Kharif* season under acid saline soil management practices in Sundarbans delta

| Treatments | 2017 | 2018 | 2019 |
|---|---|---|---|
| Control | 3.96e | 3.79c | 3.83c |
| Green manuring (GM) | 4.45d | 5.27ab | 5.39b |
| Lime (L)[a] | 4.23e | 4.68b | 5.11b |
| Rock phosphate (RP)[b] | 4.17de | 4.91ab | 5.13b |
| L + RP | 4.68c | 5.39ab | 5.28b |
| GM + L | 5.74b | 5.47ab | 5.93a |
| GM + RP | 6.04b | 5.38ab | 5.90a |
| GM + L + RP | 6.17a | 5.66a | 5.97a |

[a]Lime @ 1.5 t $ha^{-1}$
[b]RP @ 0.25 t $ha^{-1}$
Data with the same letters in a column are not significantly different
Source: ICAR-CSSRI, CSI4CZ (2020)

### 7.3.7 Site-Specific Nutrient Management Approach

The site-specific nutrient management (SSNM) approach was developed in Asian rice-producing countries through partnerships of the Irrigated Rice Research Consortium (IRRC). It emphasizes "feeding" rice with nutrients as and when needed. SSNM strives to enable farmers to dynamically adjust fertilizer use to optimally fill the deficit between the nutrient needs of a high-yielding crop and the nutrient supply from naturally occurring indigenous sources such as soil, organic amendments, crop residues, manures, and irrigation water (Dobermann and White 1998). The SSNM method does not precisely aim to either decrease or increase fertilizer use. It aims to emphasize the application of nutrients from the right source, at optimal rates, at the right time, and at the proper place to achieve high yield and high efficiency of nutrient use by the rice crop, leading to higher income from the harvest per unit investment on fertilizer.

### 7.3.8 Cropping System Approach to Improve Productivity and Profitability

To boost productivity and profitability in rice-based systems in the Sundarbans delta, the use of crop residues of the previous crop as input in the subsequent crop is essential. Recycling of 40% crop residues in rice-maize system increased rice grain yield by about 16%, reduced irrigation water requirement of *Rabi* maize by 37%, and saved 40 kg N $ha^{-1}$ (Sarangi et al. 2020c). In rice-potato system, zero tillage planting of potato with paddy straw mulching reduced the cost of cultivation of potato by 27% compared to conventional tillage intensive planting (US$ 1211.6 $ha^{-1}$) and increased net return to US$ 1779.2 $ha^{-1}$ from US$ 589.6 $ha^{-1}$ under conventional system (Sarangi et al. 2021).

## 7.4 Epilogue

Rice farming is synonymous to agriculture in coastal regions in general and Sundarbans delta region in particular. Farmer's livelihood, culture, and lifestyle are deeply linked with rice farming. In the recent years, due to changing climate, salinity intrusion, and population pressure, there is a necessity for sustainable rice farming in this highly populated region. Therefore, strategies involving genetic improvement and agronomic interventions are needed to make the avocation profitable in perpetuity. A breeding approach involving tolerant characteristics (to abiotic and biotic stresses) of local landraces, high yielding, and good grain quality of modern cultivars is needed to meet the requirement. Agronomic approaches are needed to increase cropping intensity and improve soil health and profitability of the farming system as a whole.

## References

Burman D, Maji B, Singh S, Mandal S, Sarangi SK, Bandyopadhyay BK, Ismail AM (2018) Participatory evaluation guides the development and selection of farmers' preferred rice varieties for salt-and flood-affected coastal deltas of South and Southeast Asia. Field Crop Res 220:67–77

Dobermann A, Fairhurst T. (2000) Rice: Nutrient disorders & nutrient management. International Rice Research Institute, p 188

Dobermann A, White PF (1998) Strategies for nutrient management in irrigated and rainfed lowland rice systems. Nutr Cycl Agroecosyst 53:1–18

Government of India (2020) Agricultural statistics at a glance, 2020. In: Ministry of Agriculture and Farmers Welfare, Department of Agriculture, Cooperation and Farmers Welfare, Directorate of Economics and Statistics, p 44

Gregorio G, Singh R, Islam R, Sajise A, Vergara G (2014) Salt-tolerant varieties developed to increase Rice self-sufficiency in salt-affected farmlands. In: 4th International Rice Congress, p 27

ICAR-CSSRI, CSI4CZ (2020) Final Report of ACIAR, Australia funded project on: Cropping systems intensification in the salt affected coastal zones of Bangladesh and West Bengal, India (CSI4CZ). In: Sarangi SK (ed) Indian Council of Agricultural Research (ICAR)-Central Soil Salinity Research Institute, Regional Research Station, (CSSRI, RRS), Canning Town - 743 329, South 24 Parganas, West Bengal, p 94

Ismail AM, Singh S, Sarangi SK, Srivastava AK, Bhowmick MK (2022) Agricultural system transformation for food and income security in coastal zones. In: Transforming coastal zone for sustainable food and income security. Springer, Cham, pp 3–22

Krishnamurthy SL, Sharma PC, Sharma DK, Ravikiran KT, Singh YP, Mishra VK, Burman D, Buddheswar M, Mandal S, Sarangi SK, Gautam RK, Singh PK, Manohara KK, Marandi BC, Padmavathi G, Vanve PB, Patil KD, Thirumeni S, Verma OP, Khan AH, Tiwari S, Geetha S, Shakila M, Gill R, Yadav VK, Roy SKB, Prakash M, Bonifacio J, Abdelbagi MI, Gregorio GB, Rakesh KS (2017) Identification of mega-environments and rice genotypes for general and specific adaptation to saline and alkaline stresses in India. Sci Rep 7:1–14

Sarangi SK, Islam MR (2019) Advances in agronomic and related management options for Sundarbans. In: Sen H (ed) The Sundarbans: a disaster-prone eco-region. Coastal research library, vol 30. Springer, Cham, pp 225–260. https://doi.org/10.1007/978-3-030-00680-8_8

Sarangi, SK, Maji B (2017). Sustainable rice cultivation in coastal saline soils: a case study, pp. 69-103. Sasaki, Takuji (ed.), Achieving sustainable cultivation of rice volume 2: cultivation, pest and disease management, Burleigh Dodds Science Publishing Limited, Cambridge.(ISBN: 978 1 78676 028 9; www.bdspublishing.com), 406

Sarangi SK, Maji B, Singh S, Sharma DK, Burman D, Mandal S, Haefele SM (2014) Crop establishment and nutrient management for dry season (boro) rice in coastal areas. Agron J 106(6):2013–2023

Sarangi SK, Maji B, Singh S, Burman D, Mandal S, Sharma DK, Haefele SM (2015) Improved nursery management further enhances the productivity of stress-tolerant rice varieties in coastal rainfed lowlands. Field Crop Res 174:61–70

Sarangi SK, Maji B, Singh S, Sharma DK, Burman D, Mandal S, Haefele SM (2016) Using improved variety and management enhances rice productivity in stagnant flood-affected tropical coastal zones. Field Crop Res 190:70–81

Sarangi SK, Singh S, Maji B, Sharma PC, Srivastava AK, Burman D et al (2020a) SUB1 varieties increased rice (*Oryza sativa*) yield in flood-prone rainfed lowlands of coastal regions. Indian J Agri Sci 90(11):2064–2070

Sarangi SK, Maji B, Singh S, Srivastava AK, Ismail AM (2020b) Effect of neem-coated urea and application methods on nitrogen uptake and use efficiency of rice (*Oryza sativa*) in the coastal rainfed lowlands. Indian J Agron 65(4):404–411

Sarangi SK, Singh S, Srivastava AK, Choudhary M, Mandal UK, Lama TD, Ismail AM (2020c) Crop and residue management improves productivity and profitability of rice–maize system in salt-affected rainfed lowlands of East India. Agronomy 10(12):2019

Sarangi SK, Maji B, Sharma PC, Digar S, Mahanta KK, Burman D et al (2021) Potato (*Solanum tuberosum* L.) cultivation by zero tillage and paddy straw mulching in the saline soils of the Ganges Delta. Potato Res 64(2):277–305

Singh B, Craswell E (2021) Fertilizers and nitrate pollution of surface and ground water: an increasingly pervasive global problem. SN Appl Sci 3(4):1–24

Smith CW, Dilday RH (eds) (2002) Rice: origin, history, technology, and production. John Wiley & Sons, p 3

Stephen J, Shabanamol S, Rishad KS, Jisha MS (2015) Growth enhancement of rice (*Oryza sativa*) by phosphate solubilizing Gluconacetobacter sp.(MTCC 8368) and Burkholderia sp.(MTCC 8369) under greenhouse conditions. 3. Biotech 5(5):831–837

Tahir RMA, Afzal M, Sarwar G, Noorka IR (2019) Smart nutrition Management of Rice Crop under climate change environment. In: Protecting Rice Grains in the Post-Genomic Era. Intech Open. Available from: https://www.intechopen.com/chapters/66907. https://doi.org/10.5772/intechopen.86094

Tuong TP, Bouman BA (2003) Rice production in water-scarce environments. In: Water productivity in agriculture: Limits and opportunities for improvement, vol 1, pp 13–42

Vinod, K. K, & Heuer S. (2012) Approaches towards nitrogen- and phosphorus-efficient rice. AoB PLANTS 2012: pls028. https://doi.org/10.1093/aobpla/pls028

Zhang Z, Chu G, Liu L, Wang Z, Wang X, Zhang H, Zhang J (2013) Mid-season nitrogen application strategies for rice varieties differing in panicle size. Field Crop Res 150:9–18

# Genetic and Agronomic Approaches for Higher Rice Productivity in Sodic Lands of Indo-Gangetic Plains

8

Y. P. Singh and R. K. Singh

**Abstract**

Regaining the agricultural potential of sodic soils in the Indo-Gangetic plains necessitates the development of suitable salt-tolerant rice varieties to provide an entry for other affordable agronomic and soil manipulation measures. Thus, selection of high-yielding rice varieties across a range of sodic soils is central. Evaluation of breeding lines through on-station and on-farm farmers' participatory varietal selection (FPVS) approach resulted in the identification of a high-yielding, salt-tolerant rice genotype "CSR-89IR-8," which was later released as "CSR43." Several agronomic traits coupled with good grain quality and market value contributed to commercialization and quick adoption of this variety in the sodic areas of the Indo-Gangetic plains. Matching management practices for rice production in salt-affected soils are evidently different from those in normal soils and practices. Experiments were conducted at the Indian Council of Agricultural Research-Central Soil Salinity Research Institute (ICAR-CSSRI), Regional Research Station, Lucknow, Uttar Pradesh, India, to test the hypothesis that combining matching management practices (Mmp) with an improved genotype would enhance productivity and profitability of rice in sodic soils. Technological advancements for sustainable reclamation and harnessing productivity potential of salt-affected soils like synergy of gypsum and salt-tolerant varieties, conjoint use of press mud, a sugarcane industry by-product with gypsum or phosphogypsum, and combined use of municipal solid waste compost with gypsum prove cost-effective technologies for resource-poor farmers having

Y. P. Singh (✉)
ICAR-Central Soil Salinity Research Institute, Regional Research Station, Lucknow, Uttar Pradesh, India

R. K. Singh
International Center for Biosaline Agriculture, Dubai, United Arab Emirates





R. K. Singh et al. (eds.), *Genetic Improvement of Rice for Salt Tolerance*,
https://doi.org/10.1007/978-981-99-3830-8_8

salt-affected soils and enhancing rice productivity. Improved nursery management practices like lower seed density, combined application of organic amendments and inorganic fertilizers in the seed bed, and optimum age of seedlings enhanced seedling growth resulting in higher grain yield and minimized production cost. Thirty-day-old seedlings raised using improved nursery management practices yielded 8.4% and 4.0% higher over 25- and 40-day-old seedlings, respectively. Transplanting 4 seedlings $hill^{-1}$ at a spacing of 15 × 20 cm produced significantly higher yield over other treatments. The results of on-farm evaluation trials showed that matching management practices (Mmp) increased yield by 8% over existing best management practices (Bmp) recommended by ICAR-CSSRI for sodic soils and by 16% over framers' management practices; however, combining Mmp with salt-tolerant variety resulted in 35% higher yields over farmers' current varieties and management. This approach of combining cost-effective crop and nutrient management options and a salt-tolerant variety can maximize the productivity and profitability of sodic soils in the alluvial Indo-Gangetic plains and in neighboring salt-affected areas of the Ganges mega delta in South Asia.

**Keywords**

Cost-effective reclamation technologies · Matching management practices · *Oryza sativa* · Participatory varietal selection · Salt-tolerant varieties · Crop productivity

## 8.1 Introduction

Uttar Pradesh, with a total area of 243,290 square kilometers, is India's fourth-largest state in terms of land area. It is situated on the northern spout of India. It is blessed with the fertile Indo-Gangetic plains varying from semiarid to subhumid and given the size of the state's geographical area. This Indo-Gangetic plain region is divided into three subregions, viz., eastern, central, and western regions. Uttar Pradesh has a humid subtropical climate. Summers are extreme with temperatures fluctuating anywhere between 35 °C and 50 °C. There is a low relative humidity of around 20% and dust-laden winds blow throughout the season. The mean annual rainfall ranges from 650 mm to 1000 mm. The rainfall varies from an annual average of 170 cm in hilly areas to 84 cm in western part. Given the concentration of most of this rainfall in the 4 months of the monsoon, excess rain can lead to floods and shortage to droughts.

Agriculture is the prime mover of economic growth in Uttar Pradesh. The land available for agriculture is **16.8 million ha**. A vast majority of the population (60%) in the state virtually relies on agriculture for its livelihood. Rice is the second most important crop cultivated in about 5.63 million ha area with a production of 11.94 million tons contributing 12% of the total produce and covering 13.5% of the total area under rice cultivation. The average productivity of rice in Uttar Pradesh is only

2.12 t $ha^{-1}$. It is notable that the **grain yield** of rice has not been harvested to its existing genetic potential in almost all rice-growing areas in Uttar Pradesh.

Salt-affected lands are estimated at about 955 million ha worldwide (FAO 2008), affecting over 6% of the world's total arable land (Flowers et al. 1997), of which sodic soils constitute 581 million ha. The Indo-Gangetic plains in India ($21^0 55'$ to $32^0 39'$ N and $73^0$ 45′ to $80^0$ 25′ E) has about 2.7 million ha of salt-affected lands, consisting mostly of centuries-old barren sodic soils with no land use opportunities (National Remote Sensing Application and Associates (NRSA and Associates) 1996; Singh et al. 2010a). These soils have high pH (>8.5) and exchangeable sodium percentage (ESP) >15, and the presence of soluble carbonate and bicarbonate that produce alkaline hydrolysis products such as $Na_2CO_3$ and $NaHCO_3$ adversely affect the soil physicochemical and biological properties resulting in significant reduction in yields of most crops (Shaw et al., 1998). Uttar Pradesh is having 1.37 million ha salt-affected soils. The productivity of rice in salt-affected areas in the state is very low (< 2.0 t $ha^{-1}$). One of the major reasons behind this is nonavailability of high-yielding, salt stress-tolerant varieties of rice. Population growth is the major concern that necessitates to increase the production and productivity of crops. It is expected that the world population by the end of 2025 will be about 8 billion and 9.3 billion by 2050 (US Census Bureau, 2008). The crop productivity is not increasing to a similar proportion as world population. Because of urbanization and water scarcity, the expansion of land use for food production has almost stagnated and is showing a downward trend for many important crops with high input requirements, and rice is one of them. Thus, the challenge to produce more food for a rapidly increasing population using either the same or shrinking cultivated area is an uphill task.

To produce more rice, only three options remain: (1) expand irrigated areas or favorable environments with high productivity, (2) harness unfavorable areas with potential for food production, and (3) develop high-yielding, salt-tolerant varieties with matching management practices to harness the productivity potential of salt-affected soils. The first two options are quite improbable due to shortage of irrigation water, increasing urbanization, and industrialization. The last option, developing salt-tolerant, high-yielding varieties and their matching management practices are imperative.

Certain high-yielding, salt-tolerant varieties, like CSR10, CSR13, CSR23, CSR27, CSR36, and Narendra Usar Dhan 3, which can tolerate sodicity up to pH 9.8 were developed through conventional plant breeding and have made good impacts in salt-affected areas of India (Mishra et al. 1992). However, all these varieties are of medium to long duration (130–140 days) and often do not fit well in the rice-wheat system predominating the Indo-Gangetic plains. Integrating farmers' participatory varietal selection (FPVS) approach within the evaluation process during breeding strengthens and accelerates the selection of varieties with characteristics desired by farmers and expedites their adoption when released. Using this approach, field evaluation of salt-tolerant breeding lines was conducted through the network of the "Stress-Tolerant Rice for Africa and South Asia (STRASA)" project for 5 years under varying sodic soil conditions in Uttar Pradesh, India, culminating in the release of the salt-tolerant variety "CSR43" (Singh et al. 2013, 2014; Gautam et al. 2014).

Grain yield of rice in salt-affected soils is much lower because of its high sensitivity to salt stress (Flowers and Yeo 1989; Gao et al. 2007; Ismail et al. 2007). Rice is exceptionally sensitive to salinity and sodicity at early seedling stage (Aslam et al. 1988; Aslam et al. 1993; Singh et al. 2010b) and high yield losses have been observed because of high mortality and poor crop establishment. Modern high-yielding varieties require considerable investment to ameliorate these soils to ensure reasonable yields, but this investment is beyond the capabilities of the resource-limited smallholder farmers living in these salt-affected areas. Increasing and sustaining yields in these areas will require a system that integrates salt-tolerant varieties with effective and affordable crop and nutrient management practices. Conventionally, farmers are using high seed rates (60–80 kg $ha^{-1}$) and little fertilizer in the form of urea only in their nurseries leading to poor seedling growth and crop establishment upon transplanting in sodic soils, with consequent yield losses (Sarangi et al. 2015). Chowdhury et al. (1993) observed that the number of seedlings $hill^{-1}$ and plant spacing were important factors determining plant population per unit area for optimum nutrient uptake and for accessing sufficient light for photosynthesis, which ultimately determines grain yield. It has already been established that mortality in sodic soils is high when young seedlings are transplanted, adversely affecting plant establishment and growth (Dargan and Gaul 1974; Singh et al. 2019a, 2019b). Older seedlings survive better and establish earlier, but they produce less tillers and exhibit poor growth (Mandal et al. 1994, Singh et al. 2020a). The effects of seedling age become more noticeable in short-duration varieties (Chandrakar and Chandravanshi 1988, Singh et al. 2015, Singh et al. 2019a; Singh et al. 2020b).

Sodic soils are inherently low in organic matter (<0.1%), and available N, and hence more responsive to N application. These soils are more prone to N losses due to higher N volatilization caused by high pH, further aggravating N deficiency. Microbial activity, which influences N mineralization, is restricted by salt stress (Abrol et al. 1988; Liu and Kang 2014; Wong et al. 2010). Therefore, the requirement of added N in sodic soils is higher than in normal soils and salt-tolerant varieties appear to respond better to higher N application than sensitive varieties (Dixit and Patro 1994; Meena et al. 2003; Singh et al. 2008; Singh et al. 2016).

Apparently, the lack of high-yielding, salt-tolerant varieties and good management strategies specific for sodic soils is the main reason for the low and unstable productivity of these salt-affected soils. Suitable management practices for salt-affected soils are obviously different than the normal soils. Salinity-tolerant breeding network (STBN) for screening of a large number of genotypes under different sodicity environments and developing their matching management practices was started at ICAR-Central Soil Salinity Research Institute, Regional Research Station, Lucknow. In this chapter, we limit the discussion on development of farmer's preferred high-yielding, short-duration, salt-tolerant variety of rice through farmer's participatory varietal selection approach, developing advanced cost-effective technologies for reclamation of sodic soils and improved agronomic matching management practices to ensure better establishment and higher productivity of developed salt-tolerant varieties of rice.

## 8.2 Development of High-Yielding, Salt-Tolerant Variety Through Farmer's Participatory Varietal Selection Approach

A set of 126 geographically and genetically diverse rice genotypes was screened through researcher's managed on-station and on-farm trials with soil pH between 9.6 and 9.8. This set includes local genotypes, advanced salt-tolerant breeding lines, and salt-tolerant, high-yielding varieties. The recommended package of growing practices for rice in sodic soils was followed. Cultivars were harvested at maturity with yield for each recorded. On the basis of initial screening for sodicity tolerance and grain yield, 18 genotypes were selected and subsequently tested at four locations (two on-stations and two on-farms) (Table 8.1). All the locations belong to the soil order Inceptisols with pH ranging from 9.5 to 9.7. Preference analysis was conducted at maturity following the PVS protocol (Witcombe et al. 1996). On the basis of preference ranking/voting, 12 cultivars were chosen and planted at five locations (three on-stations and two on-farms). PVS analysis was repeated following the above procedure. Nine top selections, based on grain yield and preference rankings, were planted during next year in six locations (three on-station and three on-farm), covering three highly sodic (pH 9.2–10.3) affected districts (Unnao,

**Table 8.1** PVS process for screening, evaluation, and selection of rice genotypes across years

| Year | Process | Number of genotypes planted in sodic soil | Number of genotypes selected and planted in sodic soils | Number of trial sites in sodic soil |
|---|---|---|---|---|
| 1st | Screening of genotypes | 126 ➡ | 18 | 2 |
| 2nd | To conduct PVS | 18 ➡ | 12 | 5 |
| 3rd | To conduct PVS | 12 ➡ | 9 | 6 |
| 4th | To conduct PVS | 9 ➡ | 6 | 6 |
| 5th | Final selection | 6 ➡ | 2 | 6 |
| 6th | Upscaling | 2 + 1 (farmer variety) ➡ | One genotype "CSR89-IR-8" was selected and released as variety "CSR43." | 32 |

**Table 8.2** Preference scores (mean ± SE) of desirable traits and on-station and on-farm grain yield of six rice genotypes evaluated at six locations in Uttar Pradesh, India

| Traits | CSR-2 K 239 ($n = 42$) | NDR359 ($n = 42$) | CSR13 ($n = 42$) | CSR-89IR-8 ($n = 42$) | CSR36 ($n = 42$) | CSR-2 K-262 ($n = 42$) |
|---|---|---|---|---|---|---|
| Plant height | 6.0 ± 0.5 | 6.0 ± 0.4 | 6.0 ± 0.4 | 8.0 ± 0.5 | 6.0 ± 0.5 | 7.0 ± 0.4 |
| Productive tillers $hill^{-1}$ | 6.0 ± 0.5 | 7.0 ± 0.4 | 6.0 ± 0.6 | 8.0 ± 0.5 | 8.0 ± 0.8 | 7.0 ± 0.8 |
| Days to maturity | 6.0 ± 0.4 | 6.0 ± 0.6 | 7.0 ± 0.5 | 9.0 ± 0.4 | 7.0 ± 0.6 | 7.0 ± 0.7 |
| Lodging tolerance | 6.0 ± 0.7 | 6.0 ± 0.5 | 6.0 ± 0.6 | 8.0 ± 0.6 | 7.0 ± 0.7 | 7.0 ± 0.5 |
| Sodicity tolerance | 6.0 ± 0.7 | 7.0 ± 0.8 | 7.0 ± 0.7 | 8.0 ± 0.5 | 8.0 ± 0.6 | 8.0 ± 0.7 |
| Grain quality | 6.0 ± 0.7 | 7.0 ± 0.7 | 7.0 ± 0.7 | 8.0 ± 0.7 | 7.0 ± 0.7 | 8.0 ± 0.8 |
| Cooking quality | 6.0 ± 0.7 | 6.0 ± 0.7 | 6.0 ± 0.8 | 7.0 ± 0.8 | 7.0 ± 0.8 | 7.0 ± 0.8 |
| Grain yield (t $ha^{-1}$) on-station | 4.9 ± 0.2 | 5.0 ± 0.4 | 5.0 ± 3.0 | 5.2 ± 0.04 | 5.1 ± 0.4 | 5.1 ± 0.3 |
| Grain yield (t $ha^{-1}$) on-farm | 4.8 ± 0.3 | 4.9 ± 0.2 | 4.9 ± 0.3 | 5.1 ± 0.4 | 5.1 ± 0.3 | 5.0 ± 0.3 |
| Preference score | −0.05 | 0.12 | 0.09 | 0.20 | 0.18 | 0.04 |

Scoring scale (1–10), 1 = least preferred; 10 = most preferred.

Pratapgarh, and Raebareli) of Uttar Pradesh, lying in the Indo-Gangetic river basin. From these nine, six genotypes found suitable were planted again at six (three on-station and three on-farm) different sodic environments and evaluated, taking into account traits desired by men and women farmers, including grain yield and quality. To assess the postharvest preference for cooking and eating qualities of the genotypes identified during PVS, sensory analysis was conducted by involving 26 women and 21 men farmers. The complete process of varietal selection is given in Table 8.2. Through this process, the genotype that consistently ranked first based on both farmers' preference ranking and grain yield was further evaluated in farmers' managed trials and compared with popular high-yielding varieties and then subsequently submitted to release as variety to the state variety release committee (Singh et al. 2013, 2014).

**Participation of women in varietal selection**

Participation of women in varietal selection

**A view of the participatory varietal selection exercise at site**

A view of the participatory varietal selection exercise at site

**Table 8.3** Analysis of variance among varieties and environments for grain yield

| Source of variations | D.f | Sum of squares | Mean squares | F ratio | Probability |
|---|---|---|---|---|---|
| Varieties | 5 | 1.92205 | 0.38441 | 31.989 | 0.00000 *** |
| Env. + (Var.* Env.) | 120 | 8.17273 | 0.06811 | 5.667 | 0.00000 *** |
| Environments | 20 | 5.59786 | 0.27989 | 23.291 | 0.00000 *** |
| Var.* Env. | 100 | 2.57487 | 0.02575 | 2.143 | 0.00005 *** |
| Environments (Lin.) | 1 | 5.59786 | 5.59786 | 465.829 | 0.00000 *** |
| Var.* Env. (Lin.) | 5 | 1.20494 | 0.24099 | 20.054 | 0.00000 *** |
| Pooled deviation | 114 | 1.36994 | 0.01202 | 1.000 | 0.50000 |
| Total | 125 | 10.09479 | 0.08076 | | |

D.f: degree of freedom

**Table 8.4** Sensory analysis of three rice varieties during 2005

| | Acceptability | | Rank[a] | | |
|---|---|---|---|---|---|
| | | | Female | Male | All farmers |
| Genotype/variety | Yes | No | (*n* = 26) | (*n* = 21) | (*n* = 47) |
| CSR36 | 29 | 18 | 3 | 3 | 3 |
| CSR43 | 28 | 19 | 2 | 2 | 2 |
| Samba Mahsuri | 44 | 3 | 1 | 1 | 1 |

[a]1: best, 2: very good, 3: good

Farmers ranked CSR-89IR-8, CSR36, and NDR359 as their first, second, and third preferred genotypes, respectively, with a strong correlation between the preference scores of male and female farmers (r = 0.91**) and between farmers and researchers (r = 0.74*). CSR-89IR-8 ranked first in all desired traits. CSR-2 K-239 and CSR13 scored lowest for yield and other traits indicating they are not suitable for sodic soils. CSR36 and CSR-2 K-262 showed high sodicity tolerance and grain yield but their duration is longer. Farmers' ranking of CSR-89IR-8 as highly preferred was due to several traits, including its good taste, color, and noncohesiveness when cooked and high grain yield (Singh et al. 2013, 2014) (Table 8.2).

The stability analysis of the six varieties/genotypes across a total of 21 locations revealed significant differences among varieties/genotypes and highly significant interaction between cultivars and environments (Table 8.3).

To elicit the postharvest preference for cooking and eating qualities to help identify consumer acceptance of new salt-tolerant genotypes/cultivars identified, sensory analysis comparison of two farmer-preferred salt-tolerant (CSR-89IR-8 and CSR36) varieties/genotypes and one local variety (Samba Mahsuri) was conducted by involving 26 women and 21 men farmers. Based on farmers' ranking (1 = excellent, 2 = very good, 3 = good) for taste, aroma, color, and noncohesiveness of the cooked rice, Samba Mahsuri, a local cultivar, was highly preferred by the consumers followed closely by CSR-89IR-8 (Table 8.4).

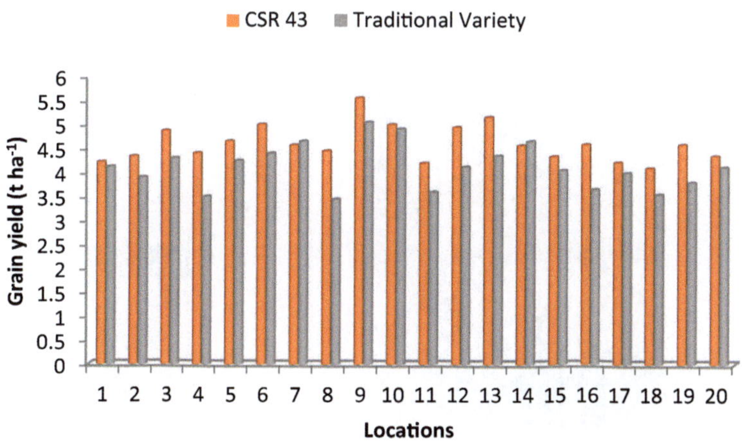


**Fig. 8.1** Grain yield of CSR43 as compared to traditional varieties on farmer's field

On the basis of this PVS exercise, one most preferred salt-tolerant genotype (CSR-89IR-8) was selected by the farmers. Later, it was released as salt-tolerant variety "CSR43." This variety is highly preferred by the farmers and is being grown in >57,000 ha salt-affected soil in Uttar Pradesh adopted by a large number of farmers. The yield advantage over locally cultivated varieties at 20 locations shows the superiority of the variety (Fig. 8.1).

## 8.3 Technological Advancements for Higher Rice Productivity in Salt-Affected Soils

### 8.3.1 Low-Cost Reclamation Technology

Gypsum (CaSO4, 2H2O), which is generally recommended for the reclamation of sodic soils, is a costly input and costs about Rs. 4000/ton. The share of gypsum amendment cost increased with severity of degradation. Under these situations, adoption of such costly technology for reclaiming the sodic soils by the resource poor farmers may not be a viable proposition. This, therefore, calls for alternative low-cost technology for the reclamation of sodic soils for harnessing their productivity potential.

Study conducted with four levels of gypsum (0, 15, 25, and 50% of GR) and two rice varieties ($V_1$) salt-tolerant variety CSR13 and ($V_2$) non-salt-tolerant traditional variety Pant 4 on highly sodic (pH 10.5, EC 1.43 dS $m^{-1}$, ESP 89, organic carbon 0.80 g $kg^{-1}$, and gypsum requirement (G.R) 15.4 t $ha^{-1}$) revealed that the increasing level of gypsum up to 25% GR significantly increased the growth, yield attributing characters, and yields of rice over the control and 15% GR (Table 8.5). Growth characters, viz., plant height, number of tillers, and dry matter accumulation, were found increased with increasing levels of gypsum up to 50% GR, but the difference between 25% and 50% GR rates was not significant. All the growth parameters of salt-tolerant rice variety, CSR13, were significantly superior over the traditional high-yielding non-salt-tolerant variety Pant 4. Similar trend was observed in yield attributing characters like panicle length, number of grains $panicle^{-1}$, and test weight. Salt-tolerant variety "CSR13" gave significantly higher panicle length, number of grains $panicle^{-1}$ over the traditional high-yielding, non-salt-tolerant variety "Pant 4" (Table 8.5). Grain yield increased significantly with increasing levels of gypsum up to 50% GR, but the difference between 25 and 50% GR were at par but significantly superior over control and 15% GR levels. Chhabra and Abrol (1977), Singh et al. (2008), and Singh et al. (2009) have also reported nonsignificant differences in rice grain yield between 25 and 50% GR. Straw yield increased significantly only up to 25% GR that in turn was at par with 50% GR. Salt-tolerant rice variety "CSR13" gave significantly higher grain yield over the nontolerant "Pant 4," the magnitude of difference being 30%.

The interaction between gypsum levels and rice varieties was significant for grain yield. Salt-tolerant variety "CSR13" significantly excelled Pant 4 in grain yield at gypsum levels from 15 to 50% GR. However, the magnitude of combined effect of reduced dose of gypsum and salt-tolerant variety of rice was reflected as 17% increase in grain yield over the traditional variety. Interaction between gypsum levels and salt-tolerant variety on grain yields indicated that the gypsum at 25% GR with sodicity-tolerant variety could save about 43% of the total initial

**Table 8.5** Effect of gypsum and varieties on growth parameters of rice and wheat

| Treatments | Plant height (cm) | Effective tillers $hill^{-1}$ | Dry matter (g $hill^{-1}$) | Panicle length (cm) | Grains $panicle^{-1}$ | Test weight (g) | Grain yield (t $ha^{-1}$) | Straw yield (t $ha^{-1}$) |
|---|---|---|---|---|---|---|---|---|
| *Gypsum levels (%GR)* | | | | | | | | |
| 0 | 62.10 | 3.80 | 19.70 | 15.81 | 29.62 | 17.60 | 0.36 | 2.35 |
| 15 | 93.30 | 6.96 | 44.10 | 21.96 | 92.37 | 22.81 | 2.99 | 6.87 |
| 25 | 98.20 | 8.16 | 57.60 | 24.34 | 126.25 | 24.88 | 4.35 | 8.68 |
| 50 | 104.20 | 9.86 | 61.80 | 25.64 | 137.37 | 25.79 | 4.75 | 944 |
| Sem± | 5.2 | 0.34 | 1.67 | 1.06 | 1.76 | 0.67 | 0.16 | 0.32 |
| CD ($P = 0.05$) | 16.52 | 1.09 | 5.29 | 3.36 | 5.72 | 1.91 | 0.50 | 1.05 |
| *Varieties* | | | | | | | | |
| $V_1$ | 94.20 | 8.36 | 55.85 | 20.73 | 103.06 | 20.73 | 3.40 | 7.47 |
| $V_2$ | 84.60 | 6.02 | 47.93 | 24.80 | 89.75 | 24.80 | 2.82 | 6.22 |
| Sem± | 1.7 | 0.22 | 0.87 | 0.53 | 1.54 | 0.60 | 0.07 | 0.16 |
| CD ($P = 0.05$) | 4.69 | 0.61 | 2.69 | 1.65 | 4.71 | 1.65 | 0.23 | 0.51 |

**Table 8.6** Interaction effect of gypsum levels and Varieties on grain yield (t $ha^{-1}$) of rice and wheat

| | 1st year | | 2nd year | | 3rd year | |
|---|---|---|---|---|---|---|
| Gypsum levels (% GR) | CSR13 | Pant 4 | CSR13 | Pant 4 | CSR13 | Pant 4 |
| 0 | 0.32 | 0.07 | 0.55 | 0.36 | 0.63 | 0.42 |
| 15 | 2.73 | 1.57 | 3.01 | 2.77 | 3.28 | 2.81 |
| 25 | 4.09 | 3.48 | 4.38 | 3.76 | 4.52 | 4.11 |
| 50 | 4.36 | 4.05 | 4.79 | 4.36 | 4.96 | 4.75 |
| Sem± | 0.09 | | 0.03 | | 0.08 | |
| CD (P = 0.05) | 0.27 | | 0.12 | | 0.28 | |

expenditure for the reclamation of sodic soils over the recommended dose of gypsum (50% GR) with non-sodicity-tolerant traditional high-yielding variety. These observations indicate saving of gypsum to the tune of 25% by using salt-tolerant variety in sodic soils (Table 8.6).

Thus, for low cost-effective technology, treating the sodic soils with reduced dose of gypsum (25% GR) and growing salt-tolerant variety of rice can save 25% gypsum and proved highly economical and sustainable technology for increasing rice productivity. With this approach, the same quantity of gypsum could be utilized for reclamation of double the area, thus accelerating the reclamation process.

### 8.3.2 Synergy of Gypsum, Press Mud, and Salt-Tolerant Varieties: An Alternate Approach for Enhancing Rice Productivity

India produces about 12.0 million tons of press mud per year (Gupta et al. 2011), a low-cost sugar industry by-product containing sizable quantities of macro- and micronutrients, high calcium sulfate, and organic matter-supplied Ca directly to the soil that replaces excess $Na^+$ from the soil exchange complex and sulfur converting into sulfuric acid that lowers down the soil pH and improves the physical, chemical, and biological properties of sodic soils (Gharaibeh et al. 2009; Dhanushkodi and Subrahmaniyan 2012; Negim 2015, Thimappa et al. 2015; Singh et al. 2019b; Singh et al. 2021). Reclamation of sodic soils using press mud with gypsum and cultivation of salt-tolerant varieties can be an alternate approach for reclamation of sodic soils and enhancing rice productivity.

Field study conducted with four sodicity levels and two reclamation practices, i.e., the existing practice (EP) including the application of gypsum at 50% GR followed by cultivation of traditional varieties of rice and the improved practice (IP) consisting of the application of reduced dose of gypsum at 25% GR + 10 Mg $ha^{-1}$ press mud followed by cultivation of salt-tolerant varieties (STV) of rice, revealed that the sodicity levels and reclamation practices had significant ($p \leq 0.05$) effects on rice growth. Successive increases in sodicity levels decreased the number of hills $m^{-2}$; dry matter accumulation by 33.5, 46.6, and 51.7%, respectively; number of panicles $m^{-2}$; and leaf area index (LAI) significantly

**Table 8.7** Effect of different sodicity levels ($S_1$–$S_4$) and reclamation practices (EP and IP) on crop growth, yield indicators, and yield of rice

| Treatments | Number of hills $m^{-2}$ | Number of panicles $m^{-2}$ | Dry matter ($g\ m^{-2}$) | LAI |
|---|---|---|---|---|
| Sodicity levels | | | | |
| $S_1$ | 31.9 | 290.7 | 1385.5 | 2.3 |
| $S_2$ | 26.8 | 199.0 | 919.2 | 2.0 |
| $S_3$ | 22.1 | 102.9 | 490.6 | 1.4 |
| $S_4$ | 17.7 | 50.1 | 236.7 | 1.2 |
| Sem± | 0.83 | 5.05 | 5.63 | 0.03 |
| LSD ($p = 0.05$) | 2.14 | 17.48 | 16.9 | 0.12 |
| Reclamation practices | | | | |
| EP | 25.1 | 130.9 | 677.7 | 1.6 |
| IP | 28.1 | 190.5 | 851.4 | 1.8 |
| Sem± | 0.67 | 2.66 | 5.83 | 0.018 |
| LSD ($p = 0.05$) | 1.86 | 8.77 | 18.20 | 0.059 |

*$S_1$* slightly sodic, *$S_2$* moderately sodic, *$S_3$* sodic, *$S_4$* severely sodic, *LSD* least significant difference, *LAI* leaf area index, *SEm±* = standard error of mean

(Table 8.7). This is because of high soil pH that reduces the availability of many nutrients, leading to reduced plant growth (Alka et al. 2012). The growth reduction in increasing sodicity levels can also be the result of toxic effects of accumulated $Na^+$ ions in the soil solution (Akhtar et al. 2001; Munns and Tester 2008; Prasad et al. 2006; Tavakkoli et al. 2010), and poor soil organic matter content is often related to less available nitrogen, and lower cation exchange capacity can cause a decrease in root growth (Onemli 2004). Significant reduction in LAI may be due to reducing the supply of carbohydrates and/or growth hormone reduction in the photosynthesis rate and excessive uptake of salts that affect the production of metabolites (Mazher et al. 2007; Munns 1992). The decomposition of organic matter in the press mud and enhanced growth of the salt-tolerant varieties increases organic acid exudates in soil that mobilizes dissolution of soil calcium (Van Asten et al. 2004). Several previous studies revealed that organic amendments can improve the root growth of crops. This may be associated with the enhancement of biological activities in crop rhizosphere by amino acid and some physiologically active substances in the organic fertilizers such as humic acid (Gondek and Filipek-Mazur 2005). The incorporation of organic matter into soil can bring beneficial effects on crop root growth by improving the physical and chemical environments of rhizosphere (Sidiras et al. 2002). Across the sodicity levels, IP significantly increased hills $m^{-2}$ (11.9%), panicles $m^{-2}$ (45.6%), dry matter g $m^{-2}$ (25.6%), and LAI (10.9%) over EP (Table 8.7).

Yield and yield-related attributes were significantly ($p \leq 0.05$) affected by increasing sodicity levels (Table 8.8). The reductions in spikelets per panicle from successive increase in sodicity levels were 3.5, 1.3, and 18.6%, respectively.

**Table 8.8** Effect of different sodicity levels ($S_1$–$S_4$) and reclamation practices (EP and IP) on yield indicators and yield of rice

| Treatments | Number of spikelets panicle$^{-1}$ | Floret fertility (%) | 1000-grain weight (g) | Straw yield (t ha$^{-1}$) | Grain yield (t ha$^{-1}$) |
|---|---|---|---|---|---|
| *Sodicity levels* | | | | | |
| $S_1$ | 121.6 | 85.2 | 25.4 | 8.7 | 5.0 |
| $S_2$ | 115.8 | 74.7 | 23.5 | 7.6 | 4.4 |
| $S_3$ | 117.3 | 59.6 | 21.5 | 4.8 | 2.8 |
| $S_4$ | 94.30 | 54.4 | 21.7 | 2.5 | 1.5 |
| Sem± | 2.12 | 0.79 | 0.11 | 0.04 | 0.02 |
| LSD (p = 0.05) | 7.1 | 2.76 | 0.37 | 0.11 | 0.06 |
| *Reclamation practices* | | | | | |
| EP | 107.5 | 64.9 | 22.8 | 5.0 | 2.9 |
| IP | 111.9 | 70.7 | 23.2 | 6.8 | 3.9 |
| Sem± | 0.64 | 0.45 | 0.22 | 0.05 | 0.03 |
| LSD ($p = 0.05$) | 2.13 | 1.49 | Ns | 0.14 | 0.08 |

Correspondingly, floret fertility and 1000 grain weight decreased by 8.5 to 20.0% and 7.4 to 9.2%, respectively. A significant reduction in 1000 grain weight was also recorded due to poor growth as a result of restricted water movement, nutrient translocation, and osmotic stress because of the toxic effect of sodium in the rhizosphere. Based on the soil data, it can be assumed that the better soil characteristics caused by both reclamation practices improved crop growth and related crop characteristics. All yield components except 1000 grain weight were significantly higher with IP, as compared with EP. The IP recorded 4.1% and 9.0% higher spikelets panicle$^{-1}$ and floret fertility, respectively, over the EP. This was attributed to the incorporation of organic amendment in combination with gypsum into soil that has been reported to have additional beneficial effects on the soil physicochemical properties as well as the rhizosphere environment (Cha-Um et al. 2011; Sidiras et al. 2002). Press mud as an organic source of amendment seems to have helped the leaching of excessive ions to deeper soil layers, alleviating the concentration of salts in the top soil (plow layer), which favored plant growth and ultimately increased grain and straw yields (Muhammad and Khattak 2011; Singh et al. 2015). A recent study has shown a highly significant correlation ($p \leq 0.01$) between grain yields and soil organic carbon content after application of organic amendments (Mikanová et al. 2012). The combined effect of sodicity levels and reclamation practices had significant effects on straw and grain yields (Table 8.8). The maximum straw and grain yields were recorded at lower pH (S1) level and the minimum with pH 10.4 (S4). This was caused by a similar response of all yield attributes to sodicity. The grain yield reduction with a progressive increase in the sodicity level from S1 to S2, S2 to S3, and S3 to S4 was 12.0%, 36.4% and 46.4%, respectively. The maximum yield reduction was recorded when the sodicity level increased from pH 10.0 to 10.4 (Table 8.8). This must be attributed to a combination of the reduced assimilate source and a partial destruction of the assimilate sink with increasing sodicity. Across all sodicity levels, IP produced 36.8% and 37.1% higher straw and grain yield over the EP. The IP had significantly excelled in straw and grain yields at all sodicity levels except S1. But the relative advantage increased with increasing levels of sodicity. Therefore, the IP caused increasing yield enhancements with increasing sodicity levels. These observations indicate that particularly, crops on sodic and severely sodic soils (S3 and S4) profit more from the synergistic effect of reduced dose of gypsum, press mud, and salt-tolerant varieties, making it a sustainable reclamation and yield maximizing technology for sodic soils.

### 8.3.3 Integration of Organic and Inorganic Amendments for Higher Rice Productivity

Generally, application of gypsum ($CaSO_4 \cdot 2H_2O$) and leaching with ponding of water for about 10 days is a standard practice used for the reclamation of sodic soils. However, gypsum-based reclamation approach has failed to improve soil physical and biological properties. Application of organic amendments in conjunction with reduced dose of gypsum improves soil physicochemical and biological properties (Naeini and Cook 2000; Singh et al. 2018; Singh et al. 2019b). The decaying organic matters enhance soil $CO_2$ concentration and release $H^+$ when it dissolves in water. The released $H^+$ increases $CaCO_3$ dissolution and liberates more calcium (Ca) for sodium (Na) exchange (Ghafoor et al. 2008). The municipal solid waste (MSW), which is imperative source of organic matter and available in abundant quantity, if converted into resources can be a good source of organic matter for the reclamation of sodic soils.

Study conducted with combined use of organic (MSW compost, press mud) and inorganic amendments (gypsum and phosphogypsum) at research farm of Central Soil Salinity Research Institute (CSSRI), Regional Research Station, Lucknow, Uttar Pradesh, India (26° 47′ 58" N, 80° 46′ 24″ E), on highly sodic soils (pH 9.8, ESP 78, OC 1.30 g $kg^{-1}$, and GR15.2 t $ha^{-1}$) revealed that the crop growth attributed by plant height, panicle density, dry matter accumulation, leaf area development, and crop growth rate was significantly higher with combined use of organic and inorganic amendments (Table 8.9).

The yield and yield component as influenced by application of organic and inorganic amendments are presented in Table 8.10. The spikelet/panicle and spikelet fertility were significantly affected with addition of organic amendments over the use of inorganic sources of amendments. Significantly higher spikelet/panicle and spikelet fertility and test weight were recorded with combined use of organic and inorganic amendments over the sole application of inorganic amendments (Shivay and Singh 2003; Singh et al. 2018; Singh et al. 2021). Rice grain and straw yields with combined application of MSW compost at 10 t $ha^{-1}$ and gypsum at 25% GR was statistically at par with application of gypsum or phosphogypsum at 50% GR. The results showed that gypsum at 25% GR may be saved by using MSW composts (Table 8.10).

**Table 8.9** Combined effect of organic and inorganic amendments on crop growth of rice in sodic soil

| Treatments | Plant height (cm) | Panicle density ($m^{-2}$) | Dry matter ($g\ m^{-2}$) | CGR ($g\ day^{-1}$) | LAI |
|---|---|---|---|---|---|
| $T_1$ | 105.77 | 336.00 | 1696.20 | 0.82 | 5.63 |
| $T_2$ | 115.93 | 389.10 | 1367.40 | 0.93 | 5.22 |
| $T_3$ | 118.37 | 432.00 | 1812.60 | 0.91 | 6.30 |
| $T_4$ | 112.66 | 330.00 | 1082.10 | 0.61 | 6.15 |
| $T_5$ | 109.40 | 359.10 | 1493.70 | 0.69 | 4.63 |
| $T_6$ | 113.99 | 345.00 | 1392.60 | 0.94 | 4.36 |
| $T_7$ | 103.97 | 387.90 | 1641.60 | 0.78 | 5.21 |
| $T_8$ | 106.55 | 325.80 | 1419.60 | 0.77 | 5.43 |
| $LSD_{0.05}$ | ns | 12.79 | 5.07 | 0.03 | 0.10 |

*ns* nonsignificant, *CGR* crop growth rate, *LAI* leaf area index, $T_1$ G at 50% GR, $T_2$ PG at 50% GR, $T_3$ G at 25% GR + on-farm MSW compost at 10 Mg $ha^{-1}$, $T_4$ PG at 25% GR + on-farm MSW compost at 10 Mg $ha^{-1}$, $T_5$ G at 12.5% GR + on-farm MSW compost at 10 Mg $ha^{-1}$ + press mud at 10 Mg $ha^{-1}$, $T_6$ PG at 12.5% GR + on-farm MSW compost at 10 Mg $ha^{-1}$ + press mud at 10 Mg $ha^{-1}$, $T_7$ G at 25% GR + industrial-processed MSW compost at 10 Mg $ha^{-1}$, $T_8$ PG at 25% GR + industrial-processed MSW compost at 10 Mg $ha^{-1}$

**Table 8.10** Combined effect of organic and inorganic amendments on yield attributes of rice in sodic soil

| Treatments | Spikelets $panicle^{-1}$ | Spikelet fertility (%) | 1000 grain wt. (g) | Grain yield ($t\ ha^{-1}$) | Straw yield ($t\ ha^{-1}$) | HI (%) |
|---|---|---|---|---|---|---|
| $T_1$ | 112.88 | 62.00 | 23.40 | 4.64 | 9.27 | 33.36 |
| $T_2$ | 120.37 | 71.72 | 25.80 | 4.34 | 13.70 | 24.06 |
| $T_3$ | 109.37 | 95.09 | 26.97 | 4.83 | 13.59 | 26.22 |
| $T_4$ | 104.28 | 76.29 | 23.97 | 4.42 | 9.50 | 31.75 |
| $T_5$ | 110.76 | 55.06 | 22.93 | 3.96 | 13.42 | 22.78 |
| $T_6$ | 109.99 | 70.41 | 25.27 | 3.71 | 12.72 | 22.58 |
| $T_7$ | 106.99 | 65.42 | 22.87 | 4.65 | 11.26 | 29.23 |
| $T_8$ | 107.93 | 79.78 | 22.73 | 4.40 | 10.14 | 30.26 |
| $LSD_{0.05}$ | 6.73 | 3.64 | 1.12 | 0.30 | 1.13 | 2.21 |

## 8.4 Improved Management Practices for Enhancing Rice Productivity

### 8.4.1 Nursery Management for Optimizing Seed Density, Fertility Levels, and Age of Seedlings for Transplanting

#### 8.4.1.1 Effect of Nursery Management Practices on Seedling Growth

Field experiment consisting of two seed densities, five nursery nutrient doses, and four seedling ages was conducted at research farm of Central Soil Salinity Research Institute (CSSRI), Regional Research Station, Lucknow, Uttar Pradesh, India. Under farmers' practices, farmers are generally using a small amount of N through urea and FYM in nursery. However, under improved management practices, 5 t FYM ha$^{-1}$ was applied uniformly in all the treatments at the time of nursery sowing. Fifty percent of the N and full dose of $P_2O_5$ and $K_2O$ were applied as basal, and the rest of the N was applied in two equal splits:12 and 22 days after sowing. Before sowing, seed was soaked for 24 h in fresh water.

Shoot and root lengths with low seeding density (25 g m$^{-2}$) were 2% and 7% higher over higher seeding density (40 g m$^{-2}$). Number of green leaves and leaf area seedling$^{-1}$ was 10.4% and 14.3% higher with low seed density than the higher seeding density. Root and shoot dry weight seedling$^{-1}$ did not show significant variation with varying seed densities. Seedling vigor index (SVI) was recorded higher with low than high seed density. Maximum SVI was recorded with 125–50-50 kg N-$P_2O_5$-$K_2O$ ha$^{-1}$; however, it was statistically at par with treatment 100–50-50 kg N-$P_2O_5$-$K_2O$ ha$^{-1}$. Forty-day-old seedlings had a higher vigor index than 25-day, 30-day, and 35-day-old seedlings, but the difference between 35- and 40-day-old seedlings was statistically not significant. Application of 5 t FYM ha$^{-1}$along with 125–50-50 kg N-$P_2O_5$-$K_2O$ ha$^{-1}$ produced significantly higher root and shoot length, number of green leaves, leaf area, and root and shoot dry weight. Shoot and root length, number of green leaves plant$^{-1}$, and shoot and root dry weight seedling$^{-1}$ were higher in 40-day-old seedling. However, the difference in these attributes between 35- and 40-day-old nurseries were not significant. Root as well as shoot dry weight increased with age (Table 8.11.).

#### 8.4.1.2 Effect of Nursery Management Practices on Crop Growth and Yield

Nursery grown with low seed density produced 3.3%, 9.4%, and 6% more productive tillers, grains panicle$^{-1}$, and panicle weight, respectively. However, 1000 grain weight and grain yield were higher in higher seed density nursery but the differences were nonsignificant (Table 8.12). Plant height did not differ significantly due to increased nitrogen levels.

Maximum number of productive tillers hill$^{-1}$, panicle length, grains panicle$^{-1}$, panicle weight, 1000 grain weight, and straw and grain yield were recorded from the nursery grown with 100–50-50 kg N-$P_2O_5$-$K_2O$ ha$^{-1}$ + 5 t FYM ha$^{-1}$ but the difference in these parameters were statistically at par with 125–50-50 kg N-$P_2O_5$_$K_2O$ ha$^{-1}$ + 5 t FYM ha$^{-1}$. Maximum plant height, 1000 grain weight, and grain and straw yields were recorded with 30-day-old seedlings. The interaction results revealed that higher nutrient levels and higher seeding density did not exhibit much effect on productive tillers hill$^{-1}$ but the combined effect of nutrient levels and seedling age had significant effect on productive tillers hill$^{-1}$ (Table 8.13).

Maximum productive tillers hill$^{-1}$were recorded in 30-day-old seedlings grown with 40 g m$^{-2}$ seeding density. Younger seedlings produced higher number of productive tillers hill$^{-1}$ than older seedlings. Interaction effect of seedling age and nutrient management was found significant on grain yield (Table 8.14). Thirty-day-old seedlings grown with 100 N-50$P_2O_5$-50$K_2O$kg ha$^{-1}$+ 5 t FYM ha$^{-1}$ recorded 8.4%, 1.22%, and 4.0% higher grain yield over 25-, 35-, and 40-day-old seedlings.

**Table 8.11** Effect of nursery management practices on seedling growth

| Treatments | Shoot length (cm) | Root length (cm) | Green leaves $seedling^{-1}$ | Leaf area ($cm^2$ $seedling^{-1}$) | Shoot dry weight $seedling^{-1}$ (g) | Root dry weight $seedling^{-1}$ (g) | SVI |
|---|---|---|---|---|---|---|---|
| Seeding density | | | | | | | |
| 25 g $m^{-2}$ | 28.2 | 12.3 | 5.1 | 41.4 | 1.5 | 0.4 | 3767 |
| 40 g $m^{-2}$ | 27.7 | 11.4 | 4.6 | 36.2 | 1.4 | 0.3 | 3636 |
| $LSD_{0.05}$ | ns[a] | 0.36 | 0.24 | 1.35 | Ns | Ns | 113.2 |
| Nutrient levels (kg N-$P_2O_5$-$K_2O$ $ha^{-1}$+ 5 t FYM $ha^{-1}$) | | | | | | | |
| 25–0-0 | 21.3 | 8.6 | 4.3 | 34.2 | 1.0 | 0.2 | 2781 |
| 50–50-50 | 23.5 | 9.2 | 4.4 | 36.4 | 1.1 | 0.3 | 3041 |
| 75–50-50 | 29.1 | 10.3 | 4.9 | 41.2 | 1.3 | 0.4 | 3664 |
| 100–50-50 | 34.9 | 11.2 | 5.9 | 43.3 | 2.4 | 0.5 | 4287 |
| 125–50-50 | 35.2 | 11.4 | 5.9 | 42.6 | 2.6 | 0.6 | 4334 |
| $LSD_{0.05}$ | 1.23 | 0.50 | 0.42 | 1.43 | 0.35 | 0.16 | 212.4 |
| Seedling age | | | | | | | |
| 25-d | 22.8 | 9.3 | 4.1 | 38.6 | 1.1 | 0.3 | 2985 |
| 30-d | 26.5 | 11.2 | 4.2 | 42.5 | 1.4 | 0.3 | 3506 |
| 35-d | 29.5 | 12.4 | 4.9 | 41.3 | 2.2 | 0.4 | 3897 |
| 40-d | 31.2 | 12.6 | 5.4 | 41.2 | 2.6 | 0.4 | 4073 |
| $LSD_{0.05}$ | 4.6 | 0.56 | 0.26 | 1.24 | 0.23 | 0.12 | 372.4 |

[a]*ns* nonsignificant, *SVI* Seedling vigor index

The seedlings grown with 100 N-50$P_2O_5$-50$K_2O$ kg $ha^{-1}$+ 5 t FYM $ha^{-1}$ recorded 19%, 14.7%, 8.4%, and 5.8% higher grain yield over 25-0-0, 50–50-50, 75–50-50, and 125–50-50 kg N-$P_2O_5$-$K_2O$ $ha^{-1}$+ 5 t FYM $ha^{-1}$.

### 8.4.2 Optimizing Number of Seedlings $Hill^{-1}$ and Spacing

Study conducted with two salt-tolerant rice varieties CSR36 and CSR43 in a sodic soil (pH 9.2), electrical conductivity (EC) 0.61 d S $m^{-1}$, and organic carbon (OC) 4.30 g $kg^{-1}$ at Shivri farm of Central Soil Salinity Research Institute, Regional Research Station, Lucknow, comprising of seedling ages and three spacings revealed that the growth, grain yield, and yield attributes were significantly higher when four seedlings were transplanted per hill compared with two seedlings per hill (Table 8.15). Productive tillers $hill^{-1}$, grain, and straw yields with four seedlings $hill^{-1}$were 15.9%, 22.7% and 30.3% higher than those with two seedlings $hill^{-1}$. Spacing had significant ($P < 0.05$) effect on plant height, filled grains $panicle^{-1}$, and grain yield. Using 15 × 20 cm spacing resulted in about 11% higher filled grains per panicle over both 15 × 15 and 20 × 20 cm spacing. It also resulted in 11% higher grain yield and 16% higher straw yields over the values obtained with 20 × 20 cm spacing (Fig. 8.2a,b).

**Table 8.12** Effect of seed density and nutrient management on yield attributing characters and yield

| Treatments | Plant height (cm) | Productive tillers $hill^{-1}$ | Panicle length (cm) | Grains $panicle^{-1}$ | Panicle weight (g) | 1000 grain weight (g) | Grain yield ($t\ ha^{-1}$) | Straw yield ($t\ ha^{-1}$) | HI (%) |
|---|---|---|---|---|---|---|---|---|---|
| *Seeding density* | | | | | | | | | |
| 25 $g\ m^{-2}$ | 100.4 | 12.0 | 24.1 | 116.0 | 3.3 | 24.8 | 4.13 | 5.00 | 45.23 |
| 40 $g\ m^{-2}$ | 93.4 | 11.6 | 23.9 | 106.0 | 3.2 | 25.0 | 4.22 | 5.24 | 44.60 |
| $LSD_{0.05}$ | ns[a] | Ns | Ns | Ns | Ns | Ns | Ns | Ns | Ns |
| *Nutrient levels (kg N-$P_2O_5$-$K_2O$ $ha^{-1}$+ 5 t FYM $ha^{-1}$)* | | | | | | | | | |
| 25–0-0 | 103.1 | 10.0 | 21.2 | 99.0 | 3.1 | 24.9 | 3.94 | 4.82 | 44.47 |
| 50–50-50 | 102.3 | 10.7 | 22.5 | 110.0 | 3.3 | 23.7 | 4.06 | 5.04 | 44.61 |
| 75–50-50 | 103.1 | 12.4 | 24.0 | 112.0 | 3.9 | 24.9 | 4.30 | 5.02 | 46.13 |
| 100–50-50 | 105.0 | 14.1 | 25.3 | 124.3 | 4.7 | 25.9 | 4.50 | 5.22 | 46.29 |
| 125–50-50 | 110.1 | 13.9 | 25.4 | 122.0 | 4.3 | 25.9 | 4.40 | 5.13 | 46.16 |
| $LSD_{0.05}$ | Ns | 1.43 | 1.12 | 6.43 | 0.46 | Ns | 0.16 | Ns | 0.43 |
| *Seedling age* | | | | | | | | | |
| 25-d | 108.8 | 9.6 | 24.2 | 115.0 | 2.31 | 27.9 | 3.84 | 4.41 | 46.65 |
| 30-d | 108.0 | 12.3 | 24.4 | 125.0 | 2.38 | 28.0 | 4.10 | 4.68 | 46.69 |
| 35-d | 106.5 | 10.5 | 23.3 | 110.5 | 2.35 | 26.9 | 3.84 | 4.39 | 46.54 |
| 40-d | 101.0 | 10.6 | 22.1 | 112.8 | 2.30 | 27.7 | 3.63 | 4.13 | 45.69 |
| $LSD_{0.05}$ | 1.43 | 1.14 | 1.40 | 10.40 | 0.78 | 0.27 | 0.28 | 0.86 | 0.52 |

[a]*ns* nonsignificant, *HI* Harvest index

**Table 8.13** Interaction effect of nutrient management, seed density, and age of seedlings on number of productive tillers hill$^{-1}$

| Nutrient levels (kg N-$P_2O_5$-$K_2O$ ha$^{-1}$) + 5 t FYM ha$^{-1}$ | Productive tillers hill$^{-1}$ | | | | | | | | | |
|---|---|---|---|---|---|---|---|---|---|---|
| | Seeding density | | | | | | | | | |
| | 25 g m$^{-2}$ | | | | | 40 g m$^{-2}$ | | | | |
| | Age of seedlings (days) | | | | | | | | | |
| | 25-d | 30-d | 35-d | 40-d | Mean | 25-d | 30-d | 35-d | 40-d | Mean |
| 25–0-0 | 9.1 | 9.2 | 9.0 | 8.4 | 8.9 | 9.1 | 9.5 | 9.0 | 8.3 | 9.0 |
| 50–50-50 | 10.6 | 11.2 | 9.5 | 9.0 | 10.1 | 11.0 | 11.5 | 10.0 | 9.2 | 10.4 |
| 75–50-50 | 11.1 | 11.6 | 10.2 | 10.0 | 10.7 | 11.4 | 12.1 | 10.4 | 10.1 | 11.0 |
| 100–50-50 | 11.3 | 12.1 | 11.2 | 10.2 | 11.0 | 12.1 | 12.3 | 11.0 | 10.5 | 11.5 |
| 125–50-50 | 11.5 | 12.1 | 11.3 | 10.5 | 11.3 | 12.1 | 12.5 | 11.1 | 11.0 | 11.7 |
| Mean | 10.7 | 11.2 | 10.2 | 9.6 | | 11.1 | 11.6 | 10.7 | 9.9 | |
| $LSD_{0.05}$ (seedling age) | | | | 1.21 | | | | | | |
| $LSD_{0.05}$ (nutrient levels) | | | | 0.24 | | | | | | |
| $LSD_{0.05}$ (seedling age x nutrient levels) | | | | 0.14 | | | | | | |

**Table 8.14** Interaction effect of nutrient levels and age of seedlings on grain yield

| Nutrient levels (kg N-$P_2O_5$-$K_2O$ha$^{-1}$) + 5 t FYM ha$^{-1}$ | Grain yield (t ha$^{-1}$) | | | | |
|---|---|---|---|---|---|
| | Seedling age (days) | | | | |
| | 25-d | 30-d | 35-d | 40-d | Mean |
| 25–0-0 | 3.4 | 3.9 | 3.7 | 3.6 | 3.7 |
| 50–50-50 | 3.6 | 3.9 | 3.9 | 3.8 | 3.8 |
| 75–50-50 | 3.8 | 4.1 | 4.2 | 4.0 | 4.0 |
| 100–50-50 | 4.1 | 4.5 | 4.5 | 4.3 | 4.4 |
| 125–50-50 | 4.1 | 4.2 | 4.0 | 4.1 | 4.1 |
| Mean | 3.8 | 4.1 | 4.1 | 4.0 | 4.0 |
| $LSD_{0.05}$ (seedling age) | | | ns[a] | | |
| $LSD_{0.05}$ (nutrient levels) | | | ns | | |
| $LSD_{0.05}$ (seedling age x nutrient levels) | | | 0.63 | | |

[a]*ns* nonsignificant

### 8.4.3 Optimizing Nitrogen Requirement of Crop

Plant height and productive tillers hill$^{-1}$ were significantly increased by increasing N levels. A 200 kg N ha$^{-1}$ produced 70%, 53.8%, 21.5%, 9.7%, and 3.5% higher number of productive tillers hill$^{-1}$ than control (no N application), 100, 125, 150, and 175 kg N ha$^{-1}$, respectively. Dry matter increased significantly with increasing N doses, which may be attributed to higher number of productive tillers hill$^{-1}$ and plant height. Maximum leaf area index was recorded with 200 kg N ha$^{-1}$, whereas minimum with control, but the difference between 175 kg N ha$^{-1}$ and

**Table 8.15** Effect of number of seedlings hill$^{-1}$ and spacing on growth and yield attributing characters of rice in an on-station field trial

| Treatments | Plant height (cm) | Productive tillers hill$^{-1}$ | Dry matter (g hill$^{-1}$) | No. of spikelets panicle$^{-1}$ | Total grains panicle$^{-1}$ | Filled grains panicle$^{-1}$ | Floret fertility (%) | 1000 grain weight (g) |
|---|---|---|---|---|---|---|---|---|
| *Number of seedlings hill$^{-1}$* | | | | | | | | |
| 2 | 98.3 | 9.4 | 52.0 | 10.77 | 124.73 | 106.9 | 85.7 | 25.9 |
| 4 | 96.7 | 10.9 | 63.7 | 10.77 | 142.87 | 120.3 | 84.2 | 25.0 |
| $LSD_{0.05}$ | ns[a] | 1.2 | 2.6 | Ns | 7.33 | 2.2 | Ns | Ns |
| *Spacing (cm)* | | | | | | | | |
| 15 x 15 | 97.8 | 9.3 | 56.8 | 10.33 | 129.21 | 109.7 | 84.9 | 24.8 |
| 15 x 20 | 98.9 | 11.1 | 58.6 | 11.33 | 138.26 | 121.4 | 87.8 | 25.3 |
| 20 x 20 | 95.7 | 10.1 | 58.2 | 10.66 | 133.17 | 109.6 | 82.3 | 26.3 |
| $LSD_{0.05}$ | 1.9 | ns | ns | ns | ns | 2.7 | 3.36 | ns |

[a]*ns* not significant

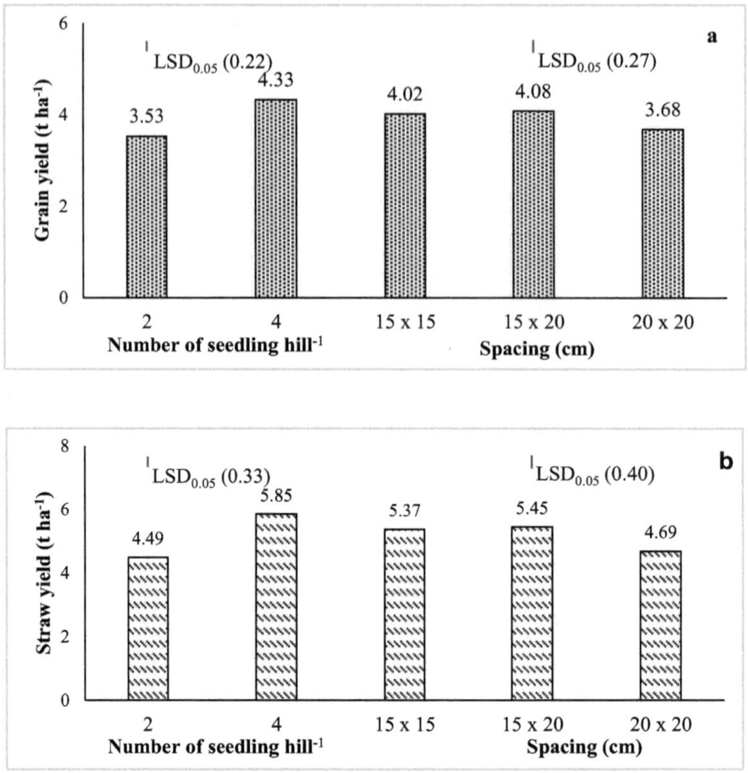


**Fig. 8.2** Grain (**a**) and straw (**b**) yields as influenced by the number of seedlings hill$^{-1}$ and spacing

200 kg N ha$^{-1}$ was not significant. Increasing nitrogen levels significantly increased days to maturity because of more vegetative growth (Table 8.16).

The yield-attributing characters (panicle length, panicle weight, number of grains panicle$^{-1}$, and 1000 grain weight) showed a significant response with increasing levels of N. Longest panicle was recorded with 200 kg N ha$^{-1}$ but the difference in this attribute was not significant between 175 kg N ha$^{-1}$ and 200 kg N ha$^{-1}$. The similar trend was recorded with panicle weight and number of grains panicle$^{-1}$. Maximum grain yield was obtained from 175 kg N ha$^{-1}$, which was 125.2%, 21.1%, 9.7%, 2.0%, and 3.3% higher over control, 100 kg, 125 kg, 150 kg, and 200 kg N ha$^{-1}$, respectively (Table 8.17). Trend analysis using curve expert 32 model revealed that response of N was recorded up to 169 kg ha$^{-1}$, and after that, there was no response to further N application. Therefore, 169 kg ha$^{-1}$ is the optimum dose of N for CSR43 (Fig. 8.3.).

**Table 8.16** Effect of N levels on growth parameters of salt-tolerant variety "CSR43"

| N levels ($kg\ ha^{-1}$) | Plant height (cm) | Productive tillers $hill^{-1}$ | Dry matter ($g\ hill^{-1}$) | Days to maturity | Leaf area index |
|---|---|---|---|---|---|
| 0 | 96.5 | 8.6 | 41.6 | 105 | 4.6 |
| 100 | 100.6 | 9.1 | 43.7 | 108 | 5.2 |
| 125 | 103.3 | 12.1 | 49.4 | 110 | 5.4 |
| 150 | 105.9 | 13.4 | 52.1 | 112 | 5.6 |
| 175 | 106.7 | 14.2 | 53.4 | 115 | 6.0 |
| 200 | 108.3 | 14.7 | 55.3 | 120 | 6.1 |
| $LSD_{0.05}$ | ns[a] | 1.23 | 1.5 | 3.2 | 0.1 |

[a]*ns* non-significant

**Table 8.17** Effect of N levels on yield-contributing characters and yield of salt-tolerant variety

| N levels ($kg\ ha^{-1}$) | Panicle length (cm) | Panicle weight (g) | Number of grains $panicle^{-1}$ | 1000 grain weight (g) | Grain yield ($t\ ha^{-1}$) | Straw yield ($t\ ha^{-1}$) |
|---|---|---|---|---|---|---|
| 0 | 21.0 | 2.3 | 83 | 24.6 | 2.50 | 3.80 |
| 100 | 23.4 | 2.3 | 111 | 27.6 | 4.65 | 5.08 |
| 125 | 24.5 | 2.7 | 121 | 26.7 | 5.13 | 6.13 |
| 150 | 26.2 | 3.0 | 133 | 28.1 | 5.53 | 7.70 |
| 175 | 27.9 | 3.3 | 142 | 28.1 | 5.63 | 7.65 |
| 200 | 28.3 | 3.4 | 145 | 28.4 | 5.45 | 7.20 |
| $LSD_{0.05}$ | 1.52 | 0.33 | 8.7 | 1.05 | 0.52 | 0.58 |

### 8.4.4 On-Farm Evaluation of Sal-Tolerant Varieties Using Improved Agronomic Approaches

Salt-tolerant varieties had an average yield advantage of 0.62 $t\ ha^{-1}$ over farmers' varieties across locations (range of 0.39 to 0.82 $t\ ha^{-1}$). However, combined effect of improved variety with matching management practices resulted in a 35% increase in grain yield over farmers' current practices. The average yield enhancement due to these best management practices across locations was 16%. Studies by Ismail et al. (2012) highlighted the benefits of combining good management practices with genetic tolerance for better crop establishment in less favorable areas, as tolerant genotypes respond better to these agronomic manipulations than sensitive ones. Purushottam et al. (2012) reported that grain yield of a rice-wheat system in on-farm trials did not increase significantly with the use of improved varieties alone, but a significant increase over farmers' practices was observed when improved varieties were combined with line sowing/transplanting. Sarangi et al. (2015) and Singh et al. (2015) also reported higher yields when farmers use improved varieties with improved nursery management practices. Our study highlighted the importance of developing and validating management options for new salt-tolerant varieties to maximize their attainable yields in farmers' fields (Fig. 8.4).

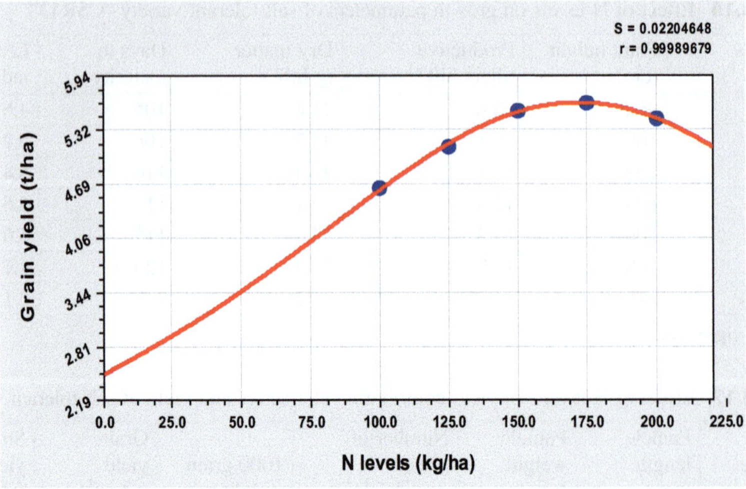


**Fig. 8.3** Trend analysis for grain yield of CSR43 as affected by different N levels

**Fig. 8.4** Performance of salt-tolerant and traditional varieties under farmers managed baby trials

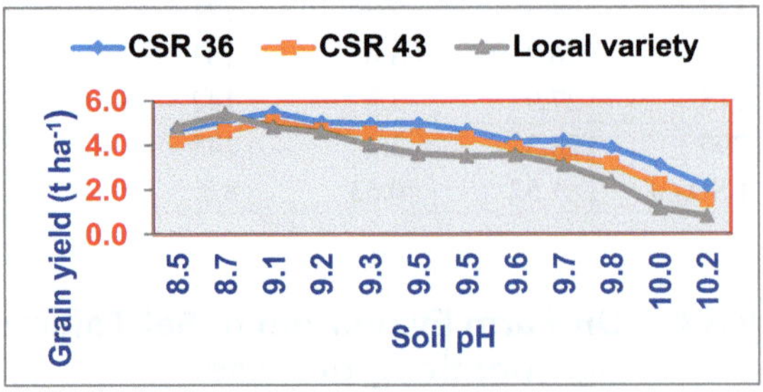

Before Reclamation

After Reclamation

Overall impact of participatory reclamation and management approaches on landscape after reclamation

## 8.5 Conclusions

Salt-tolerant varieties with matching management practices produced about 0.5 t $ha^{-1}$ high grain yields than farmers' current varieties and their management practices across locations. This yield advantage, together with tolerance of sodicity, short duration, good grain quality, and market value contributed to the selection and commercialization of the varieties and their quick adoption in the sodic soils of the Indo-Gangetic plains. Improved technological advancement for reclamation of sodic soils and matching management practices increased crop yield of salt-tolerant varieties significantly compared with traditional varieties and agronomic practices used by local farmers. Suggested recommendations including seed rate, nutrient management and age of seedling in nursery, number of seedlings $hill^{-1}$, and spacing and fertilizer dose for salt-tolerant varieties if adopted by the farmers can boost rice productivity in salt-affected soils. On-farm validation trials of these practices using salt-tolerant varieties showed that the matching management practices alone increased grain yield by 8% over existing recommendations and by 16% over management strategies being practiced by local farmers. Synergy of genetical and agronomical approaches is essential for harnessing the full potential of a tolerant variety when grown in targeted stress-prone conditions. These cost-effective management options for sodic soils can further be refined to match other new tolerant varieties as they become available. The benefits can then be extended to larger areas including the salt-affected areas of the Ganges mega deltas in India and Bangladesh, in addition to the Indo-Gangetic plains of eastern Uttar Pradesh and Bihar in India.

## References


Abrol IP, Yadav JSP, Massoud FI (1988) Salt-affected soils and their management (No. 39). Food & Agriculture Org., p 119

Akhtar J, Naseem A, Mahmood K, Nawaz S, Qureshi RH, Aslam M (2001) Response of some selected wheat (*Triticum aestivum* L.) genotypes to salinity: growth and ionic relations. Pakistan J Soil Sci 19:1–7

Alka U, Satyendra T, Pandey SN (2012) Effects of soil sodicity on growth, nutrients uptake and bio-chemical responses of *Ammi majus* L. Res J Soil Biol 4(3):69–80

Aslam M, Qureshi RH, Ahmad N (1993) Mechanisms of salinity tolerance in rice (*Oryza sativa* L.). In: Leith H, Massoum AA (eds) Towards the Rational Use of High Salinity Tolerant Plants. Kluwer Academic Publisher, Netherlands, pp 135–138

Aslam M, Qureshi RH, Ahmed N, Muhammed S (1988) Response of rice to salinity shock at various growth stages and type of salinity in the rooting medium. Pakistan J Agric Sci 25:199–205

Chandrakar BL, Chandravanshi BR (1988) Effect of age and number of seedling, spacing and fertilizer on tall indica rice. Indian J Agron 33(2):131–134

Cha-Um S, Pokasombat Y, Kirdmanee C (2011) Remediation of salt-affected soil by gypsum and farmyard manure-Importance for the production of Jasmine rice. Austr J Crop Sci 5(4):458–465

Chhabra R, Abrol IP (1977) Reclaiming effect of rice grown in sodic soils. Soil Sci. 124(1):49–55

Chowdhury MJU, Sarker AU, Sarkar MAR, Kashem MA (1993) Effect of variety and number of seedlings hill-1 on the yield and its components on late transplanted Aman rice. Bangladesh J. Agril. Sci 20(2):311–316

Dargan KS, Gaul BL (1974) For paddy in alkali soil seedling age and planting date vital. Indian Farming 245:11–12

Dhanushkodi V, Subrahmaniyan K (2012) Soil management to increase rice yield in salt affected coastal soil-a review. Int J Res Chem Environ (IJRCE) 2(4):1–5

Dixit UC, Patro N (1994) Effect of Levels of NPK, Zinc and Plant Density on Yield Attributes and Yield of Summer Rice. Environ Ecol 12:72–72

FAO, Land and plant nutrition management service. 2008. http://www.FAO.org/ag/agl/agll/spush.

Flowers TJ, Garcia A, Koyama M, Yeo AR (1997) Breeding for salt tolerance in crop plants-the role of molecular biology. Acta Physiologiae Plantarum 19(4):427–433

Flowers TS, Yeo AR (1989) Effects of salinity on plant growth and crop yield, in environmental stress in plants. In: Cherry JH (ed) Springer-Verlag, Berlin, pp 101–119

Gao JP, Chao DY, Lin HX (2007) Understanding abiotic stress tolerance mechanisms: recent studies on stress response in rice. J Integr Plant Biol. 49(6):742–750

Gautam RK, Singh YP, Sharma DK, Nayak AK, Mishra VK, Singh R, Verma CL, Singh RK, Mishra B, Sharma SK (2014) Variety notification "CSR 43". Indian J Genet 74(2):265–269

Ghafoor A, Murtaza G, Ahmad B, Boers TM (2008) Evaluation of amelioration treatments and economic aspects of using saline sodic soils for rice and wheat production on salt affected soils under arid land conditions. Irrigation Drainage 57:424–434

Gharaibeh MA, Eltaif NI, Shunnar OF (2009) Leaching and reclamation of calcareous saline-sodic soil by moderately saline and moderate-SAR water using gypsum and calcium chloride. J Plant Nutr Soil Sci. 172(5):713–719

Gondek K, Filipek-Mazur B (2005) The effects of mineral treatment and the amendments by organic and organomineral fertilisers on the crop yield, plant nutrient status and soil properties. Plant, Soil and Environ 51(1):34–45

Gupta N, Tripathi S, Balomajumder C (2011) Characterization of pressmud: A sugar industry waste. Fuel 90(1):389–394

Ismail AM, Heuer S, Thomson MJ, Wissuwa M (2007) Genetic and genomic approaches to develop rice germplasm for problem soils. Plant Mol. Biol. 65(4):547–570

Ismail AM, Johnson DE, Ella ES, Vergara GV, Baltazar AM (2012) Adaptation to flooding during emergence and seedling growth in rice and weeds, and implications for crop establishment. AoB Plants 2012:pls019. https://doi.org/10.1093/aobpla/pls019

Liu SH, Kang YH (2014) Changes of soil microbial characteristics in saline-sodic soils under drip irrigation. J.Soil Sci Plant Nutr. 14(1):139–150

Mandal BK, Sainik TR, Ray PK (1994) Effect of age of seedling and level of nitrogen on the productivity of rice. Oryza 21(4):225–232

Mazher AA, El-Quesni EF, Farahat MM (2007) Responses of ornamental and woody trees to salinity. World J. Agric. Sci 3(3):386–395

Meena SL, Singh S, Shivay YS (2003) Response of hybrid rice (*Oryza sativa*) to nitrogen and potassium application in sandy clay-loam soils. Indian J Agric Sci. 73(1):8–11

Mikanová O, Šimon T, Javůrek M, Vach M (2012) Relationships between winter wheat yields and soil carbon under various tillage systems. Plant, Soil and Environ 58(12):540–544

Mishra B, Singh RK, Bhattacharya RK (1992) CSR10, a newly released dwarf rice for salt affected soils. IRRN 17(1):19

Muhammad DOST, Khattak RA (2011) Wheat yield and chemical composition as influenced by integrated use of gypsum, pressmud and FYM in saline-sodic soil. J Chem Soc Pakistan 33(1): 82–89

Munns R (1992) A leaf elongation assay detects an unknown growth inhibitor in xylem sap from wheat and barley. Funct Plant Biol. 19(2):127–135

Munns R, Tester M (2008) Mechanisms of salinity tolerance. Annu Rev Plant Biol. 59:651

Naeini SM, Cook HF (2000) Influence of municipal waste compost amendment on soil water and evaporation. Commun Soil Sci Plant Anal. 31(19):3147–3161

National Remote Sensing Application and Associates (NRSA & Associates) (1996) Mapping of Salt Affected Soils of India.1:250000 map sheets, legend. NRSA, Hyderabad

Negim O (2015) Effect of addition pressmud and gypsum by product to reclamation of highly calcareous saline sodic soil. Am Assoc Sci Technol J Environ I 1:76–84

Onemli F (2004) The effects of soil organic matter on seedling emergence in sunflower (*Helianthus annuus* L.). Plant Soil Environ 50(11):494–499

Prasad A, Chattopadhyay A, Chand S, Naqvi AA, Yadav A (2006) Effect of soil sodicity on growth, yield, essential oil composition, and cation accumulation in rose-scented geranium. Commun Soil Sci Plant Anal. 37(13):1805–1817

Purushottam K, Singh DK, Bharadwaj AK (2012) On-farm evaluation of agronomic management practices under rice–wheat cropping system. In: Extended Summaries. Third International Agronomy Congress, New Delhi, pp 1295–1296

Sarangi SK, Maji B, Singh S, Burman D, Mandal S, Sharma DK, Haefele SM (2015) Improved nursery management further enhances the productivity of stress-tolerant rice varieties in coastal rainfed lowlands. Field Crop Res 174:61–70

Shivay YS, Singh S (2003) Effect of planting geometry and nitrogen level on growth, yield and nitrogen-use efficiency of scented hybrid rice (*Oryza sativa*). Indian J Agron 48(1):42–44

Sidiras N, Bilalis D, Vavoulidou E (2002) Effects of tillage and fertilization on some selected physical properties of soil (0–30 cm depth) and on the root growth dynamic of winter barley (*Hordeum vulgare* cv. Niki). J Agronomy Crop Sci 187(3):167–176

Singh RK, Redoña E, Refuerzo L (2010b) Varietal improvement for abiotic stress tolerance in crop plants: special reference to salinity in rice. In: Abiotic stress adaptation in plants. Springer, Dordrecht, pp 387–415

Singh YP, Arora S, Mishra VK, Bhardwaj AK (2021) Regaining the agricultural potential of sodic soils and improved smallholder food security through integration of gypsum, pressmud and salt tolerant varieties. Agroecol Sustain Food Syst 46(3):410–431

Singh YP, Arora S, Mishra VK, Dixit H, Gupta RK (2018) Conjoint use of chemical amendments and municipal solid waste compost for amelioration of degraded sodic soil. J Indian Soc Soil Sci 66(4):392–398

Singh YP, Arora S, Mishra VK, Dixit H, Gupta RK (2019b) Plant and soil responses to the combined application of organic amendments and inorganic fertilizers in degraded sodic soils of indo-gangetic plains. Commun Soil Sci Plant Anal 50(20):2640–2654

Singh YP, Mishra VK, Arora S, Sharma DK, Singh S (2015) Barren to bountiful: a successful livelihood security option for resource poor farmers. STRASA News 8

Singh YP, Mishra VK, Bharadwaj AK, Arora S, Singh AK, Singh S, Haefele SM (2019a) Synergy of reduced gypsum and pressmud–a cost effective approach for sustainable reclamation of degraded sodic lands. Indian J Agric Sci 89(6):1027–1032

Singh YP, Mishra VK, Gupta RK (2020a) Optimizing seedling number and hill spacing: a way forward to harness productivity potential of salt-tolerant rice cultivars in salt-affected soils. Agric Res 9(1):77–84

Singh YP, Mishra VK, Singh D (2020b) Enhancing rice productivity in sodic soils of Indo-gangetic plains through improved nursery management practices. Indian J Agric Sci 90(5):995–999

Singh YP, Mishra VK, Singh S, Sharma DK, Singh D, Singh US, Ismail AM (2016) Productivity of sodic soils can be enhanced through the use of salt tolerant rice varieties and proper agronomic practices. Field Crop Res 190:82–90

Singh YP, Nayak AK, Sharma DK, Gautam RK, Singh RK, Singh R, Ismail AM (2014) Farmers' participatory varietal selection: a sustainable crop improvement approach for the twenty-first century. Agroecol Sustain Food Syst 38(4):427–444

Singh YP, Nayak AK, Sharma DK, Gautam RK, Singh RK, Singh R, Mishra VK, Paris T, Ismail AM (2013) Varietal selection in sodic soils of Indo-gangetic plains through farmer's participatory approach. Afr J Agric Res *8(23)*:2849–2860

Singh YP, Singh R, Gautam AK (2008) Effect of nitrogen levels on yield and nutrient uptake of salt tolerant rice and wheat cultivars in gypsum amended alkali soils. J Indian Soc Soil Sci 56(1): 86–91

Singh YP, Singh R, Sharma DK (2009) Combined effect of reduced dose of gypsum and salt tolerant varieties of rice (*Oryza sativa*) and wheat (*Triticum aestivum*) on rice-wheat cropping system in sodic soils. Indian J Agron 54(1):24

Singh YP, Singh R, Sharma DK (2010a) Determination of time frame for substitution of salt tolerant varieties of rice (*Oryza sativa*) and wheat (*Triticum aestivum*) through crop diversification in sodic soils. Indian J Agric Sci 80(7):245–250

Tavakkoli E, Rengasamy P, McDonald GK (2010) High concentrations of Na + and Cl–ions in soil solution have simultaneous detrimental effects on growth of faba bean under salinity stress. J Exp Bot 61(15):4449–4459

Thimappa K, Singh YP, Raju R, Kumar S, Tripathi RS, Pal G, Reddy AA (2015) Reducing farm income loss through land reclamation: a case study from Indo-Gangetic Plains. J Soil Sali Water Qual 7(1):68–76

US Census Bureau, 1998 (2008). www.census.gov/ipc/www/wp98001.html. Accessed 30 July 2008.

Van Asten PJ, Van't Zelfde JA, Van der Zee SEATM, Hammecker C (2004) The effect of irrigated rice cropping on the alkalinity of two alkaline rice soils in the Sahel. Geoderma 119(3):233–247

Witcombe JR, Joshi A, Joshi JK, Sthapit BR (1996) Farmer participatory crop improvement. I. Varietal selection and breeding methods and their impact on biodiversity. Exp Agric 32:445–460

Wong VN, Greene RSB, Dalal RC, Murphy BW (2010) Soil carbon dynamics in saline and sodic soils: a review. Soil Use Manage. 26(1):2–11

# Genetic Enhancement of Rice for Alkalinity-Affected Area of Central Uttar Pradesh, India

9

Vijay Kumar Yadav

**Abstract**

Uttar Pradesh is a significant contributor of food security of the nation. The rice-wheat cropping system is a major production system of the state. About 28% of India's wheat and 12% of rice is produced by the state. It is blessed with the fertile Indo-Gangetic plains. The problem of salt affected soils in the state was earliest observed in1876 by Mr. David Robarts, a landlord of Sikandara Rao. The central part of Uttar Pradesh of Ganga-Yamuna Doab is severely affected by alkalinity which reflected high pH, EC, ESP and preponderance of sodium bicarbonate and carbonate. Rice is one of the most important crops of Uttar Pradesh on which research work was initiated in 1924 at Kanpur and onward more rice research station established at Nagina where famous variety Nagina 22 developed, Basti, Gorakhpur, Panchperwa, Tissuhin, Surha Tal to address the problems of rice in eastern Uttar Pradesh. Ghaghraghat research station was established in 1959 to tackle the problems of deep water rice in flood affected area of the eastern Uttar Pradesh developed number of famous varieties for deep water. Kanpur centre was reoriented in 1971–1972 to develop rice variety & technology (s) for saline—alkaline soil of the state. Salt tolerance research work was more focused on alkalinity tolerant rice with the assistance of STRASA project operated since 2013.

Famous rice varieties viz., N22, Chakiya 59, Sudha, Majhera3, Jalmgna, Madhukar, Uasar 1, Ashwani for various ecosystems were developed and some of which are being used as transgression work. The technology for reclamation of alkali soils like application of RSC water passing through 15 cm phospho-gypsum bed for soil properties and improving productivity of paddy and wheat

V. K. Yadav (✉)
C S A University of Agriculture and Technology, Kanpur, Uttar Pradesh, India
e-mail: vkyadav@csauk.ac.in





R. K. Singh et al. (eds.), *Genetic Improvement of Rice for Salt Tolerance*,
https://doi.org/10.1007/978-981-99-3830-8_9

crop; gypsum for reclamation of sodic soil @ 50% G.R replaced with 25 % G.R. if supplemented with press mud @ 5 t $ha^{-1}$; Standardized Resource Conservation Technologies for alkali soils and Identified suitable tree species for alkali land plantations. Efforts are being made to developed inbuilt salt stress tolerant varieties, suitable production technology to enhance production, productivity and profitability of farmers of Uttar Pradesh.

**Keywords**

Rice · Alkalinity · Salt stress · Tolerance · Genetic · Uttar Pradesh

## 9.1 Introduction

Uttar Pradesh is spread over an area of 240,928 sq. km and has 75 districts. It is an agrarian economy, where 47% of the population is directly dependent on agricultural activities for their livelihood. The net cultivated area of Uttar Pradesh is 16.42 million hectares, constituting 70 per cent of the total geographical area of the state. The area under forest and fallow land is 6.90% and 7.00%, respectively. Other uncultivated land excluding fallows is 3.70% and area not available for cultivation stood at 14.30% of total geographical area. The irrigated area is over 73% and cropping intensity is 153%. This state has very wide agroclimatic conditions. It is one of the major producers of food grains in the country. The major crops of the state are wheat, rice, sugarcane, maize, chickpea, pigeon pea, lentil and potato. Within cereals, wheat and rice are the most important crops accounting for 38% and 12% of gross cropped area, respectively. Uttar Pradesh is largely dominated by small and marginal farmers with 93% of agricultural households operating 65% of land. The rice-wheat cropping system (RWCS) is a major production system of the state.

C. S. Azad University of Agriculture and Technology, Kanpur, is situated between 27.24° N latitude and $77.5^0$ E longitudes and at an altitude of 178 metres above the mean sea level in the Gangetic plain of central Uttar Pradesh. The climate of district of Kanpur is semi-arid with hot summer and cold winter. Nearly 80 per cent of total rainfall occurs during the monsoon (only up to September) with a few showers in the winter.

## 9.2 Salt Stress: Historical Perspective

Salt-affected soils have been known to occur in Uttar Pradesh from time immemorial. Their reference is made in the ancient Indian scriptures, and it is also recorded in the British memoirs of the nineteenth century that some of the wastelands in Uttar Pradesh were covered with vegetation till 1800. Continuous deforestation to meet the demand of firewood and food grain production led to ecological imbalance resulting in excessive surface evaporation and finally extension of salt-affected lands. As early as 1876, Mr. David Roberts, a substantial *Zamindar* of *Pargana*

*Sikandra Rao*, Aligarh, filed a petition to the Board of Revenue for relief on the grounds of disastrous spread of *Reh* in parts of Meerut, Aligarh, and throughout of *Kaali Nadi* valley areas of his land owing to the appearance of salt efflorescence as a result of the introduction of canal irrigation. This culminated into the constitution of the historical Indian *Reh* Committee in 1877 with the objectives of investigating into the causes of the deterioration of land by salt efflorescence. Voelcker (1897) reported that in the State of Uttar Pradesh (then United Provinces) alone, an area of about 10,000–12,000 $km^2$ was affected by soil salinity.

Such lands in local parlance are known by a variety of names in different parts of the subcontinent and the most common term used is '*usar*', which is derived from the Sanskrit word '*ushtora*' meaning barren or sterile. The names like *Reh*, *Chhar*, *Khari*, *Lona*, *Shera*, etc. are also used in some places. These are addressed for all kind of saline alkaline soils in North India, particularly in Uttar Pradesh. Most of these lands were uncultivated and served as village grazing grounds.

Scientific study of salt-affected soils in India was first started in Uttar Pradesh in the second half of the last century. Leather (1914), the then Imperial Agricultural Chemist, initiated systematic study on these soils in Uttar Pradesh. Considerable research has been done on the salt-affected soils during the last 90 years and different packages of practices have been developed for successful reclamation of such precarious lands. However, no systematic efforts were undertaken for amelioration of salt-affected soils in farmers' field prior to 1947.

## 9.3 Type and Extent of Salt Stress in Uttar Pradesh

Salinity and alkalinity are growing problems worldwide. It can either be natural or primary or secondary or man-made salinity. Primary salinity of soil and groundwater is due to the weathering of naturally saline rocks, while secondary salinity is due to human activities like unreliable irrigated source of water.

## 9.4 Extent and Distribution of Salt-Affected Soils in Uttar Pradesh

In Uttar Pradesh, Usar Land Reclamation Committee (1938–1939) constituted by the state government assessed 0.84 million hectares under salt-affected soils, which in subsequent survey (Agrawal et al. 1979) revised it to 1.37 million hectares. The division worst affected by salt stress are Allahabad, Lucknow, Kanpur and Agra, which together constitute nearly 54% of the total area of salt-affected soils in the state. The districts worst hit by the ravages of salt stresses were listed as Kanpur, Allahabad, Mainpuri, Raebareli, Etah, Farrukhabad, Aligarh, Etawah, Sultanpur, Fatehpur, Hardoi, Pratapgarh, Shahjahanpur, Moradabad, Azamgarh and Faizabad. Canal-irrigated area land of state has already been affected by salinity/water logging while the low rainfall areas, viz., Aligarh, Agra and Mathura districts where exploitation of saline groundwater coupled with the intensive use of canal water have created serious problem of soil deterioration. In Ramganga, Sharda and Gandok command projects, a large area has already become saline due to water logging. The total area under wastelands in these 23 districts is estimated about 11.23 lakh ha. The area under salt-affected wastelands is 5.21 lakh ha.

## 9.5 What Are Salt-Affected Soils?

All the soils contain some amount of salts that though in small quantities are not harmful for the growing plants, their excessive presence may either inhibit or deteriorate plant growth resulting in conspicuous reduction in crop yields, the extent depending on the kind and amount of salts present, the species and varieties of plants grown and the environmental factors. Soils that contain excess soluble salts that affect plant growth adversely are called *salt-affected soils.*

These problematic soils comprise of 'saline soils' that contain sufficient soluble salts to interfere with the growth of plants and 'alkali soils' (also called sodic-form sodium) that contain excessive exchangeable sodium to adversely affect crop production. Salt-affected soils is a general term that includes soils having an excess of soluble salts or an excess of exchangeable sodium, or both, to an extent that their presence or action can adversely modify growth or completely inhibit production of most crops. Saline soils are known as *'reh'* or *'thur'* and alkali as *'usar'* or *'kallar'* in village parlance.

Salt-affected soils of Uttar Pradesh may be divided into three categories:

1. Saline soils.
2. Sodic soils/alkaline soils.
3. Saline sodic soils.

### 9.5.1 Saline Soils

Soils that contain sufficient neutral soluble salts in the root zone that adversely affect the crop growth and production is called 'saline soils'. Soluble salts most commonly present are the chlorides and sulphates (anions) of sodium, calcium and magnesium (cations). In saline soils, the electrical conductivity of the saturation extract exceeds 4 $dSm^{-1}$ and the pH is less than 8.2. When the salinity level is low, the proportion of sodium is less and calcium and magnesium are more but sodium predominates in highly saline soils.

In Uttar Pradesh, such soils are found in the districts of Mathura and Agra and *Khadirs* (riverside tracts) of rivers Ganga and Yamuna. The saline soils are recognized by patchy growth of crops and sometimes visible signs of salt injury such as tip burn of leaves and chlorosis (pale yellow colour) of leaves. White salt crusts on the surface can be seen during post monsoon period. In *Khadir* or riverine areas of rivers Ganga and Yamuna, salt crusts may not be visible at the surface during early winters due to higher soil moisture content and better flushing of salts through underground circulation in the vicinity of the rivers. But salt efflorescence becomes well marked on convex spots during drier parts of the year even in these areas. In *Yamuna Khadirs*, saline soils have a tendency to undergo alkalization in the subsoil rapidly. Secondary salinization is also evident sporadically with the expansion of canal system under command project areas owing to inadequate provision of drainage. With the large-scale expansion of surface irrigation systems in the country, this problem is likely to assume bigger dimension (Joshi 1990). The chlorides and sulphates are the dominant anions followed by bicarbonates. Due to the preponderance of neutral salts, these soils remain flocculated and possess good internal drainage. The textures of the soils usually vary from loamy sand to loam, and there is usually no hard pan or *Kankar* layer (calcium carbonate nodules) in the subsoil. The groundwater table remains high during most part of the year especially in Agra and Mathura districts.

### 9.5.2 *Sodic* or Alkali Soils

These soils are formerly termed as 'saline-alkali' and alkali soils in scientific literature and 'usar' in common parlance; these soils represent the most extensive and most problematic soils of the state located mainly in the semi-arid heartland of *Gangetic* plains protruding gradually to sub-humid eastern plains of river Ganga. These soils have developed under impeded drainage due to nearly impermeable, clay subsoils usually underlain by hard indurate *Kankar* pans and high fluctuating groundwater table. The soil surface has extensive white salt efflorescence during drier parts of the year turning into pellets of black to brownish grey flakes owing to dispersion and dissolution of *humic* substance at high alkalinity (Agarwal and Gupta 1968).

### 9.5.3 Saline-Alkali

These soils do not occupy extensive areas but are located in small pockets within the vast stretches of salt-affected soils. Excessive leaching leading to dealkalization of surface horizon has resulted in lowering of pH values to nearly 5.2, while the adjacent sodic soils have pH values above 10. In eastern districts of Uttar Pradesh, paddy cultivation on non-calcareous *sodic* soils for a long parried has also led to degradation or solodization of these soils. The groundwater table is within 2 m even during the summer months and greying in subsurface horizon is well marked. The degradation process leads to the replacement of exchangeable $Na^+$ by exchangeable $H^+$ and $A1^{+++}$ in the surface layer along with the accumulation of crystalline, amorphous and water-soluble silica. The soil pH is strongly acidic declining to about 5.2. However, the disruption of clay complex is not conspicuous since exchangeable $A1^{+++}$ is much less than exchangeable $H^+$. The dark colour of surface horizon owes its origin to accumulation of organic matter under submerged conditions prevailing up to late winters. In subsurface horizons, the soil pH gradually tapers to neutrality with depth but magnesiumisation of these horizons is very conspicuous, recording values of EM2P ranging from 60 to 74, which imparts prismatic structure of these soils breaking into medium blocks. The ESP is, however, below 15 in these soils. High exchangeable magnesium imparts poor physical properties to these soils. Drainage is the major problem of these soils and paddy or sugarcane can thrive well after making adequate provisions for drainage.

Out of 6.40 million ha wasteland in Uttar Pradesh, 1.37 million hectares was reported to be *sodic* land. The sustainable reclamation of *sodic* land and prevention of further increase in sodicity definitely contributed in developing food security of state. On-farm development of *sodic* lands with provision of assured irrigation and reclamation through introducing new technology, tolerant cultivar and better cultivation practices would definitely add to farmer's food basket.

## 9.6 Status of Rice Area, Production and Productivity in the State

Rice is the major crop in Uttar Pradesh that is grown in about 5.70 million hectares, sharing 13% of total rice area in India. Uttar Pradesh has favourable and suitable climate, fertile soil, adequate water resources and sunshine hours. The state ranks second in county's rice production. Rice is cultivated mainly in *Kharif* season in around 5.52 million hectares followed by *Zaid* covering 0.035 million ha. The *Boro* rice is grown in deeply flooded area of around 0.003 million ha mainly in eastern districts of Uttar Pradesh. Critical analysis of productivity of rice in Uttar Pradesh state revealed that it gradually increased to 10.90 qt/ha by 1981 and finally rose to 27.04 qt/ha by 2019–2020 (Agricultural Statistics at a Glance 2020).

The average rice productivity of the state is around 27 qt/ha. Surprisingly, 32% of the irrigated rice area of Uttar Pradesh is giving low yields. Yield gap analysis reveals that 30–40% of the potential yield is yet to be tapped with available high-yielding varieties sown on highly productive irrigated soils. This gap is due to degraded and less fertile soils, pockets of endemic pests and diseases, low input use, faulty cropping systems and a less adoption of improved technology by the farmers. Saline and *sodic* soils are widespread in inland areas and are progressively expanding because of improper water management. Rice is suitable for rehabilitating salt-affected soils because it can grow under flooded conditions and has a high potential for genetic improvement. Rice productivity in salt-affected areas could be increased by 10–20 q/ha, providing food for millions of the poorest people and making use of some of the least exploited land and water resources. With adoption of salt-tolerant varieties and advanced production technologies, diagnostic and corrective measures will substantially increase yield levels.

## 9.7 Background of Rice Research at C. S. Azad University of Agriculture and Technology, Kanpur, Uttar Pradesh

- Rice is one of the most important crops of Uttar Pradesh. Research work on rice was initiated in 1924 with the existing staff of economic botanist (Rabi cereals) at Kanpur.
- Later on, a post of research assistant was created specifically for rice research with its headquarter at Nagina, Bijnor.
- In 1932, the Indian (the then Imperial) Council of Agricultural Research also provided financial assistance for expanding the research work on rice, and a post of assistant paddy botanist was created, who worked under the overall supervision of economic botanist (sugarcane and cotton).
- During 1938, a research substation was established in district of Basti to study the problems of rice in the eastern districts of Uttar Pradesh and it was later on shifted to Gorakhpur.
- In year 1942, rice research work was separated from economic botanist (Rabi cereals), and up to 1945, the rice research remained the sole responsibility of state government. This centre was provided only financial assistance and advice to rice research.
- During 1946, mega change occurred in respect of rice research in the country and government of India also took up rice research responsibility by establishing Central Rice Research Institute (CRRI) at Cuttack, Orissa, and later on, two additional substations were also created in eastern districts of the Uttar Pradesh to study the problems of late-sown paddy. Consequently, research work on late-condition paddy was started at Pachperwa in districts of Gonda and Tissuhin, Mirzapur in 1949. In the same time, assistant paddy botanist was given an independent status and responsibility for research in the state.
- In 1957–1958, Indian Council of Agricultural Research, New Delhi sanctioned a research scheme for deep water rice, which was started to work on the problems in famous *Surha Tal* with an area of 30 sq. km located at Bansdih district in Ballia.
- A new separate research station was established at Ghaghraghat in district of Bahraich to tackle the problems of flood-affected area of the eastern Uttar Pradesh in 1959. The major mandate of this research station was related to evolve flood-resistant rice varieties and develop suitable agro-technology for deep water rice. Subsequently, a hill rice scheme at Majhera (district of Nainital) was started to work on rice husbandry relevant to hill areas of Uttar Pradesh, now in Uttarakhand, and another scheme to work on rainfed upland paddy was started at Balchandpur in district of Bahraich.
- The research work on rice was geared up in 1958 with creation of post of economic botanist with full financial and administrative power, and subsequently, a new rice research farm was established at Masodha, Faizabad. The systematic rice research work in state has been started from 1964 by economic botanist (rice) with establishment of its headquarter at Masodha, Faizabad.
- The rice research in the state was further strengthened and closely linked up with the national programme during 1968 and 1969 with set-up of zonal and regional

research centres of the All India Coordinated Rice Improvement Project at Masodha in Faizabad and Majhera in Nainital, respectively.

- Later in 1961, G. B. Pant University of Agriculture and Technology (GBPUAT) was established at Pant Nagar, Nainital, with an aim to give new direction to agricultural research, education and extension. Rice research work was initiated in 1963 and the most effective control for '*Khaira*' disease was worked out in 1966. The work on varietal improvement began in 1969 adopting multidisciplinary approach to rice improvement.
- Further, the government of Uttar Pradesh sanctioned a new Rice Research Station at Kanpur during 1971–1972 to develop *suitable rice technology(s) for saline–alkaline soil of the state and to develop postharvest technology of rice.*
- In 2005, AICRP on Rice was established in Kanpur to strengthen research on saline and alkaline soils while at Nagina for Basmati rice.
- ICAR Network Project on Development of New Plant Type varieties with higher yield and inbuilt resistance to major insect and pest operated from March 2010 to 2014. Segregating and advance breeding lines were also screened out under alkaline stress.
- Stress Tolerance Rice for Africa and South Asia (STRASA), a programme of IRRI funded by Bill and Melinda Gates Foundation, was operated since April 2013 to 2018 for evaluation of salt stress tolerance line at Bhoja Crop Research Farm, Daleepnagar, Kanpur.

## 9.8 Rice Varietal Development Work at Kanpur

### 9.8.1 Introduction from Other Countries

The introduction method of breeding was also utilized and materials were introduced from different countries like China, Russia, Italy, etc., and from these materials, China 4, China 12, Russian 2652 and Kachsuing 22 (VL 8) became very popular in Uttar Pradesh.

### 9.8.2 Selection and Purification

With the view to help farmers to raise their rice productivity, the indigenous and wild races were evaluated by scientists and purification and selection work was done based on pure line breeding method adopted by the plant breeder. As a result, several improved varieties in different maturity groups (*Bhadain, Kuwari and Agahani*) were developed and popularized in Uttar Pradesh. Some of these varieties are T1, T3, T9, T17, T22, T22A, T23, T25, T26, T29, T36, T43, T65, T88, T96, T100, T110, T119, N23, N27, N32, Chakiya 59, Sudha, T2099, Majhera3, Cross 116, Jalmgna and Madhukar for various ecosystems.

### 9.8.3 Recombination Breeding (Hybridization and Selection)

Inspired by breeding achievement using hybridization technique in sugarcane, maize and few other crops, similar work was started in rice from 1930 onwards. This enables to combine certain traits dispersed in different varieties. The varieties so evolved were T 136 from cross (T1xT100), T137 (T1xT27) and Kashti (T136 x Russ. 2652).

### 9.8.4 Introduction of Japonica Varieties

Japonica, having better plant type and response to high level of management, were introduced for direct cultivation. But it could not succeed due to climate variation.

### 9.8.5 *Indica* x *Japonica* Hybridization Programme

To combine high responsiveness to Japonica rice, the crosses were undertaken between *Indica* and Japonica but were not successful due to large amount of sterility problems for free recombination of characters. Under this programme, the varieties like NSJ 200 (Oshu x T m136), NSJ 98 (Oshu x China 4), Saket 1 (Norin 8 x T21), Dhaneshwar (Norin 80x21), Ruhela, Bundela, Varuna and Naina were bred and released for general cultivation, but being tall like indica, it could not succeed as expected (Tables 9.1 and 9.2).

### 9.8.6 Technology Developed for Reclamation and Management of Sodic Soils and RSC Water

C. S. Azad University of Agriculture and Technology, Kanpur, carried research on reclamation and management of heavy textured soils of alluvial and black soil regions and established remarkable findings in association with AICRP on water management since 1969.

- The detailed technology for reclamation of alkali soils, which constitute a major area under salt-affected soils, and its utilization in agriculture production, has been developed. The technology would essentially consist of proper bunding and levelling of the land, application of pyrites or gypsum as per requirement and by the right method based on soil test, judicious use of fertilizers including zinc, growing of salt-tolerant variety of paddy and wheat during kharif and rabi followed with green manuring (*Dhaincha*) in summer. Wheat can be replaced by barley, Egyptian clover during rabi.
- Groundwater sampling and analysis for different quality parameters were conducted for Unnao, Kanpur, Fatehpur, Raebareli, Kannauj and Kanpur Dehat districts. On the basis of analysis, samples were grouped into good, marginally

**Table 9.1** Names, designation, year, parentage and salient traits of salt-tolerant rice varieties (CVRC/SVRC) developed

| Variety name/ IET no. | Cross combination | Year of release | Ecosystem | Maturity (d) | Yield q/ha | Grain type | Released by | State/area recommended | Notification details | Special features |
|---|---|---|---|---|---|---|---|---|---|---|
| Type 100 | Self from Bhainsalot | 1978 | Irrigated saline and alkaline | 150 | 25–30 | SB | SVRC | UP | 13(E) dated 19.12.1978 | Tall and tolerant to sodicity |
| Usar-1 | Jaya x Getu | 1985 | Irrigated saline and alkaline | 140–150 | 32–38 | LB | SVRC | UP | No 540 (E) 24.7.85 | Coarse grain, suitable for saline and alkaline condition |
| Ashwini | N 22x Chauvery | 1987 | Rainfed | 110 | 45–50 | LB | SVRC | Up | No. 662 (E) 17.09.1997 | Semi dwarf, early maturity, drought tolerance |

**Table 9.2** Entries showed higher yield under Stress Tolerance Rice Varieties for Africa and South Asia (STRASA) programme

| S. no | Designation | Cross combination | Remarks |
|---|---|---|---|
| 1 | CSAR 1604 | CST7-1/IRGC-11010// FL478 | Good yield performance, plant height of 88 cm, DF 107, grain SB |
| 2 | CSAR1610 | CST 7-1/IRGC-69861// PUSA 1601 | Plant height 92 cm, DF 104, grain LS |
| 3 | CSAR1611 | CSR-36/IRGC-69708// DSB-3 | Good yield performance, plant height 92 cm, DF 102 |
| 4 | CSAR 1620 | CST7-1/IRGC-11010// RP bio 226 | Good yield performance, plant height 93 cm, DF 99, grain MS |
| 5 | CSAR1628 | Swarna/IRGC-26940// FL478 | Plant height 92 cm, DF 102, grain MS |
| 6 | CSAR 17135 | CST7-1/IRGC-1891// CSR 36 | Good yield performance, plant height 99 cm, DF 109, grain LS |
| 7 | CSAR 17817 | CSR 36/IRGC-11010// AVT1 BT2410 | Plant height 92 cm, DF 104, grain MS |

saline, saline, high SAR-saline, marginally alkali, alkali and highly alkali categories.

- The numbers of technologies of immense utility have been developed from studies, and it was inferred that application of RSC water passing through 15 cm phospho-gypsum bed proved by far superior to equivalent amount of gypsum in controlling the effect of bed quality (RSC) water on soil properties and improving productivity of paddy and wheat crop.
- Experiments showed that the amount of gypsum for reclamation of sodic soil, which was hitherto recommended at 50% GR, could be reduced to half, i.e., 25% GR if supplemented with press mud at 5 t $ha^{-1}$.
- Yield obtained in sodic soil with 90 kg N $ha^{-1}$ supplemented with G.M. (*Dhaincha*) or FYM (15 t $ha^{-1}$) were comparable with inorganic added through urea at 120 kg or 150 kg $ha^{-1}$. Studies have clearly revealed usefulness of organic manuring in rice cropping system in sodic soils.
- Identified suitable tree species for alkali land plantations.
- Resource conservation technologies are being standardized for alkali soils.
- On the basis of field study, it was found that crop rotations like rice-wheat-*Dhaincha* and rice-Egyptian clover are better suited crops under alkali soil conditions.

## 9.9 Infrastructure for Salt-Tolerance Research Work

C. S. Azad University of Agriculture and Technology, Kanpur, has Daleepnagar, Bojha, farm of 278-hectare area with various gradient sodic or alkaline soils with 9.0 to 10.8 pH value. It is considered a natural hotspot to conduct various types of salt tolerance research work. The site characterization where rice research work carried is given below.

### 9.9.1 Site Characterization (Tables 9.3 and 9.4)

Salt tolerance research work at Daleepnagar Bojha farm monitored by ICAR Team

## 9.10 Conclusion

Rice culture have a variety of complex factors that influence from water access to changing climate, wind, soil quality, fertility, presence of weeds and insects, disaster and more. Besides climate changes, malnutrition is now viewed as a global challenge and is being considered as one of the major millennium development goals with improvement of standards of health by reducing child mortality and improving maternal health. Efforts will be made to improve germplasm adaptation to cope up with the prevailing unfavourable rice environments. Development of aerobic rice, drought-tolerant and efficient problem soil-tolerant genotypes and hybrids will be

**Table 9.3** Soil attributes: Bojha Farm, CSAUAT, Kanpur, UP (0–15 cm depth)

| Parameters | Unit | Result | Guide Low | Guide High | Low | Optimum | High |
|---|---|---|---|---|---|---|---|
| $pH_{(1:2)}$ | | 10.03 | 6.0 | 6.8 | | | |
| $pH_s$ | | 8.97 | | >8.2 | | | |
| $EC_{(1:2)}$ | dS $m^{-1}$ | 0.93 | | | | | |
| $EC_e$ | dS $m^{-1}$ | 3.16 | <4.0 | > 4 | | | |
| Organic C | % | 0.19 | <0.5 | >0.75 | | | |
| Cation exchange capacity | $cmol_{(P+)}$ $kg^{-1}$ | 15.6 | <10.0 | >15.0 | | | |
| Exchangeable sodium per cent | % | 90.4 | | >15 | | | |
| Sand | % | 49.6 | 30.0 | 55.0 | | | |
| Silt | % | 33.0 | 30.0 | 50.0 | | | |
| Clay | % | 17.4 | 20.0 | 55.0 | | | |
| Textural Class | | Sandy loam | | | | | |
| Available N | kg $ha^{-1}$ | 135.5 | <280.0 | >560.0 | | | |
| Available P (Olsen) | kg $ha^{-1}$ | 18.5 | <10.0 | >25.0 | | | |
| Available K | kg $ha^{-1}$ | 206.1 | <120.0 | >280.0 | | | |
| Available S | mg $kg^{-1}$ | 20.1 | <10.0 | >20.0 | | | |
| Available Fe | mg $kg^{-1}$ | 1.4 | <2.4 | >5.0 | | | |
| Available Mn | mg $kg^{-1}$ | 8.0 | <1.0 | >2.0 | | | |
| Available Zn | mg $kg^{-1}$ | 0.15 | <0.25 | >2.0 | | | |
| Available Cu | mg $kg^{-1}$ | 1.6 | <0.12 | >2.5 | | | |
| Available B | mg $kg^{-1}$ | 0.61 | <0.1 | >2.0 | | | |
| Per cent Exchangeable K | % | 3.8 | 3.0 | 8.0 | | | |
| Per cent Exchangeable Ca | % | 3.5 | 60.0 | 70.0 | | | |
| Per cent Exchangeable Mg | % | 2.3 | 10.0 | 16.0 | | | |
| Ex Ca: Ex Mg | | 1.5 | 4.0 | 7.0 | | | |

**Table 9.4** Soil attributes: CSAUAT, Kanpur, UP (15–30 cm depth)

| Parameters | Unit | Result | Guide Low | Guide High | Low | Optimum | High |
|---|---|---|---|---|---|---|---|
| $pH_{(1:2)}$ | | 10.04 | 6.0 | 6.8 | | | |
| $pH_s$ | | 9.13 | | >8.2 | | | |
| $EC_{(1:2)}$ | $dS\ m^{-1}$ | 1.06 | | | | | |
| $EC_e$ | $dS\ m^{-1}$ | 2.43 | <4.0 | > 4 | | | |
| Organic C | % | 0.15 | <0.5 | >0.75 | | | |
| Cation exchange capacity | $cmol_{(P+)}\ kg^{-1}$ | 16.99 | <10.0 | >15.0 | | | |
| Exchangeable sodium per cent | % | 92.49 | | >15 | | | |
| Sand | % | 58.63 | 30.0 | 55.0 | | | |
| Silt | % | 25.46 | 30.0 | 50.0 | | | |
| Clay | % | 15.91 | 20.0 | 55.0 | | | |
| Textural Class | | Sandy loam | | | | | |
| Available N | $kg\ ha^{-1}$ | 98.56 | <280.0 | >560.0 | | | |
| Available P (Olsen) | $kg\ ha^{-1}$ | 15.84 | <10.0 | >25.0 | | | |
| Available K | $kg\ ha^{-1}$ | 595.20 | <120.0 | >280.0 | | | |
| Available S | $mg\ kg^{-1}$ | 3.27 | <10.0 | >20.0 | | | |
| Available Fe | $mg\ kg^{-1}$ | 0.87 | <2.4 | >5.0 | | | |
| Available Mn | $mg\ kg^{-1}$ | 6.33 | <1.0 | >2.0 | | | |
| Available Zn | $mg\ kg^{-1}$ | 0.02 | <0.25 | >2.0 | | | |
| Available Cu | $mg\ kg^{-1}$ | 1.15 | <0.12 | >2.5 | | | |
| Available B | $mg\ kg^{-1}$ | 0.55 | <0.1 | >2.0 | | | |
| Per cent Exchangeable K | % | 3.17 | 3.0 | 8.0 | | | |
| Per cent Exchangeable Ca | % | 3.04 | 60.0 | 70.0 | | | |
| Per cent Exchangeable Mg | % | 1.30 | 10.0 | 16.0 | | | |
| Ex Ca: Ex Mg | | 2.33 | 4.0 | 7.0 | | | |
| **Sub soil is more sodic than the surface soils, deficient in Available N, Fe and Zn; very low in exch. Ca** | | | | | | | |

the main focus of rice improvement research in the future, following both conventional and marker-assisted backcross breeding techniques. In order to develop salt-tolerant genotype for inland salinity areas, efforts are being made to transfer 'Saltol' gene in popular varieties following marker-assisted backcross breeding. Phenotyping of these introgressed lines developed by marker-assisted backcross (MABC) breeding will be done in target environments to identify tolerant varieties to various abiotic stresses. IPM, IDM and INM approach with focussing on soil health are very essential components for enhancing production, productivity and profitability of rice farming community.

## 9.11 Future Outlook

The main constraints faced by the farmers generally focus on three issues, viz., yield gap between research farm and farmer's field, technological constraints and socio-economic constraints. Analysing these aspects will help to understand the real situations in rice and immensely useful to frame policies for overcoming the constraints and enhancing rice productivity. The yield gap can be minimized by making the inputs available to the farmers in sufficient quantity, good quality and at right time. The major constraints responsible for yield gap were late sowing/transplanting, higher prices of high-yielding variety seed, non-availability of fertilizer at sowing time, lack of funds with farmers and infestation of pest and disease. The following areas have a great potential to enhance rice yield in Uttar Pradesh:

- By incorporating Dhaincha as green manure in the crop cycle.
- In non-monetary inputs, timely completion of the planting and adoption of suitable technology/techniques that are specified for stress environment.
- Improvement of soil health by use of green manure and compost for sustainability of rice productivity of the state.

In order to develop salt-tolerant genotype for inland salinity areas, efforts are being made to transfer 'Saltol' gene in popular varieties following marker-assisted backcross breeding. Phenotyping of these introgressed lines developed by MABC will be done in target environments to identify tolerant varieties to various abiotic stresses. Screening and development of high content of Fe and zinc genotypes for nutritional security will be given more emphasis. IPM and RCT need emphasis in the future in order to increase production and productivity.

**Acknowledgement** The author would like to acknowledge the support provided in 2010–2012 from project on Rice Knowledge Management Portal funded by ICAR-IIRR, Hyderabad, for the development of rice knowledge content of Uttar Pradesh, in 2010–2014 from the project on Development of New Plant Type varieties with higher yield and inbuilt resistance to major insect and pest funded by ICAR-IIRR, Hyderabad, for the development of salt-tolerant genetic material and in 2013–2018 from the international project on Stress Tolerance Rice for Africa and South Asia (STRASA), IRRI and Bill and Melinda Gates Foundation for giving the opportunity to evaluate various types of lines for alkaline tolerance, capacity building and exposure and interaction with salt-tolerant breeding expert of international repute. We are grateful to our honourable vice chancellor of this university to support for all necessary facilities, our fellow scientists and field assistants working in rice.

## References

Agarwal RR, Gupta RN (1968) Saline alkali soils in India. ICAR, Tech. Bull. (Agri. Series) No. 15, New Delhi, pp. 1–65

Agrawal RR, Yadav JSP, Gupta RN (1979) Saline and alkali soils in India. New Delhi, ICAR, p 286

Agricultural Statistics at a Glance (2020) Govt, Of India, Ministry of Agriculture & farmers welfare. Directorate of Economics and Statistics, New Delhi

Joshi PK (1990) Influence of surface irrigation on crop productivity. Indian J Agric Sci 60:91–106

Leather JW (1914) Investigation on *Usar* land in the united provinces. Govt. Press, U. P

Voelcker JA (1897) Report on improvement of Indian agriculture. Govt. Press United Provinces

# Genetic Enhancement of Rice for Salt-Affected Areas of Bangladesh

# 10

Mirza Mofazzal Islam, Shamsun Nahar Begum,
and Mst. Sifate Rabbana Khanom

**Abstract**

Soil salinity causes significant adverse impact on agriculture since Bangladesh is one of the seaside countries. Significant progress has been made by Bangladesh Institute of Nuclear Agriculture in the last decade for developing salt-tolerant varieties. Development of Binadhan-8 and Binadhan-10 for dry season was a new dimension for breeding salt-tolerant rice in Bangladesh. The salinity tolerance and yield of the varieties ranged from electrical conductivity (EC) 8 to 12 dS $m^{-1}$ and 5.5 to 8 t/ha, respectively, in different saline-prone areas. Using marker-assisted selection (MAS) and marker-assisted backcrossing (MABC) as an advanced tool provided opportunities to introgress of *Saltol* QTL (quantitative trait loci) to popular varieties in Bangladesh. Considerable progress was made to introgress *Saltol* QTL into Binadhan-7, a popular short-duration aman variety in Bangladesh through marker-assisted backcrossing (MABC). Those introgressed lines are tested in the farmers' field in coastal region. The main focus was to pyramid multiple stress-tolerant genes (salinity, submergence) into mega-varieties of rice to ensure better adaptability in salt-affected areas.

**Keywords**

Breeding · Salinity tolerance · Rice · Marker-assisted selection · Submergence tolerance

M. M. Islam (✉)
Bangladesh Institute of Nuclear Agriculture (BINA), Mymensingh, Bangladesh

S. N. Begum · M. S. R. Khanom
Plant Breeding Division, BINA, Mymensingh, Bangladesh


R. K. Singh et al. (eds.), *Genetic Improvement of Rice for Salt Tolerance*,
https://doi.org/10.1007/978-981-99-3830-8_10

## 10.1 Introduction

In Bangladesh, there are about 2.86 million hectares of land at the coastal regions, of which almost 1.05 million hectares are arable, but these areas are affected by varying degrees of soil salinity (SRDI 2010). Due to environmental changes, sea levels are gradually rising and therefore, many areas in the coastal regions of Bangladesh will experience a significant increase in soil salinity during the coming decades (Hasan et al. 2020; Dasgupta et al. 2014). For these reasons, coastal farming is very vulnerable to climate change. As the coastal zone of Bangladesh constitutes 32% of the country's area and supports 28% livelihood of the population (Islam 2004), a decrease in coastal agricultural production, mainly rice, would also have significant consequences for food security as well as socioeconomic conditions for that population, which is living in those coastal regions (SRDI 2010).

Spread of salinity has the potential to reduce national rice production significantly as rice is one of the most salt-sensitive crops (Maas and Hoffman 1977; Munns and Tester 2008). During dry season, salinity is a serious constraint for rice production in coastal regions. It is also important during the transplanted Aman (T. Aman) season where salinity is high during seedling establishment particularly in areas where rice is grown after shrimp cultivation. The problem has been aggravated due to tidal storm like Sidr, Aila, etc. and waterlogging caused by the cyclones and other natural calamities.

## 10.2 Methodology of Rice Breeding for Salinity Tolerance

During the last decade, BINA released salinity-tolerant, dual-tolerant (salinity and submergence) rice varieties along with some advanced rice lines through several studies and experiments to develop rice that can grow in coastal regions. In this section, the methodology for breeding salinity-tolerant rice at BINA is described.

### 10.2.1 Salt-Tolerant Rice Varieties Developed by Conventional Breeding

Through collaboration between BINA and the International Rice Research (IRRI), some advanced breeding lines have been developed through hybridization of two genetically different parents using conventional breeding approach. Genetically fixed lines are developed through repeated selection in consecutive generations. Those homozygous lines were evaluated for salinity tolerance both at seedling and reproductive stages with salt stress of EC 12 dS $m^{-1}$ and 6 dS $m^{-1}$, respectively, using IRRI standard protocol (Gregorio et al. 1997). Based on the IRRI modified Standard Evaluation System (SES) (IRRI 1996) with a ranking of 1–9 (Table 10.1), the visual symptoms of salt stress injury of the seedlings are evaluated. In order to select promising salt-tolerant varieties, selected salt-tolerant genotypes were screened in natural salt-affected areas as well as in farmers' fields. As like as the

**Table 10.1** Modified IRRI Standard Evaluation Score (SES) based on the visual symptoms of saline injury at seedling stage of rice (*Oryza sativa* L.)

| Score | Observation | Tolerance |
|---|---|---|
| 1 | Normal growth, no leaf symptoms | Highly tolerant |
| 3 | Nearly normal growth, but leaf tips or few leaves whitish and rolled | Tolerant |
| 5 | Growth severely retarded; most leaves rolled; only a few are elongating | Moderately tolerant |
| 7 | Complete cessation of growth; most leaves dry; some plants dying | Susceptible |
| 9 | Almost all plants dead or drying | Highly susceptible |

Source: Gregorio et al. (1997)

"Mother and Baby" system trial of participatory variety selection (PVS), all these genotypes were grown for evaluation (Islam et al. 2008). PVS trials have been performed at various sites in the salt-affected coastal area. Selected PVS trial entries were then recommended to the National Seed Board (NSB) for release as varieties. Consequently, BINA successfully released two salt-tolerant varieties, namely, Binadhan-8 and Binadhan-10, following the same procedures.

### 10.2.2 Introgression of the Salinity Tolerance to a Target Rice Genotype

To initiate the development of a suitable short-duration, high-yielding, salt-tolerant variety, the experiment was conducted at BINA and IRRI through introgression of salt-tolerant QTL, *Saltol,* into salt-sensitive, Bangladeshi, short-duration popular aman rice variety Binadhan-7 by MABC. FL478 was used as donor of *Saltol* locus, which was introgressed into recipient Binadhan-7. Then Binadhan-7 was crossed with the introgression and selection of salt-tolerant lines. Those were carried out through MABC scheme in several cycles to the recurrent parent followed by selfing as illustrated in the flowchart.

### 10.2.3 Marker-Assisted Backcrossing (MABC) Scheme

To obtain $F_1$ seeds, Binadhan-7 was used as the female parent and FL478 was the male parent. This crossing was carried out during aman season in 2012 at the Plant Breeding Division of BINA. The true $F_1$ seeds were backcrossed to Binadhan-7 to obtain $BC_1F_1$ seeds. In the $BC_1F_1$ generation, a total of 141 $BC_1F_1$ plants were used for genotyping and 47 plants were obtained heterozygous at *Saltol* locus for the foreground markers. Foreground selection was carried out by tightly linked *Saltol* markers RM3412b (11.5 Mb) and AP3206f (11.2 Mb). The selected plants were then backcrossed with the recurrent parent to produce $BC_2F_1$ seeds.

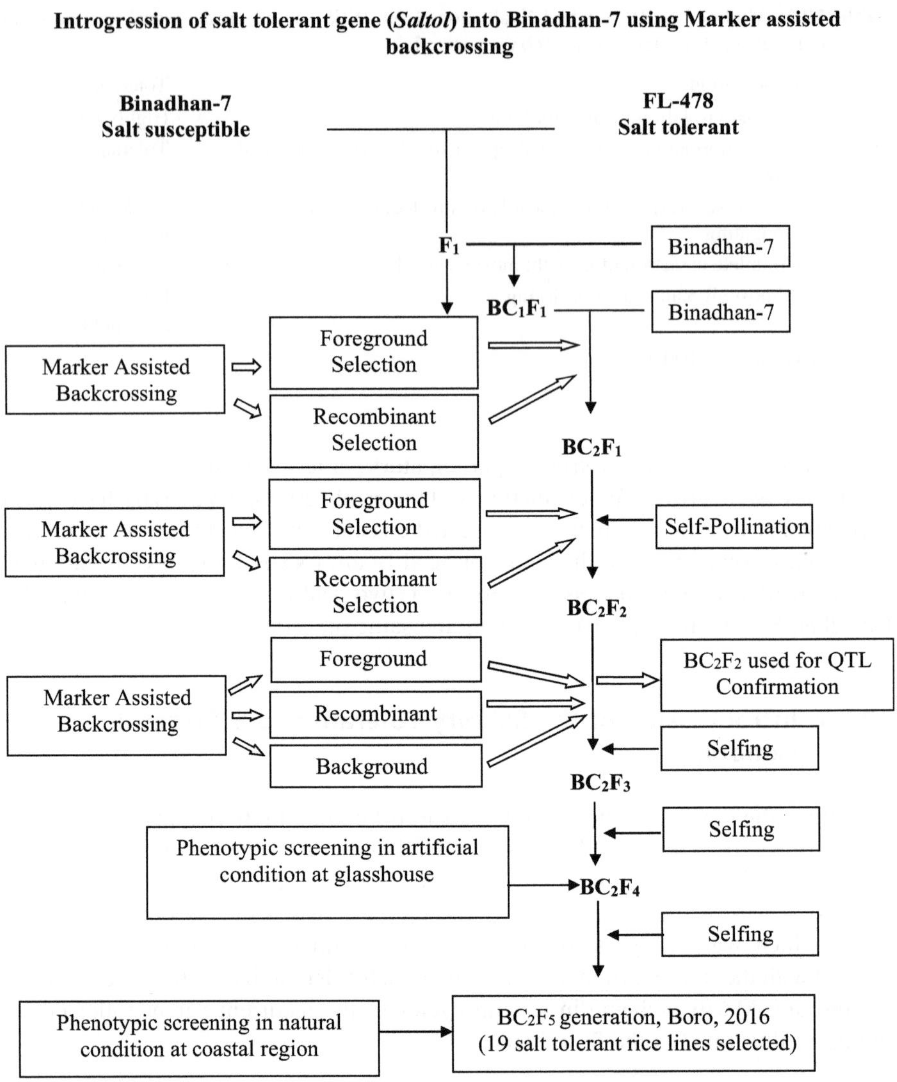


1. Flowchart entitled "Foreground, recombinant, and phenotypic selection approach with selection details in each generation and development of salt-tolerant rice varieties".

Plants that were heterozygous for *Saltol* and homozygous for the recipient allele were then surveyed using the markers RM10793 (distal end) flanking the *Saltol* locus (i.e., recombinant selection) (Collard and Mackill 2008). Selected $BC_2F_1$ plants were allowed to self-pollinate to obtain $BC_2F_2$ for further analysis. In $BC_2F_2$ generation, plants were selected based on foreground and recombinant selection with the same markers used in $BC_1F_2$ generation. Plants homozygous for the recurrent parent alleles at the flanking markers were used for genotyping by using markers unlinked to *Saltol* and covering all the chromosomes including the *Saltol* carrier chromosome 1 to recover the recurrent parent background or recurrent selection (Tanksley et al. 1989; Young and Tanksley 1989; Hospital and Charcosset 1997; Neeraja et al. 2007; Thomson et al. 2010).

The individuals homozygous at the target *Saltol* locus and with maximum background of Binadhan-7 were selected in the $BC_2F_2$ generation. $BC_2F_3$ seeds were produced by selfing of the selected $BC_2F_2$ plants. In $BC_2F_3$ generation, five plants were selected from each line (a total of 40 plant selections) based on earliness and plant type. $BC_2F_4$ seeds were produced by selfing of the selected $BC_2F_3$ plants. Selection was done on $BC_2F_3$ based on earliness and similarity to recurrent parent Binadhan-7 to produce $BC_2F_4$. Selected $BC_2F_4$ seeds were used to screen for salinity tolerance at seedling stage to verify the tolerance level of the newly developed Binadhan-7-*Saltol* lines and to compare them with their parents and salt-sensitive check, IR29. Based on phenotypic screening, selected plants were evaluated for salt tolerance in natural condition at coastal region of Satkhira district.

### 10.2.4 Phenotypic Selection

$BC_2F_4$ seeds of selected plants were used to screen in artificial condition for salinity tolerance at the seedling stage to confirm the tolerance level of the newly developed Binadhan-7-*Saltol* lines as compared with their parents and a tolerant check using Binadhan-10. Selected 19 plants were evaluated for salt tolerance in natural condition at coastal region (Satkhira). $BC_2F_4$ lines were also investigated for salinity tolerance at the seedling stage in hydroponic culture with EC 12 dS $m^{-1}$, and a glasshouse study was conducted to evaluate the response of backcross population at salt stresses up to 6 dS $m^{-1}$ at reproductive stage following IRRI standard protocol (Gregorio et al. 1997).

### 10.2.5 Identification of QTLs Associated with Salinity Tolerance.

The experiment was carried out at the glasshouse and genotyping services laboratory (GSL) of plant breeding, Genetics and Biotechnology Division, Los Baños, IRRI, Philippines, from February 2015 to July 2015. To identify QTLs for salinity tolerance at seedling stage, the following activities were carried out. $BC_2F_2$ backcross population developed from Binadhan-7 × FL478 were used as mapping population in this study. The Java-based Qgene 4.0 software (Joehanes and Nelson 2008) is used to identify the major QTLs for salinity tolerance at seedling stage. The analyses were performed using the 281 $BC_2F_{2:3}$ individuals of Binadhan-7/FL478 population based on the available information of the genotype data from the genetic linkage map. The QTLs were identified using single marker analysis (SMA), interval mapping (IM), and composite interval mapping (CIM). To assess the relationship between individual marker loci and putative QTLs, these methods were used, as well as to discover the marker loci that can be similar to the QTLs, single marker analysis has been used. Analysis of IM and CIM was carried out to detect accurate locations of the detected QTLs. For the CIM, the default forward cofactor selection command was used (Zeng 1994). A technique that can identify the existence of a putative QTL was used with an LOD (logarithm of the odds) score of 3.0 for the given genomic area. Using 10,000 permutation checks, the minimum LOD value for CIM was calculated (Churchill and Doerge 1994) for each trait to declare a QTL as significant. For each phenotype, the percentages of the total phenotypic variation explained by individual QTLs as $R^2$ value ($R^2$ = PVE, phenotypic variation explained by the QTL) and additive effects (the effects and sources of parent-contributed alleles) were calculated. The structures suggested by McCouch (2008) were followed by the nomenclature of the described QTLs.

### 10.2.6 Introgression of *Saltol-Sub1* Genes Into Binadhan-10 and Binadhan-11

Binadhan-10 as recurrent parent was crossed with selected four donor lines, namely, RC-251, RC-222, RC-291, and RC-249, and the same way Binadhan-11 was crossed with RC-193 and RC-292. $F_1$ hybrids were screened for heterogeneity using simple sequence repeats (SSRs) markers linked to the salinity-tolerant QTL *Saltol* and submergence-tolerant QTL *Sub1*. $F_1$ progenies derived from Binadhan-10 were grown and backcrossed with the same recurrent parent to obtain $BC_1F_1$ progenies. The same strategies were followed in case of Binadhan-11 × RC-193 and Binadhan-11 × RC-292. Foreground selection was performed using tightly linked SSR markers. $BC_1F_1$ lines were grown and backcrossed with recurrent parents following the backcross strategy and harvested $BC_2F_1$ seeds. A total of 3703 $F_1$ seeds were harvested. These were cultivated in entirety 2134 $BC_1F_1$ and also a total of 3613 $BC_2F_1$ seeds. All the $BC_2F_1$ lines were genotyped with linked and flanking markers of *Saltol* and *Sub1* locus. All the selected populations were allowed for selfing to develop $BC_2F_2$ and $BC_2F_3$ generations to fix the genes in the advanced lines.

## 10.3 Achievement of BINA in Developing of Salt-Tolerant Rice

BINA started their research for salinity-tolerant rice development since 2007. A short description about the varieties and advanced lines for salinity tolerance has been mentioned below.

### 10.3.1 Binadhan-8

An advanced line IR66946-3R-149-1-1 (PBRCSTL-20) was developed from IR29/ Pokkali, which was released as Binadhan-8 in 2010 and showed an impressive result in different areas of the country in Boro season. It is a high-yielding (5.5 t/ha under salt stress), erect, stout culm, early maturing (130–135 days), coarse grain, and semi-dwarf rice variety that can tolerate salinity of 8–10 dS $m^{-1}$. Cultivation of Binadhan-8 was more profitable compared to another variety in coastal areas of Bangladesh.

### 10.3.2 Binadhan-10

In the seedling and reproductive stages, 60 rice germplasms were tested for salinity tolerance in the hydroponic system using the IRRI standard protocol. SSR markers showed salinity tolerance in tested line PBRC-37. During 2011–2012 Boro season, the Seed Certification Agency (SCA) carried out an evaluation of the salinity-tolerant line PBRC-37 in the coastal saline areas of Satkhira, Bagerhat, and Khulna and then named it as Binadhan-10. Out of the 60 rice germplasms, 24 were selected based on salt tolerance. Nine lines were found to be tolerant at the level of EC 12 dS $m^{-1}$ at seedling stage screening, while 15 lines were moderately tolerant. Three lines out of the 24 germplasms, namely, Pokkali, PBRC-37 (salt-tolerant variety, Binadhan-10), and FL-478, were found to be salinity tolerant, and four lines were found to be moderately tolerant at the EC level of 6–12 dS $m^{-1}$. Binadhan-10 gave a higher yield (6.0 t/ha) under salinity stress (10–12 dS $m^{-1}$) than the earlier released salinity-tolerant rice variety Binadhan-8 (Table 10.2). Binadhan-10 can be cultivated in 40–50% of fallow lands in both Boro (dry season) and Aman (wet season) seasons. The farmers can get potential yield of 5.5–6.5 t/ha in saline land and 7.5–8.5 t/ha in nonsaline land. The new salt-tolerant variety Binadhan-10 can be cultivated in a large part of saline-prone 13 districts in Ganges coastal zone and more 4–5 million tons of additional rice can be produced a year. Large-scale distribution of this variety will enable farmers to achieve greater productivity, ensure food security, and improve their livelihoods.

**Table 10.2** Yield and other characteristics of salt-tolerant modern varieties developed by BINA suitable for dry season

| Variety | Growth duration | Plant height (cm) | Grain shape | Amylose content (%) | Salt tolerance (dS $m^{-1}$) | Grain Yield (t/ha) |
|---|---|---|---|---|---|---|
| Binadhan-8 | 135 | 97 | MB | 25.0 | 8–10 | 5.5 |
| Binadhan-10 | 130 | 105 | MS | 26.0 | 10–12 | 6.0 |

*MS* = medium slender, *MB* = medium bold

### 10.3.3 Introgression and Mapping of Salt-Tolerant Genes in Binadhan-7

A major QTL for salinity tolerance on chromosome 1, *Saltol*, provided the opportunity to apply MABC for introgression of salt-tolerant genes into popular salt-sensitive Bangladeshi rice variety. In this study, FL478 was used as donor where *Saltol* locus was introgressed into Binadhan-7. Homozygous 19 $BC_2F_4$ individuals were selected. Seven of them were within 1.2 Mb (11.0–12.2) introgression segment containing *Saltol* locus and were 83.6%, 78.5%, 78.1%, 75.8%, 74.6%, 73.8%, and 73.4% similar to Binadhan-7, respectively. All the introgressed lines did not show any undesirable traits. Four Binadhan-7 *Saltol* introgresssed lines, namely, Binadhan-7-*Saltol*-89, Binadhan-7-*Saltol*-230, Binadhan-7-*Saltol*-63, and Binadhan-7-*Saltol*-242, were selected (Fig. 10.1).

The aim of this study was also to evaluate the role of *Saltol* QTL in the effects of salinity on plant growth and the yield components of the rice backcross population at various stages of growth. This study was conducted to identify QTLs in a $BC_2F_2$ backcross population derived from the cross between FL478 and Binadhan-7 for salinity tolerance at the seedling stage. The $BC_2F_2$ backcross population was tested at the seedling stage for 10 traits possibly associated with salinity tolerance. Phenotyping was performed in hydroponic media with salt stress of EC 12 dS $m^{-1}$.

A total of 10 QTLs were identified by SMA, IM in the range, and CIM with an LOD value in the range of 3.1–4.8 and phenotypic variance of 4.8–7.4%. The major QTLs *qRFW*1.1, *qSL*1.1, *qK*1.1, *qNa*1.1, and *qNa*1.2 were detected on the short arm of chromosome 1. Another QTLs *qSL*2.1, *qSl*7.1, *qRDW* 10.1, *qRFW* 10.2, and *qRL* 11.1, on chromosomes 2, 7, 10, and 11 are novel in this study and can be further exploited for fine mapping, gene discovery, and subsequent use in breeding by pyramiding them with other QTLs for salinity tolerance or for tolerance of other abiotic stresses (Khanom 2019).

### 10.3.4 Introgression of Dual-Tolerant Genes into Leading Rice Varieties

Adverse effects of climate change in coastal areas compel the breeders to combine multiple abiotic stress tolerance traits into one genotype (tolerance to salinity and submergence) because combined attributes are needed as such problems often coexist in one cropping season. This study motivated to unite saline and submergence tolerant QTLs from dual-tolerant advanced lines into Binadhan-10 and Binadhan-11 through MAS and to determine the genetic basis of dual QTLs of salt and submergence tolerance using SSR markers. MAS helped in the selection of recombinants with *Saltol-Sub1* introgressed into one genetic background.

The first experiment was conducted to select donor lines. For the experiment, a total of 14 advanced lines were collected from IRRI and screened with other check varieties at three separate experiments. At the first, the genotypes were screened through hydroponic system with EC 12 dS $m^{-1}$ at seedling stage. For submergence

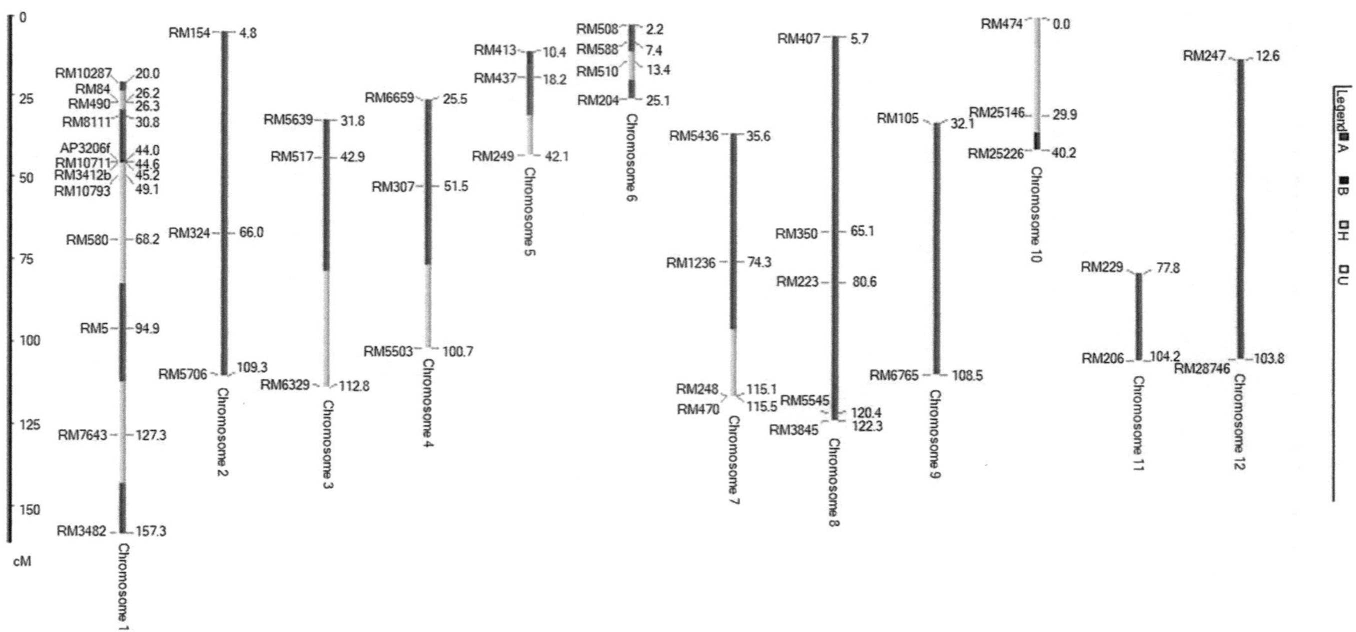


**Fig. 10.1** Graphical genotype (GGT) of selected plants of Binadhan-7-Saltol-89 (53–14–11-3) in $BC_2F_2$ generation. The red-colored regions represent the homozygous regions of the recipient genome, whereas the blue-colored regions on the short arm of chromosome 1 represent the *Saltol* introgression from FL478

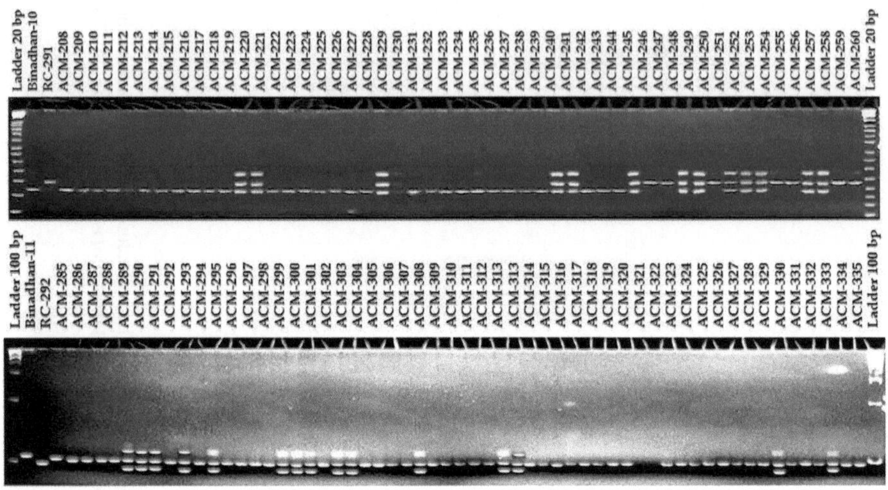


**Fig. 10.2** DNA banding profiles of the genotypes ($BC_2F_2$) of cross combination of Binadhan-10 and RC-251 for salinity-tolerant SSR markers

tolerance screening, the progenies were completely submerged for 21 days under 40 cm water. For salinity-submergence tolerance, the (Near Isogenic lines) NILs together with check varieties were submerged for 21 days at 12 dS $m^{-1}$ salinization and 40 cm depth of water. Six lines were selected out of 14 lines through combining the results of salinization, submergence, and salinized-submerged screening.

In the second experiment, the six lines selected from the previous experiment were used as donor. Binadhan-10 and Binadhan-11 were crossed with the donor lines to introgress submergence-tolerant QTL, *Sub1*, into submergence-sensitive but salt-tolerant variety, Binadhan-10, and *Saltol* QTL into salinity-sensitive but submergence-tolerant variety Binadhan-11. The submergence-tolerant QTLs, *Sub1*, was introgressed from the double-tolerant (*Saltol-Sub1*) donor lines RC-251, RC-222, RC-291, and RC-249 into Binadhan-10 and salt-tolerant QTL, *Saltol*, from the dual-tolerant (*Saltol-Sub1*) donor lines RC-193 and RC-292 into Binadhan-11 through MAS. Final genotyping of the progenies, derived from these all (6) combinations was done at $BC_2F_2$ (named as ACM) generation using the SSRs, namely, RM493, RM10793, RM7075, RM490, and AP3206, to identify *Saltol* region (Fig. 10.2). *Sub1*-linked markers Sub1BC2, RM5688, RM336, RM23662, RM216, and SC3 were used to detect donor segments in Binadhan-10 and Binadhan-11 (Fig. 10.3). Less NILs were selected from the combinations, namely, Binadhan-10 × RC-222, Binadhan-10 × RC-291, Binadhan-10 × RC-249, and Binadhan-11 × RC-292, from various generations. In total, 10 plants were identified as heterozygous at both *Saltol* and *Sub1* locus in Binadhan-10 × RC-222 combination. In Binadhan-10 × RC-291 combination, only eight plants were identified as heterozygous while 10 plants were selected with heterozygous alleles at both loci *Saltol-Sub1* based on foreground selection in Binadhan-10 × RC-249 combination. In the

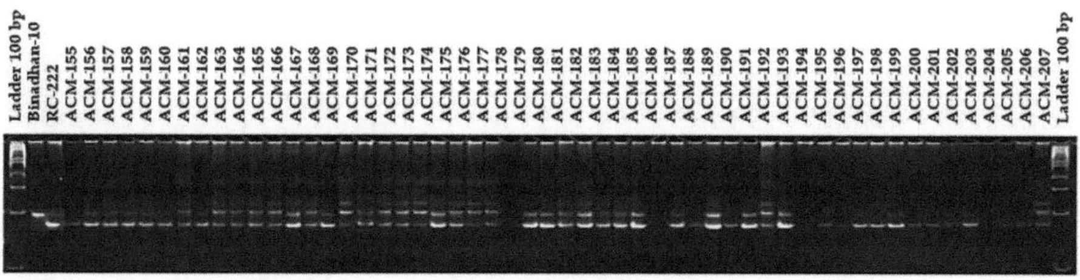


**Fig. 10.3** DNA banding profiles of the genotypes ($BC_2F_2$) of cross combination of Binadhan-11 and RC- 251 for submergence tolerant SSR markers and RC-251 for salinity tolerant SSR markers

combination Binadhan-11 × RC-292, only 12 plants were identified as heterozygous at both loci *Saltol-Sub1.* More NILs were selected from Binadhan-10 × RC-251 and Binadhan-11 × RC-193 combinations. In Binadhan-10 × RC-251, a total of 45 progenies were identified containing both *Saltol-Sub1* QTLs in the same genotype. In the combination of Binadhan-11 × RC-193, foreground selection was done on $BC_2F_2$ generation to detect 48 progenies, which indicated the presence of both *Saltol-Sub1* QTLs. The best individuals with target loci (*Saltol-Sub1*), selected through GGT analysis, were used for phenotypic evaluation. A total of 45 progenies along with parents of Binadhan-10 × RC-251 and 48 individuals of Binadhan-11 × RC-193 having both S*altol-Sub1* loci, along with parents, were tested in the field.

Correlation coefficient distance for Binadhan-10 × RC-251 and Binadhan-11 × RC-193 was constructed through UPGMA (unweighted pair group method with arithmetic mean) method based on 10 morphological traits. For Binadhan-10 × RC-251, the genotypes were grouped into two clusters. Based on the cluster analysis, 17 NILs were grouped with the recurrent parent Binadhan-10 in cluster II with 95.19% similarities. These individuals were genotypically and phenotypically selected NILs with 95.19% similarities with recurrent parent, Binadhan-10. On the other hand, the recurrent parent, Binadhan-11, along with 33 NILs were grouped with 99.55% similarities (Fig. 10.4). Forty-five progenies along with parents of Binadhan-10 × RC-251 and 48 individuals of Binadhan-11 × RC-193 with parents were included for the field evaluation. The best individuals with target loci (S*altol-Sub1)* identified through GGT analysis were screened following the protocol derived from IRRI. A total of 10 NILs with Binadhan-10 background and 24 NILs with Binadhan-11 were selected based on the germination reduction, SES (Standard Evaluation Score), survival percentage, and elongation percentage. The improved NILs with both S*altol-Sub1 QTLs* could be further exploited for fine mapping of genes/QTL and gene discovery. The main emphasis was on pyramiding multiple stress-tolerant QTL/genes into mega-varieties of rice (Sultana 2020).

### 10.3.5 Binadhan-23

Twenty-two (22) rice genotypes were screened for dual tolerance. Among these, RC-249 and RC-251 were selected as dual tolerant after vigorous phenotypic and

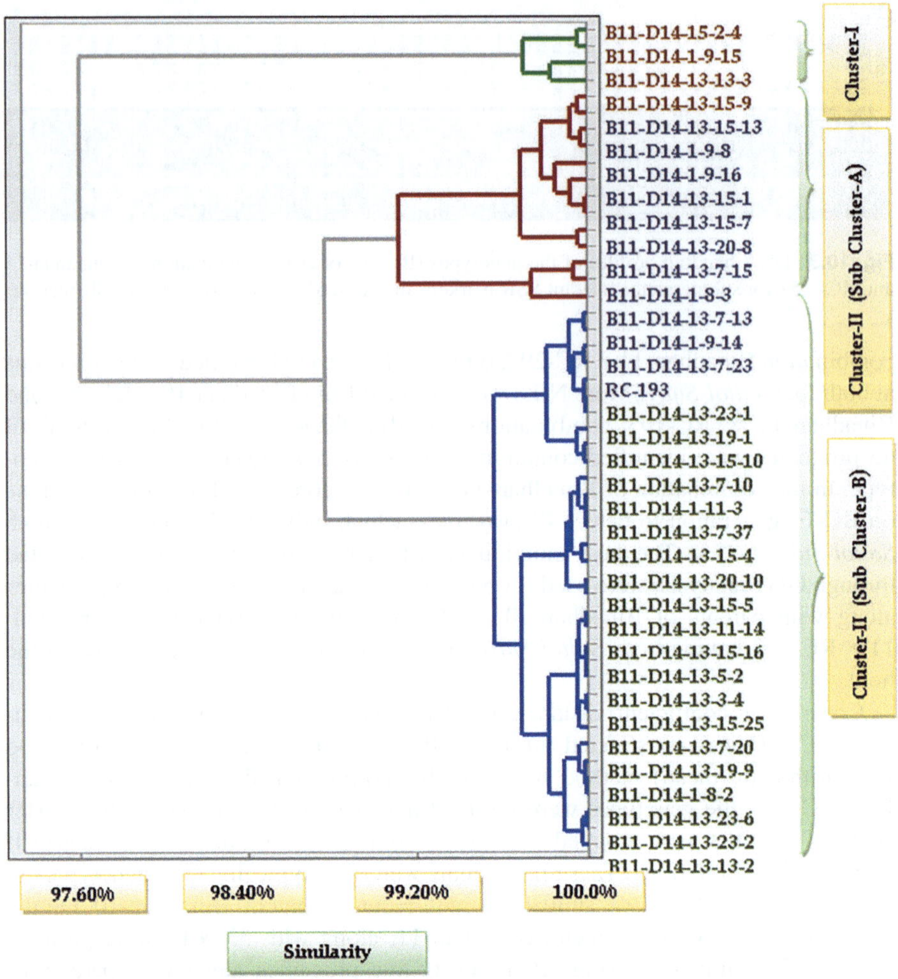


**Fig. 10.4** Complete linkage, correlation coefficient distance of 48 NILS with parents of Binadhan-11 × RC-193, constructed through UPGMA method using Minitab 17 software

genotypic screening. IRRI developed the salinity- and submergence-tolerant line through MABC technique and distributed for trial in the environment of Bangladesh at BINA, Mymensingh. Evaluation of dual-tolerant line RC-251 was done during aman season, 2018, in the coastal saline areas of Bagerhat, Barishal, Jhalokati, Bhola, Patuakhali, Barguna, and Khulna. RC-251 gave a higher yield (5.3 t/ha) and matured 8–10 days earlier than check variety BRRI dhan78, which was then released as Binadhan-23 in 2019. Binadhan-23 is a dual-tolerant variety that can tolerate salinity of EC 8 dS $m^{-1}$ and can give yield even if submerged in floodwater up to 15 days (Table 10.3)

**Table 10.3** Yield and other characteristics of dual-tolerant (salinity and submergence) rice variety, Binadhan-23, developed by BINA suitable for tidal flood-prone areas for aman season

| Variety | Growth duration | Plant height (cm) | Grain size and shape | Amylose content (%) | Salt tolerance (dS $m^{-1}$) | Yield (t/ha) |
|---|---|---|---|---|---|---|
| Binadhan-23 | 125 | 115 | Medium | 25.1 | 8 | 5.3 |
| BRRI dhan78 (check) | 135 | 114 | Medium | 25.2 | 8 | 4.4 |

## 10.4 Future Perspective of Rice Breeding for Salinity Tolerance

At present, rice productivity in salt-affected areas is very low, which is around 1 t/ha. But this can potentially increase to at least 3–4 t/ha with improved, newly developed rice varieties that can withstand soil salinity. Earlier, farmers received good yields from this variety, but nowadays, due to increased salinity intrusion in these areas, they face serious problems in getting a good harvest from these non-salt-tolerant rice varieties during the Boro season, although the coastal belt salinity intrusion trend is still rising. Scientists have undertaken comprehensive rice research and breeding programs to make climate-smart rice varieties that are more tolerant to submergence, drought, heat, and salinity conditions that are predicted to increase in frequency and severity with climate change. It is often argued that the future development of the country depends predominantly on the sustainable development of such stress-tolerant rice varieties as well as extension and utilization of such varieties in stress-prone areas.

The development of Binadhan-8 and Binadhan-10 improved the rice cultivation scenario of salt-affected area in Bangladesh in the dry season. These two varieties cover 20% of salt-affected area in Bangladesh. These varieties can tolerate EC 8 to 12 dS $m^{-1}$ salt stress. The potential yield of these varieties ranged from 5.5 to 8.5 t/ha in different saline-prone areas. More recently, Binadhan-23 with tolerance to both salinity and submergence has been developed for tidal flood-prone areas. This dual-tolerant variety can tolerate salinity of 8 dS $m^{-1}$ and can produce yield even if submerged in floodwater up to 15 days.

The application of MABC breeding to introgress *Saltol* QTL into Binadhan-7 was completed. Among the four introgression lines of Binadhan-7-*Saltol*-89, Binadhan-7-*Saltol*-63 performed better than other lines. The use of MABC is best demonstrated by the amazing achievement of introgression of dual tolerance in rice. In conclusion, 38 NILs with Binadhan-10 background and 36 individuals with Binadhan-11 background were selected through MAS and combined salinity-submergence screening. Significant morphological features with salinity-submergence tolerance were found in these NILs as compared to Binadhan-10 and Binadhan-11. Similarly, difference between selected rice lines and recurrent parents at genetic level was also observed. This may be somewhat helpful to the breeders that offer a wider option to combine superior QTLs into one genetic background through gene pyramiding. All of these introgression lines are tested in the farmers' field in coastal regions of Bangladesh. The prime focus is to pyramid multiple stress-tolerant genes, i.e., salt, submergence and water stagnation into mega-varieties for better adaptability to the climate changes faced by the coastal environments for the larger benefit of the farmers and rice consumers.

## References

Churchill GA, Doerge R (1994) Empirical threshold values for quantitative trait mapping. Genetics 138(3):963–971

Collard BC, Mackill DJ (2008) Marker-assisted selection: an approach for precision plant breeding in the twenty-first century. Philosophical Transactions of the Royal Society B: Biological Sciences 363(1491):557–572

Dasgupta S, Hossain M, Huq M, Wheeler D (2014) Climate change, soil salinity, and the economics of high-yield rice production in coastal Bangladesh. Development Research Group, Environment and Energy Team, The World Bank

Gregorio GB, Senadhira D, Mendoza RD (1997) Screening rice for salinity tolerance no, 22. International Rice Research Institute, Los Baños, Laguna

Hasan MK, Kumar L, Gopalakrishnan T (2020) Inundation modelling for Bangladeshi coasts using downscaled and bias-corrected temperature. Clim Risk Manag 27:100207

Hospital F, Charcosset A (1997) Marker-assisted introgression of quantitative trait loci. Genetics 147(3):1469–1485

IRRI (1996) Standard evaluation system for rice, 4th edn. International Rice Research Institute, Manila, pp 1–52

Islam MM (2004) Mapping salinity tolerance genes in rice (*Oryza sativa* L.) at reproduction stage. Ph.D. Dissertation,. University of Philippines Los Banos College, Laguna, pp 1–149

Islam MR, Singh RK, Salam MA, Hassan L, Gregorio GB (2008) Molecular diversity of stress tolerant rice genotypes using SSR markers. SABRAO Journal of Breeding and Genetics 40: 127–139

Joehanes R, Nelson JC (2008) QGene 4.0, an extensible Java QTL-analysis platform. Bioinformatics 24(23):2788–2789

Khanom MSR (2019) Introgression and mapping of salt tolerant QTL in early rice variety Binadhan-7. PhD. Thesis,. Bangladesh Agricultural University, Mymensingh, pp 85–89

Maas EV, Hoffman GJ (1977) Crop salt tolerance—current assessment. J Irrig Drain Div 103(2): 115–134

McCouch SR, CGSNL (Committee on Gene Symbolization, Nomenclature and Linkage, Rice Genetics Cooperative) (2008) Gene nomenclature system for rice. Rice 1:72–84

Munns R, Tester M (2008) Mechanisms of salinity tolerance. Annu Rev Plant Biol 59:651

Neeraja CN, Maghirang-Rodriguez R, Pamplona A, Heuer S, Collard BC, Septiningsih EM, Mackill DJ (2007) A marker-assisted backcross approach for developing submergence-tolerant rice cultivars. Theoretical and Applied Genetics 115(6):767–776

SRDI (2010) SRMAF Project. Soil Resource Development Institute. Ministry of Agriculture, Government of Bangladesh

Sultana A (2020) Introgression of salinity and submergence tolerant genes into leading rice varieties through marker assisted selection. PhD. Thesis,. Bangladesh Agricultural University, Mymensingh, pp 132–136

Tanksley SD, Young ND, Paterson AH, Bonierbale MW (1989) RFLP mapping in plant breeding: new tools for an old science. Bio/Technology 7(3):257–264

Thomson MJ, Ismail AM, McCouch SR, Mackill DJ (2010) Marker assisted breeding. In: Pareek A, Sopory SK, Bohnert H, Govindjee. (eds) Abiotic stress adaptation in plants, physiological, molecular and genomic foundation. Springer, pp 451–469

Young ND, Tanksley SD (1989) Restriction fragment length polymorphism maps and the concept of graphical genotypes. Theoretical and Applied Genetics 77(1):95–101

Zeng ZB (1994) Precision mapping of quantitative trait loci. Genetics 136(4):1457–1468

# Physical and Chemical Characteristics of Salt-Affected Soils in India

11

Arvind Kumar Rai, Nirmalendu Basak, Parul Sundha, Subedar Patel, Sanjay Kumar, Sandeep Bedwal, Kajal, Rajender Kumar Yadav, and Parbodh Chander Sharma


**Abstract**

Understanding on physical and chemical properties of salt-affected soils (SAS) is imperative for planning and executing reclamation programs and greening degraded salty lands for agricultural use. This issue received priority in Indian agriculture since pre-independent period and still occupies the central agenda in current policies of agriculture and the welfare of the farming community. Largely, SAS is categorized into saline due to soluble electrolyte load; sodic as it contains disproportionately large amount of sodium upon calcium and magnesium in the soil exchange sites and the soil solution with alkaline reaction; and saline–sodic with the dual problem of alkaline pH and salinity and typical categories of SAS having spatial existence. SAS often coexists with poor quality (saline, sodic, or both and associated problem) groundwater, prolonged water trajectories, erratic rainfall, and high evapotranspiration demand for crops under arid to semiarid agroecology. The physical and chemical properties of SAS demand specific reclamation strategies or management of soils to grow the crop. Further, application of mineral or organic amendment, soil types, and irrigation with varied quality water concomitantly change the chemical equilibria and may degrade soil physical health. Inherently, SAS is disturbed with the ionic environment that had impact on the different processes of nutrient cycles operating in these soils. Therefore, guidance on soil physical and chemical environment, distribution of SAS in India with their physiochemical characteristics in different agroecoregions, and the impact of saline–sodic water on SAS are highlighted.



A. K. Rai · N. Basak (✉) · P. Sundha · S. Patel · S. Kumar · S. Bedwal · Kajal · R. K. Yadav · P. C. Sharma
ICAR-Central Soil Salinity Research Institute, Karnal, Haryana, India
e-mail: Nirmalendu.Basak@icar.gov.in







R. K. Singh et al. (eds.), *Genetic Improvement of Rice for Salt Tolerance*,
https://doi.org/10.1007/978-981-99-3830-8_11

**Keywords**

Salt-affected soil · Alkalinity · Semiarid region · Poor quality water · Amendments · Soil physical health

## 11.1 Introduction

Salts are indispensable part of soil and water. Nearly all types of soil and underground water carry salts. Therefore, the quality and quantity of salts play a vital role for soil and water. As salts are the chemical compound of ions, the nature and characteristics of salts depend on the types of cations/anions that constitute the salts. The basic cation sodium when conjugated with $CO_3^{2-}$ or $HCO_3^-$ forms sodium carbonates/sodium bicarbonate, which has a potentiality of alkaline hydrolysis. Soil and water are predominant in these types of salts and show sodic properties. The chloride and sulfates when conjugated with basic cations (sodium, calcium, magnesium) and potassium forms nearly neutral salts, whereas carbonates and bicarbonate of calcium and magnesium form nearly insoluble salts in soil and water. The presence of these types of salts increases the electrolytes concentration and soils are defined as saline. The salt-affected soil (SAS) having the potentiality of alkaline hydrolysis and containing appreciable quantity of electrolytes is defined as saline–sodic soil.

India is endowed with extended geographic areas with versatile landform of the world's tallest mountain (the Great Himalaya) on his north side to the wide coastline on the southeast, south, and southwestern sides and a per-humid hilly mountain area in the east to the arid desert at west. The ancient Deccan peninsula in the central south is rich with clayey soil. The aeolian deposition by the northern river system (Ganges, Yamuna, Brahmaputra, and Sindhu and their tributaries) generates the extended alluvium region (Indo-Gangetic Plain, IGP) in the north and northwest and eastern India. The aeolian deposits carry alkali salts in some pockets of IGP and possess the occurrence of alkali/sodic soil. The high evapotranspiration demand because of arid/semiarid climate, presence of shallow underground saline water, and indiscriminate irrigation with saline water generates the problem of salinity. The development of salinity is further triggered when rainfall is scanty or fresh canal water irrigation is a crisis. Therefore, this chapter describes the distribution of SAS in India with their physiochemical characteristics in different agroecoregions and the impact of saline–sodic water on SAS.

## 11.2 Distribution of SAS in India

In India, ~3.67 Mha of geographic area is affected by salinity, which is different from previous approximations (6.74 Mha) done by NRSC, CSSRI, and NBSS and LUP based on Landsat imagery of 1996 (SAC 2016). This difference for estimation of SAS area illustrates data accuracy, followed methodology, and scalability

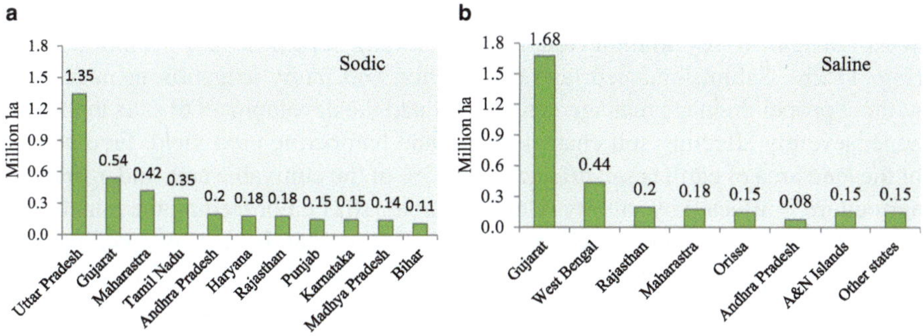


**Fig. 11.1** State-wise distribution of sodic (**a**) and saline (**b**) soils

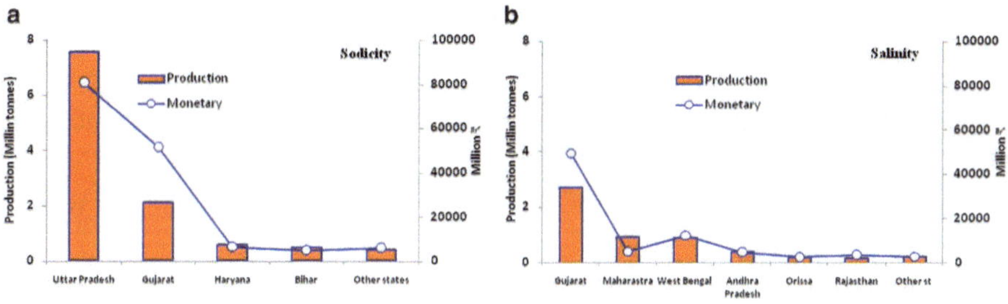


**Fig. 11.2** State-wise annual production and monetary losses due to sodicity (**a**) and salinity (**b**)

problem. The salt-affected area raised from 0.23 Mha in 1996 to 0.32 Mha in 2010–2013 of the recently updated map of Haryana, India, by CSSRI (Mandal et al. 2011) due to irrigation from the canal in impeded, drained area and use of high RSC/SAR groundwater for irrigation. Spatiotemporal distribution pattern of SAS, its factors, and incidence knowledge play a dynamic role in understanding the variations of SAS area and essential effective planning for arresting further degradation process in future climatic and other uncertainty drivers of SAS. In India, a total of 6.74 Mha area was affected by soil salinization including 3.79 and 2.95 million ha of sodic and saline, respectively. Uttar Pradesh (35.8%), Gujarat (14.4%), Maharashtra (11.2%), Tamil Nadu (9.4%), Haryana (4.9%), and Punjab (4.0%) collectively share 80% of the total sodic soils in India (Fig. 11.1a). Soil salinity, intrusion of seawater, salinity associated with waterlogging, and groundwater irrigation salinity are distributed across 13 states with largest area in Gujarat (56.8%) followed by West Bengal (14.9%), Rajasthan (6.6%), and Maharashtra (6.2%) (Fig. 11.1b). Presently, sodicity diminishes ~11 million tons of produce with a monetary value of 1904 billion US dollar, whereas salinity hampered ~5.76 million tons of produce with value of 1014 billion US dollar of annual loss. The state-wise crop production losses due to sodicity reveal that Uttar Pradesh suffered a yield penalty of 7.6 million tons, followed by Gujarat with annual production losses of 2.1 million tons (Fig. 11.2a), whereas state-wise production losses due to salinity are in order of Gujarat (2.7 million tons) followed by Maharashtra (0.92 million tons), West Bengal (0.89 million tons), and Andhra Pradesh (0.40 million tons) (Fig. 11.2b). Salinity-induced land degradation and faulty irrigation managements without proper drainage management aggravated the development of salts in the root zone, severely affecting soil characteristics and hampering crop yield. Five percent of the land area of earth is salt affected and 75% of the cultivable area under irrigated agriculture is affected by salinity (Hopmans et al. 2021). For feeding the nine billion people by 2050, it is anticipated that crop production can be met by greening SAS soil with appropriate agro-techniques. The yield reduction in wheat, rice, sugarcane, and cotton by 40, 45, 48, and 63 percent is noted in SAS in the Indo-Gangetic Plain of India, respectively (Sharma et al. 2015). Prolong irrigation with high sodic water increases the soil pH and builds ESP and reduces rice and wheat yield by 14 to 16%, respectively (Sheoran et al. 2021). However, low permeability and waterlogging in occurrence with the concretion of '*kankar* layer of $CaCO_3$' below the rhizosphere of paddy produce desirable paddy yield in some areas in India and Pakistan. But the alkalinity-induced submergence along with the toxic appearance of $HCO_3^-$, B, Al, Fe, and Mn reduced the productivity of wheat and winter crops in IGP and sodic areas in Australia (Sharma et al. 2018).

## 11.3 Classification of SAS

The classification and characterization of SAS is the first step for completion of a reclamation program of barren or underproductive land. SAS carries electrolytes of $Na^+$, $Ca^{2+}$, $Mg^{2+}$, and $K^+$ and $Cl^-$, $SO_4^{2-}$, $CO_3^{2-}$, $HCO_3^-$, $SiO_x^{-n}$, etc. Soluble, less soluble, and relatively insoluble salts are produced on combinations of cations and anions. The order of solubility of salts are follows: $K^+ > Na^+ > Mg^{2+} > Ca^{2+}$ and anions are $Cl^- > SO_4^{2-} > HCO_3^- > CO_3^{2-}$. The degree of alkaline hydrolysis is $Na^+ > Mg^{2+} > Ca^{2+} > K^+$ and $CO_3^{2-} > HCO_3^-$ and anions $Cl^-$ and $SO_4^{2-}$ produce salts of near neutral to acidic (Tavakkoli et al. 2015). The pH of the soil water environment regulates the forms of silicate in soil water solution. Decomposition of primary minerals liberates alkaline salts in the case of extreme equilibrium of hydrological stagnation with short trajectories (Jobbágy et al. 2017). Saline, sodic, and saline–sodic soils are the three categories of SAS that are normally available.

### 11.3.1 Saline Soil

The abundance of soluble electrolytes generates osmotic effects for roots when established in saline soil, and this is termed "physiological drought." This environment may create ion(s) toxicity for plant roots and soil microbes. Salts with encrustation of white color at the soil surface and poor crop stand are the common syndromes of saline soil. $Na^+$, $Ca^{2+}$, $Mg^{2+}$, $Cl^-$, and $SO_4^{2-}$ are the dominant electrolytes in this category of soil, and the occurrence of these electrolytes disrupts the chemical equilibria of soil, and crop nutrition is affected but soil physical quality is merely unaffected. The pH of the soil water saturation paste ($pH_s$) of saline soil is lower than 8.2, exchangeable sodium percent (ESP) not more than 15.0 percent, SAR (sodium adsorption ratio [$SAR_e$]) not more than 13.0 $mmol^{1/2}$ $L^{-1/2}$, and electrical conductivity of the soil water-saturated paste extract ($EC_e$) greater than 4.0 dS $m^{-1}$ at 25 °C (Ahmad et al. 1998). The genesis of saline soil is due to the existence of electrolytes in soil strata, capillary rise of underground saline water, prolonged irrigation with saline or marginal water, and drainage congestion or intrusion of the sea or brackish water at the coastal line (Singh 2009).

### 11.3.2 Sodic Soil

Sodic soils usually termed as "alkali soils" carry unduly high levels of $Na^+$ than $Ca^{2+}$ and $Mg^{2+}$ in the soil solution and exchange sites. These soils have an ESP, $SAR_e$, and $pH_s$ of more than 15 percent and 13.0 $mmol^{1/2}$ $L^{-1/2}$, > 8.2, respectively, and $EC_e$ usually varies subject to the occurrence of soluble electrolytes. The exchangeable Na causes clay dispersion and impedes soil physical structure, blocks the pore space, keeps slowing down the water and air entry, aggregates failure, and slows infiltration, and the soil organic matter remains mobile and facilitates the erosion of organic matter. In a severe alkaline environment, the concentration of Na reaches in toxic level, and Ca starts precipitating out as $CaCO_3$ and causing Na-induced nutritional deficiency. A frequent N deficiency may appear in sodic soil because of losses of soil organic matter, a wasteful conversion of supplied N, and poor symbiotic N fixation due to legumes having low sodicity-tolerant potential (Sundha et al. 2017). Soil aggregates in alkali soil are coated with SOM and show black color. The deposited $CaCO_3$ in the soil profile in arid and semiarid environments frequently release $OH^-$ ions in soil water solution; therefore, the $OH^-$ ions increase pH in calcareous sodic soils more than that in noncalcareous alkali soil. The presence of exchangeable Na without a bulk quantity of neutral soluble salts shows a high pH (Basak et al. 2015a).

## 11.4 Characteristics of SAS in Different AEZ

India is endowed with 20 agroecoregions. These agroecoregions are very diverse from humid to the hot dry arid, cold arid desert to the humid coastal line. Therefore, several provinces of the country are endowed with a wide variety of soils. Nearly, SAS covers pockets to extensive areas of fourteen states (Punjab, Haryana, Uttar Pradesh, Rajasthan, Gujarat, Bihar, West Bengal, Madhya Pradesh, Chhattisgarh, Maharashtra, Karnataka, Telangana, Andhra Pradesh, and Kerala) and one union territory (Andaman and Nicobar Island). The typical characteristics of SAS are mentioned below.

### 11.4.1 Physical Properties

Sodification promotes aggregate failure and declines structural stability; therefore, the stability of soil aggregates worsened. After irrigation or rainfall, typical convex surfaces appear in sodic soil. Wide (1–2 cm) and deep (several centimeters) cracks appear during the progress of the drying cycle. However, these cracks may disappear upon wetting. The tendencies of formation of cracks arise at similar sites on the surface each time, drying the wet sodic soil until the perturbation or mechanical disturbance is applied to this soil. The emergence of seedling and, consequently, the plant population is largely reduced because of surface sealing and hard crust formation (Mandal et al. 2008). Negative repulsion charges among clay colloids and the persistence of net negative charge on exchange sites induce swelling, aggregate slaking, dispersion, and movement of clay clog water–conducting soil pores, increase compaction, reduce pore space ultimately cause reduced water intake rates in irrigated soils (Chaganti and Crohn 2015; Levy and Shainberg 2005; Sharif et al. 2012). Swelling and dispersion of clay and their migration due to the presence of alkaline pH together with blockage of pores deteriorate the physical characteristics of soil (Dang et al. 2018) and also increase soil bulk density (Josan et al. 1998). Poor aggregate stability in sodic soils provides inadequate protection to the soil organic matter that further accentuates the degradation of structural stability (Murphy 2015). Sodic soils have low infiltration capacity and are, therefore, prone to waterlogging, the impedance of air and water entry, and the formation of anoxia (Das et al. 2022; Sharma et al. 2018).

The poor hydraulic conductivity with increment in sodicity causes swelling and leads to clogging of pores in smectite, clayey soils with extended surface area (Bhardwaj et al. 2019). Further, pore blocking in coarser soils arises because of dispersion and migration of clay (Shainberg and Letey 1984) and crops suffer during waterlogging problems due to surface crusting after rain or application of irrigation (Sharma et al. 2018). Such a situation further worsens with the beating effect of raindrops. The aggregate failure and the soil dispersion in alkaline-irrigated soils not only tend to be an obstacle to water intake but also decline water infiltration followed by water storage in the soil profile and also take a long time for redistribution of water due to drainage (Quirk et al. 1986). Sodicity on one hand reduces both the binding of the microaggregates and simultaneously increases slaking. The clay translocation and its settling depth in a high sodic condition depend upon the clay mineralogy and soil texture, which regulate the infiltration yet modifying factors like sodicity level of the soil, release of electrolytes, and quantity and quality of inherent and applied organic matter (Minhas et al. 2019; Suarez 2012). Adoption of the intensive rice–wheat system in heavy-textured soils (Surapaneni and Olsson 2002) makes them more prone to sodification and problems when irrigated with sodic waters (Minhas et al. 2019). The use of sodic groundwater for irrigation is the key issue for deterioration of soil physical condition. The hardness of water, alkalinity, SAR (>10), and appearance of residual $CO_3^{2-}$ and $HCO_3^-$ and the presence of a larger quantity of $Mg^{2+}$ than $Ca^{2+}$ also degrade soil physical condition (Balpande et al. 1996; Qadir et al. 2018; Vaidya and Pal 2002).

### 11.4.2 Chemical Properties

The supply of carbonate in the form of $Na_2CO_3$ on weathering of aluminosilicate minerals and reaction of $CaCO_3$ with neutral salts also generate $CO_3^{2-}$ (Gupta and Mathur 2011). The larger quantities of $HCO_3^-$ and $CO_3^{2-}$ ions increase alkalinity, and the Na hydrolysis produces hydroxyl ($OH^-$) ions showing a toxic effect on plant growth. Toxic appearance of these anions and the large concentration of ESP and $SAR_e$ lower the solubility and availability of Zn and Fe. The persistence of alkalinity in soil solution is the function of $CO_3^{2-}$ and $HCO_3^-$ concentration, the evaporation demand, and input of carbonate excess water as irrigation or ingression or aeolian deposition. The irrigation of RSC-born (carbonate excess) groundwater in arid and semiarid regions of northwestern India and waterlogged *vertisols* of southern and central India (Ranade and Gupta 1987; Sheoran et al. 2021, 2022) is responsible for sodicity development in sodic soil. The supply of both inorganic carbon ($HCO_3^-$ and $CO_3^{2-}$) and divalent cation on the decomposition of rock (hydrological trajectories) is one of the causes of carbonate surplus or deficit solution in nature. Sodic soil sustains precipitation and evapotranspiration demand. The ecosystem ingression of inorganic carbon is inadequate in this soil because of low biological activity (occurrence of poor vegetation and slow microbial cycling), alkaline pH, and mobility of organic matter. The microbial $CO_2$ may persist in the dissolved or (bi)carbonated form of a sodic environment. The decomposition of rock primarily supplies $HCO_3^-$ and $CO_3^{2-}$ in soil solution where system alkalinity limits the degassing of $CO_2$. The supply of divalent cation in soil–water system prevails via weathering and exchange process. The extended water trajectories system is excessively saturated with divalent cations over $HCO_3^-$ (Jobbágy et al. 2017). But in short trajectories, the chances of excess $HCO_3^-$ may occur upon weathering of calcareous rocks. Further, $Na^+$ selectivity in alkaline pH regresses $Ca^{2+}$ and slows down the Ca solubility.

$$NaAlSi_3O_8 + HOH + CO_2 \rightarrow HAlSi_3O_8 + NaHCO_3$$

$$NaHCO_3 \rightarrow Na_2CO_3 + CO_2 + H_2O$$

$$Na_2CO_3 + 2HOH \rightarrow 2Na^+ + 2OH^- + H_2CO_3$$

The highly alkaline sodium carbonate is water soluble and its hydrolysis raises the pH up to 12.0 (Bajwa and Swarup 2012). The $OH^-$ ions slow a higher pH in calcareous alkali compared to that in noncalcareous alkali soil. Therefore, the exchangeable Na of sodic soil with the inadequate quantity of neutral electrolytes ($Cl^-$ and $SO_4^{2-}$) possesses a high pH. However, the concentration of $Na_2CO_3$ and level of ESP play an edge role in soil pH. Irrigation having high residual sodium carbonate (RSC > 2.5 me $L^{-1}$) and soils with shallow sodic water tables rise the EC buildup in sodic soil. Therefore, it appears to be confusing to discriminate sodic soil from saline–sodic soil. At this stage, some ionic indices are developed. The quantities of $Na^+$, $CO_3^{2-}$, and $HCO_3^-$ in saturation extract (Chhabra 2004) or the ratio of $[Na^+]/([Cl^-] + [SO_4^{2-}])$ and $(2CO_3^{2-} + HCO_3^-) / (Cl^- + 2SO_4^{2-})$ in soil water saturation paste extract play an important role. If the ratio of these two indices is more than one, then soils are treated as nitric (sodic) and need chemical amendments for reclamation (Bajwa and Swarup 2012). If the soils show the indices value <1, then regardless of their pH and SAR, these should be coined as salic (saline). In such circumstances, both SAR and $EC_e$ lower down on leaching with good quality water. Here, we categorize SAS in India with ionic concentration in soil water saturation extract for discriminating and characterizations for reclamation strategies (Table 11.1).

Soils of central and southern India in *vertisol* categories under semiarid climatic conditions with reasonable rainfall zone (annual rain ~875 mm) favor the development of calcareous sodic soils (Table 11.2). The Purna Valley of Maharashtra state frequently reports the difficulties of poor drainage, waterlogging (slow, saturated hydraulic conductivity), and poor soil physical conditions but no indication of salt efflorescence. The irrigation is the important issue for the deterioration of physical condition of the soil (Padekar et al. 2016). Some areas having water contain variable electrolytes, besides salinity, presence of appreciable quality of Mg than Ca, high SAR, and appearance of residual $CO_3^{2-}$ and $HCO_3^{2-}$ (Table 11.3). The meta-analysis of Balpande et al. (1996) and Vaidya and Pal (2002) showed the deterioration of soil properties due to an increase of ESP and/or exchangeable Mg. The rapid deterioration of soil hydraulic conductivity (K*s*) was due to smectite soils in which 98 percent of clay is fine. An increment in Mg to Ca ratio > 1 in waters and exceeding EMP (exchangeable magnesium percent) > 25% in soils are an indicator of soil degradation, and environmental quality is deleteriously affected (Basak et al. 2015b). Therefore, the external application of Ca and removal of Mg salts are promoted to mitigate this effect (Qadir et al. 2018).

A low to medium range of soil organic C (SOC) is present in SAS while the inorganic C may vary. Poor vegetation cover and low return of rhizodeposition support a low SOC storage in SAS. Alkalinity neutralization, improvement of biomass production, and return of those to soil favor enriching SOC in reclaimed soils. In addition, leaching losses of DOC and clay richness into deeper layers may cause a low SOC content in the top layer of sodic soils. With the succession of grassland, SOC remains in physical protection in microaggregates due to continual expansion of the root system. With an increment of SOC, the closed space –COOH and –OH (phenolic) group present in organic matter increase the surface charge density (SCD) and ease selectivity toward multivalent cations. Further, an increment in SOC increases the internal and external surface area and exchange sites. The SOC increment in SAS prefers to hoard Ca over Na in exchange sites and decline the values of $K_G$. Oppositely, in sodic soil, a high-alkaline environment with the presence of Na and bicarbonate, the native SOC remains highly mobile and prone to erosion and decomposable loss (Jobbágy et al. 2017). The adverse soil physical condition, poor air, and water entry and inadequate nutrient availability hinder germination and root development. A lower amount of SOC is reported in SAS of northwest India because of sodicity (Basak et al. 2021; Datta et al. 2019) as sodic soil remains dispersed upon wetting and is prone to vulnerability for decomposition of

**Table 11.1** The chemical behaviors of SAS with reclamation strategies (recalculated from Rai et al. 2022 [personal communication])

| Sites/soil attributes | Kampasagar, Telangana | Lucknow, Uttar Pradesh | Faizabad, Uttar Pradesh | Pusa, Bihar | Machilipatnam, Andhra Pradesh | Raigad, Maharashtra | Cuttack, Odisha | Port Blair, Andaman and Nicobar | Canning Town, WB | Old Goa, Goa |
|---|---|---|---|---|---|---|---|---|---|---|
| $pH_{1:2}$ | 9.77 | 10.34 | 10.01 | 10.4 | 8.41 | 7.09 | 4.91 | 8.09 | 7.42 | 6.21 |
| $pH_s$ | 9.3 | 9.83 | 9.28 | 9.78 | 7.67 | 6.88 | 4.24 | 7.12 | 7.4 | 5.42 |
| $EC_{1:2}$ | 2.6 | 2.78 | 2.54 | 8.08 | 4.39 | 8.24 | 4.31 | 3.52 | 1.52 | 12.07 |
| $EC_e$ | 4.68 | 3.22 | 4.88 | 14.2 | 11.95 | 26.1 | 12.64 | 7.72 | 5.34 | 23.99 |
| $Na^+$ | 56.1 | 15.8 | 20.4 | 163.0 | 122.8 | 253.3 | 111.1 | 51.7 | 37.3 | 285.9 |
| $Ca^{2+} + Mg^{2+}$ | 8.9 | 4.5 | 5.2 | 6.3 | 56.5 | 181.1 | 68.9 | 112.1 | 31.1 | 145.2 |
| $K^+$ | 0.2 | 0.8 | 0.3 | 0.1 | 1.3 | 2.9 | 1.2 | 1.2 | 0.1 | 4.3 |
| $Cl^-$ | 27.1 | 8.1 | 6.1 | 19.1 | 118.3 | 220.1 | 128.1 | 6.4 | 27.3 | 269.1 |
| $SO_4^{2-}$ | 9.9 | 5.0 | 2.9 | 24.4 | 5.1 | 29.6 | 4.5 | 15.8 | 6.9 | 8.8 |
| $CO_3^{2-}$ | 0.0 | 1.0 | 0.0 | 15.1 | 0.0 | 0.0 | 0.0 | 0.0 | 0.0 | 0.0 |
| $HCO_3^-$ | 10.0 | 22.0 | 11.0 | 6.0 | 5.0 | 4.0 | 4.0 | 5.0 | 5.0 | 6.0 |
| $SAR_e$ | 26.4 | 10.5 | 12.9 | 94.1 | 23.1 | 26.6 | 18.9 | 6.9 | 9.5 | 33.6 |
| CBCS | 0.2 | 1.3 | 0.9 | 0.5 | 0.0 | 0.0 | 0.0 | 0.1 | 0.1 | 0.0 |
| SCSR | 1.2 | 0.9 | 1.7 | 2.4 | 1.0 | 0.9 | 0.8 | 0.6 | 0.9 | 1.0 |
| Reclamation strategies | Sodic soil alkaline hydrolysis, application of Ca (agricultural grade mineral gypsum) source to reduce ESP and $SAR_e$ followed by leaching with good quality water | | | | | Saline leaching with good quality water based on underground water table | | | | |

$pH_{1:2}$: pH in soil water ratio 1:2; $EC_{1:2}$: EC in soil water ratio 1:2; $pH_s$: pH soil water saturation; $EC_e$: electrical conductivity of soil water saturation paste extract, dS $m^{-1}$; $Na^+$, $Ca^{2+}$, $Mg^{2+}$, $K^+$, $CO_3^{2-}$, $HCO^-_3$, $Cl^-$, and $SO^{2-}_4$: concentration of all electrolytes in soil water saturation paste extract in me $L^{-1}$; $SAR_e$: $Na^+/[(Ca^{2+} + Mg^{2+})/2]^{1/2}$ sodium adsorption ratio of soil water saturation paste extract in $mmol^{1/2}\ L^{-1/2}$; CBCS: $(2CO^{2-}_3 + HCO^-_3) / (Cl^- + 2SO^{2-}_4)$; SCSR: $Na^+ / (Cl^- + 2SO^{2-}_4)$

**Table 11.2** Chemical properties of SAS in central and southern India

| Soil attributes | $pH_{1:2}$ | ESP | EMP | Exchangeable Ca/Mg | $SAR_e$ | $HCO_3^-$/ $Ca^{2+}$ | Recalculated from Balpande et al. (1996) |
|---|---|---|---|---|---|---|---|
| Range | 8.1–9.1 | 0.7–26.2 | 18.6–60.1 | 0.2–4.7 | 0.5–18.8 | 0.6–11.7 | |
| ESP | 0.77** | 1.0 | | | | | |
| EMP | 0.37* | 0.136 | 1.0 | | | | |
| Exchangeable Ca/Mg | 0.17 | 0.07 | −0.33 | 1.0 | | | |
| $SAR_e$ | 0.702** | 0.62** | 0.32 | 0.04 | 1.0 | | |
| $HCO_3^-/Ca^{2+}$ | 0.585** | 0.49** | 0.51** | −0.26 | 0.62** | 1.0 | |
| Soil attributes | $pH_{1:2}$ | ESP | EMP | Exchangeable Ca/Mg | $SAR_e$ | | Recalculated from Vaidya and Pal (2002) |
| Range | 8.0–8.8 | 1.4–33.9 | 11.1–36.4 | 1.2–4.8 | 2.4–41.7 | | |
| ESP | 0.65** | 1.0 | | | | | |
| EMP | 0.36 b | −0.25 | 1.0 | | | | |
| Exchangeable Ca/Mg | −0.65 | −0.04 | −0.90 | 1.0 | | | |
| $SAR_e$ | 0.58** | 0.89** | −0.29 | 0.002 | 1.0 | | |

*$P < 0.05$; ** $P < 0.01$; EMP is the exchangeable magnesium percent

**Table 11.3** Effect of organic C on Gapon's selectivity coefficient of soils collected from different agroecological regions

| Soils | AER | CEC (c mol $_{p+}$ kg$^{-1}$) | ESP (%) | Organic C (g kg$^{-1}$) | $K_G$ |
|---|---|---|---|---|---|
| Baruipur, West Bengal | 15 | 7.7 | 28.9 | 11.75 | 0.060 |
| Ela, Goa | 20 | 17.2 | 54.3 | 9.47 | 0.046 |
| Karaikal, Pondicherry | 19 | 21.5 | 22.8 | 8.25 | 0.038 |
| Panvel, Raigad, Maharashtra | 20 | 36.0 | 60.8 | 5.77 | 0.060 |
| Canning Town, WB | 19 | 8.8 | 39.7 | 5.62 | 0.063 |
| Trichy, Tamil Nadu | 9 | 13.3 | 38.9 | 5.48 | 0.036 |
| Machilipatnam, Andhra Pradesh | 19 | 13.1 | 48.9 | 5.18 | 0.041 |
| Chidambaram, Tamil Nadu | 19 | 6.9 | 5.7 | 4.02 | 0.030 |
| Trichy (fallow land), Tamil Nadu | 9 | 13.7 | 36.1 | 3.87 | 0.084 |
| Bathinda, Punjab | 2 | 2.7 | 18.6 | 3.72 | 0.081 |
| Pusa, Bihar | 13 | 6.1 | 71.6 | 3.55 | 0.122 |
| Cuttack, Orisha | 12 | 2.1 | 39.2 | 3.29 | 0.046 |
| Barwaha, Madhya Pradesh | 7 | 24.9 | 79.7 | 2.96 | 0.032 |
| Kanpur, UP | 6 | 11.3 | 82.0 | 2.70 | 0.207 |
| Kampasagar, Telangana | 8 | 23.0 | 88.0 | 2.56 | 0.263 |
| Dharamgarh, Panipat, Haryana | 4 | 12.5 | 72.3 | 2.30 | 0.110 |
| Faizabad, Uttar Pradesh | 6 | 20.5 | 93.0 | 1.92 | 1.643 |
| Kanpur, Uttar Pradesh | 6 | 15.6 | 90.4 | 1.92 | 0.735 |
| Shivari, Lucknow, UP | 6 | 20.9 | 84.0 | 1.50 | 0.029 |
| Haibatpur, Karnal, Haryana | 4 | 11.9 | 53.1 | 1.40 | 0.069 |
| Sangrur, Punjab | 4 | 27.8 | 70.3 | 1.10 | |

soil microbial biomass. Table 11.4 shows greater values of Gapon's selectivity coefficient ($K_G$) with a decrement of SOC in different SAS across the country. The Ca source and organic amendments from external supply favor the Ca enrichment in exchange sites and decrease $K_G$. The supplied organic matter produced clay–humus complex and divalent Ca is preferred for the formation of the complex (Poonia and Pal 1998). The closely spaced organic sites remain inactivated and reduce ESP and $K_G$. However, the addition of 10 Mg ha$^{-1}$ farmyard manure or city waste composts had the potential to supply both Ca and Mg and produce clay humus complex of Ca or Mg and reduce the injurious impact of sodicity.

## 11.5 Salt-Affected Soil and Poor Quality Water

Soil solution chemistry is largely influenced by water quality. Underground saline or saline–sodic irrigation water imports salt from outside, and the capillary rise of underground saline–sodic water and hydrological stagnation alters soil solution

**Table 11.4** Attributes of collected groundwater from different agroecological regions

| Sites | AER | pH | EC (dS $m^{-1}$) | $Mg^{2+}$/$Ca^{2+}$ + $Mg^{2+}$ | Cl-/$SO_4^{2-}$ | SAR ($cmol^{1/2}$ $L^{-1/2}$) | RSC ($meL^{-1}$) |
|---|---|---|---|---|---|---|---|
| Bathinda, Punjab | 2 | 7.29 | 5.60 | 0.17 | 3.66 | 15.24 | Nil |
| Nain, Panipat, Haryana | 4 | 7.99 | 16.00 | 0.45 | 0.93 | 0.23 | Nil |
| Trichy, Tamil Nadu | 9 | 8.61 | 3.90 | 0.13 | 23.02 | 10.71 | 1.40 |
| Chidambaram, Tamil Nadu | 19 | 7.71 | 1.29 | 0.33 | 7.47 | 2.16 | Nil |
| Baruipur, West Bengal | 15 | 7.56 | 1.23 | 0.17 | 10.60 | 4.92 | 0.72 |
| Karaikal, Puducherry | 19 | 7.83 | 1.78 | 0.47 | 4.73 | 8.31 | 2.00 |
| Kampasagar, Telangana | 8 | 7.09 | 1.70 | 0.37 | 3.61 | 3.65 | 2.00 |
| Panvel, Maharashtra | 20 | 7.83 | 1.37 | 0.22 | 3.71 | 6.42 | 0.83 |
| Lucknow, Uttar Pradesh | 6 | 7.58 | 1.15 | 0.48 | 7.06 | 2.26 | 1.90 |
| Machilipatnam, Andhra Pradesh | 19 | 7.88 | 1.28 | 0.20 | 5.48 | 3.25 | Nil |
| Canning, West Bengal | 19 | 7.26 | 1.36 | 0.21 | 10.56 | 5.29 | 2.40 |
| Pusa, Bihar | 13 | 7.63 | 0.99 | 1.08 | 4.52 | 3.87 | 0.37 |
| Kanpur, Uttar Pradesh | 6 | 7.72 | 1.22 | 0.23 | 9.28 | 15.12 | 7.20 |
| Faizabad, Uttar Pradesh | 6 | 5.05 | 0.77 | 0.09 | 7.06 | 1.61 | 0.40 |

chemistry (Minhas and Tyagi 1998). Chemical hydrolysis of electrolytes present in water along with dissolved organic compounds with varied functional groups play a huge role in the equilibrium of soil pH (Ulrich 1991). Nearly, soils of arid regions usually have an alkaline reaction (pH >7) as precipitation is inadequate to leach down the weathered mineral (Brady and Weil 2014). With the inadequate rainfall and occurrence of a water deficit environment in the major times of the calendar year, both the soil solution and exchange phases remain rich with mainly $Ca^{2+}$, $Mg^{2+}$, $K^{+}$, and $Na^{+}$. $Ca^{2+}$, $Mg^{2+}$, $K^{+}$, and $Na^{+}$ do not provide $OH^{-}$ and soil water solution remains nearly neutral unless certain anions ($CO_3^{2-}$ and $HCO_3^{-}$) appear in the solution (Tavakkoli et al. 2015). Therefore, in certain environmental conditions, these cations performed as the base-forming cations, and hydrolysis of these cations is expected to cause alkaline pH. Calcite on the dissolution of and dissociation of carbonic acid supply $CO_3^{2-}$ and $HCO_3^{-}$.

$$Ca^{2+}/Mg^{2+}/Na^{2+} \rightarrow \text{Neutral}(pH \le 7.0)$$

$$CO_3^{2-}/HCO_3^{-} \rightarrow [OH^{-}]_n ? pH$$

Irrigation with saline water creates the disturbance of soil chemical equilibrium, whereas sodic water makes worse the physical health of soil, and thus soil became structurally impaired and shows toxicity of soluble Na. Field observation shows a wide, special, and temporal variability of both salinity and sodicity due to the inherent presence of salts, adoption of different cultural practices, the requirement of irrigation, availability of rainfall, crop evapotranspiration demand, and climatic variability. Saline water is abundant in neutral soluble salts mostly $Cl^-$ and $SO_4^{2-}$. Some $HCO_3^-$ may be present but $CO_3^{2-}$ is absent. Therefore, irrigation with saline water shows a higher $EC_e$ of soil and a lower value of pH (Basak et al. 2015a; Chaudhari and Somawanshi 2002). The $OH^-$ generated by the hydrolysis of exchangeable bases was compensated by proton generated by the anionic hydrolysis of $Cl^-$ and $SO_4^{2-}$ (Tavakkoli et al. 2015). Similarly, several research findings show a negative relationship between soil pH and electrolytes concentration in soils (Kopittke et al. 2016). Thus, transformations of saline–sodic soils to degraded sodic soil have greater pH compared to saline–sodic soils with high ionic strength (Sundha et al. 2020). However, it should be emphasized that pH of the equilibrium solution of neutral salts increased with raising SAR of irrigation water (Basak et al. 2015a). Therefore, electrolyte concentration and composition of applied water for irrigation have an overriding influence on soil pH. Further, ionic strength and ionic hydrolysis are regulated by dilution (soil–water ratio). Like with a low soil–water ratio as characterized by the saturated paste, the pH varied from near neutral in the highly saline soil to slightly alkaline in normal soil. A slight increase in pH (7.1 to 8.0) with the increase in salinity levels may be due to anionic hydrolysis of small quantity of $CO_3^{2-}$ and $HCO_3^{2-}$. The presence of clay, organic matter, etc., shows visual disparity on the increment of soil pH in saline soil (Basak et al. 2015a; Tavakkoli et al. 2015). Chemically disturbed saline soil shows the relative preponderances of electrolytes and their indices upon changing of salinity gradient and management practices. Barren saline land upon bringing in cultivation shows relative migration of electrolytes and associated C and N fraction (dissolved organic C and nitrogen, DOC, DON, $NH_4^+$, $NO_3^-$, and total N) with shifting of salinity gradient (Rai et al. 2022; Soni et al. 2021a, 2021b; Basak et al. 2022). The multivariate analysis shows the relative migration of electrolytes and their indices with DOC, $NO_3^-$, and P in aqueous soil–water saturation paste extract. At less salinity condition, total electrolytes concentration ($EC_e$) play a dominant role, but in higher salinity, DOC, $NO_3^-$, and SCSR ($Na^+/Cl^- + 2SO^{2-}{}_4$) are the main limiting soil attributes. Management practices (mulch and deficit irrigation) in such cases had a dominant role in reducing the deleterious impact of salinity and concentration of alkaline anion (Soni et al. 2021a, 2021b Soni et al. 2023). Mulching improves soil structure and declines surface evaporation, thus enhancing salt leaching, and hence inhibits salt accumulation at the surface (Rai et al. 2022). Furthermore, organic mulch supplies substrate that induces respiration and generates $CO_2$. Week carbonic acid forms upon reacting with $CO_2$ and water and lowers soil pH overcultivation (Wong et al. 2010). The supply of electrolytes upon irrigation with saline water developed salinity; oppositely, irrigation with the best available water leaches down the most actively moved anions (chloride followed by sulfate) across the soil profile compared to relatively static bicarbonate and shows a greater value of soil pH at the surface (Choudhary et al. 2004; Jiao et al. 2013).

## 11.6 Conclusion

A comprehended knowledge of physical and chemical characteristics can economically manage the rehabilitation program of SAS. The know-how about these soil properties explores the invention of cost-effective technologies for sodicity reclamation, managing saline soil, strengthening plant nutrition, and increasing fertilizer use efficiency of SAS. This will attract the policymaker to invest money in the reclamation and management of SAS for the longtime benefit of the country. Therefore, further intensified research is needed to understand the changes in the mineral composition of SAS during reclamation. Change in variable charge and its effect on reclamations progresses the sodic *vertisols*; the poor quality irrigation affects the dissolution–precipitation cycle. Considerable research in the past has enabled the use of poor quality water in agriculture. However, with the rapid development of new irrigation techniques and the adoption of pressurized irrigation, it is important to understand the effect of alkali/sodic (residual sodium carbonated) water irrigation on global carbon fluxes, $CaCO_3$ formation, and its fate in the soil profile in soils of varied mineralogy. The effect of pH-buffering capacity on the effectiveness of different amendments, soil pH and ESP, pH, and gypsum requirement relationships need to revisit to address varied soil types. Soil organic carbon stability has received much attention in the present context of global carbon cycle. The SAS is one of the important sinks for carbon sequestration. The effect of changes in ionic strength and ionic composition of solution–solid phase on dissolved organic carbon movement and its stabilization in soil profile needs to be included in future research. Sustainable development goals, the land degradation neutrality, and associated environmental policies of globe underscore showed that from an environmental view, soil salinity is a serious concern primarily to land degradation and subsequently disturbs crop growth and productivity in several irrigated commands and arid and semiarid regions. The poor physical and chemical health of salt-affected soils needs definite amending protocols of rehabilitation for crop production.

## References

Ahmad A, Abraham G, Gandotra N, Abrol YP, Abdin MZ (1998) Interactive effect of nitrogen and sulphur on growth and yield of rape-seed-mustard (*Brassica juncea* L. Czern and Coss and *Brassica campestris* L) genotypes. J Agron Crop Sci 181(4):193–199

Bajwa MS, Swarup A (2012) Soil Salinity and Alkalinity. In: Goswami NN, Rattan RK, Dev G, Narayansamy G, Das DK, Sanyal SK, Pal DK, Rao DLN (eds) Fundamentals of Soil Science, 2nd edn, pp 329–339

Balpande SS, Deshpande SB, Pal DK (1996) Factors and processes of soil degradation in vertisols of the Purna Valley, Maharashira. India L Degrad Dev 7(4):313–324

Basak N, Chaudhari SK, Sharma DK (2015a) Influence of water quality on exchange phase-solution phase behaviour of texturally different salt-affected soils. J Indian Soc Soil Sci 63(4): 365–372

Basak N, Chaudhari SK, Sharma DK (2015b) Impact of varying ca/mg waters on ionic balance, dispersion, and clay flocculation of salt-affected soils. Commun Soil Sci Plant Anal 46(7): 827–844. https://doi.org/10.1080/00103624.2014.1003937

Basak N, Sheoran P, Sharma R, Yadav RK, Singh RK, Kumar S, Krishnamurthy T, Sharma PC (2021) Gypsum and pressmud amelioration improve soil organic carbon storage and stability in sodic agroecosystems. L Degrad Dev 32(15):4430–4444. https://doi.org/10.1002/ldr.4047

Basak N, Rai AK, Sundha P, Bedwal S, Soni PG, Patel S, Narjary B, Yadav G, Kumar S, Yadav RK (2022) Conversion of barren saline soil to cultivation changes Soil enzymes activities: effects of irrigation water quality. Journal of the Indian Society Soil Science 70:347–354. https://doi.org/10.5958/0974-0228.2022.00028.7

Bhardwaj AK, Mishra VK, Singh AK, Arora S, Srivastava S, Singh YP, Sharma DK (2019) Soil salinity and land use-land cover interactions with soil carbon in a salt-affected irrigation canal command of Indo-Gangetic plain. Catena 180:392–400. https://doi.org/10.1016/j.catena.2019.05.015

Brady NC, Weil RR (2014) The Nature and Properties of Soils, 14th edn. Pearson Publication

Chaganti VN, Crohn DM (2015) Evaluating the relative contribution of physiochemical and biological factors in ameliorating a saline–sodic soil amended with composts and biochar and leached with reclaimed water. Geoderma 259–260:45–55. https://doi.org/10.1016/j.geoderma.2015.05.005

Chaudhari SK, Somawanshi RB (2002) Effect of water quality on exchange phase — solution phase behavior of three soils. J Plant Nutr Soil Sci 165(2):229–234

Chhabra R (2004) Classification of salt-affected soils. Arid L Res Manag 19(1):61–79. https://doi.org/10.1080/15324980590887344

Choudhary OP, Josan AS, Bajwa MS, Kapur ML (2004) Effect of sustained sodic and saline-sodic irrigation and application of gypsum and farmyard manure on yield and quality of sugarcane under semi-arid conditions. F Crop Res 87(2):103–116. https://doi.org/10.1016/j.fcr.2003.10.001

Dang A, Bennett JM, Marchuk A, Biggs A, Raine SR (2018) Quantifying the aggregation-dispersion boundary condition in terms of saturated hydraulic conductivity reduction and the threshold electrolyte concentration. Agric Water Manag 203:172–178. https://doi.org/10.1016/j.agwat.2018.03.005

Das BT, Menzies NW, Dalzell SA, McKenna BA, Kopittke PM (2022) Avoiding the point of no return: Maintaining infiltration to remediate saline-sodic Vertosols in high rainfall environments. Agric Water Manag 270:107725. https://doi.org/10.1016/j.agwat.2022.107725

Datta A, Setia R, Barman A, Guo Y, Basak N (2019) Carbon Dynamics in Salt-affected Soils. In: Dagar JC, Yadav RK, Sharma PC (eds) Research Developments in Saline Agriculture. Springer, Singapore, pp 369–389. https://doi.org/10.1007/978-981-13-5832-6_12

Gupta RN, Mathur A (2011) Origin, distribution and classification of salt-affected soils of India. In: Rattan RK, Singh AK (eds) Salt-affected soils: ISSS Bulletin No, vol 28. Indian Society of Soil Science, pp 1–29

Hopmans, J. W., Qureshi, A. S., Kisekka, I., Munns, R., Grattan, S. R., Rengasamy, P., Ben-Gal, A., Assouline, S., Javaux, M., Minhas, P. S., Raats, P. A. C., Skaggs, T. H., Wang, G., De Jong van Lier, Q., Jiao, H., Lavado, R. S., Lazarovitch, N., Li, B. & Taleisnik, E. (2021). Chapter One - Critical knowledge gaps and research priorities in global soil salinity Advances in agronomy

(A. Sparks (ed.), Academic Press, 169, pp. 1–191. https://doi.org/10.1016/bs.agron.2021.03.001
Jiao F, Wen Z-M, An S-S, Yuan Z (2013) Successional changes in soil stoichiometry after land abandonment in Loess Plateau. China Ecol Eng 58:249–254. https://doi.org/10.1016/j.ecoleng.2013.06.036
Jobbágy EG, Tóth T, Nosetto MD, Earman S (2017) On the fundamental causes of high environmental alkalinity (ph $\geq$ 9): an assessment of its drivers and global distribution. L Degrad Dev 28(7):1973–1981. https://doi.org/10.1002/ldr.2718
Josan AS, Bajwa MS, Choudhary OP (1998) Effect of sustained sodic irrigations on physical properties and root growth in a Typic Ustochrept soil under two cropping systems. Agric. Res. J 35:125–131
Kopittke PM, Menzies NW, Wang P, Blamey FPC (2016) Kinetics and nature of aluminium rhizotoxic effects: a review. J Exp Bot 67(15):4451–4467. https://doi.org/10.1093/jxb/erw233
Levy GJ, Shainberg I (2005) Encyclopedia of Soils in the Environment
Mandal AK, Obi Reddy GP, Ravisankar T (2011) Digital database of salt affected soils in India using Geographic Information System. J Soil Salin Water Qual. 3(1):16–29
Mandal UK, Bhardwaj AK, Warrington DN, Goldstein D, Bar Tal A, Levy GJ (2008) Changes in soil hydraulic conductivity, runoff, and soil loss due to irrigation with different types of saline–sodic water. Geoderma 144(3):509–516. https://doi.org/10.1016/j.geoderma.2008.01.005
Minhas PS, Qadir M, Yadav RK (2019) Groundwater irrigation induced soil sodification and response options. Agric Water Manag 215:74–85. https://doi.org/10.1016/j.agwat.2018.12.030
Minhas PS, Tyagi NK (1998) Guidelines for irrigation with saline and alkali waters. Central Soil Salinity Research Institute
Murphy BW (2015) Impact of soil organic matter on soil properties—a review with emphasis on Australian soils. Soil Res 53(6):605–635
Padekar D, Bhattacharyya T, Ray SK, Tiwary P, Chandran P (2016) Influence of irrigation water on black soils in Amravati district, Maharashtra. Curr Sci 10(9):1740–1755. https://doi.org/10.18520/cs/v110/i9/1740-1755
Poonia SR, Pal R (1998) Cation-exchange equilibria in soils. In: Minhas IPS, Tyagi NK (eds) Agricultural Salinity Management in India. Central Soil Salinity Research Institute, Karnal, Haryana, pp 93–107
Qadir M, Schubert S, Oster JD, Sposito G, Minhas PS, Cheraghi SAM, Murtaza G, Mirzabaev A, Saqib M (2018) High-magnesium waters and soils: Emerging environmental and food security constraints. Sci Total Environ 642:1108–1117. https://doi.org/10.1016/j.scitotenv.2018.06.090
Quirk J, Pereira C, Tanton TW (1986) Soil permeability in relation to sodicity and salinity. Philos. Trans R Soc London Ser A Math Phys Sci 316:297–317
Rai AK, Basak N, Soni PG, Kumar S, Sundha P, Narjary B, Yadav G, Patel S, Kaur H, Yadav RK, Sharma PC (2022) Bioenergy sorghum as balancing feedback loop for intensification of cropping system in salt-affected soils of the semi–arid region: Energetics, biomass quality and soil properties. Eur J Agron 134:126452. https://doi.org/10.1016/j.eja.2021.126452
Ranade DH, Gupta RK (1987) Salt effect on hydraulic conductivity of a vertisol. Catena 14(1):45–53. https://doi.org/10.1016/S0341-8162(87)80005-X
SAC. (2016). Desertification and Land Degradation Atlas of India (Based on IRS AWiFS data of 2011–13 and 2003–05)
Shainberg I, Letey J (1984) Response of soils to sodic and saline conditions. Hilgardia 52:1. https://doi.org/10.3733/hilg.v52n02p057
Sharif A, Chaudhari SK, Dagar JC, Basak N (2012) Soil aggregates as indicator of soil health in waterlogged sodic soil. J. Soil Salin. Water Qual. 4(2):92–96
Sharma DK, Thimmappa K, Chinchmalatpure AR, Mandal AK, Yadav RK, Chaudhury SK, Kumar S, Sikka A (2015) Assessment of Production and Monetary Losses from Salt-affected Soils in India. Technical Bulletin

Sharma SK, Kulshreshtha N, Kumar A, Yaduvanshi NPS, Singh M, Prasad KRK, Basak N (2018) Waterlogging effects on elemental composition of wheat genotypes in sodic soils. J Plant Nutr 41(10):1252–1262. https://doi.org/10.1080/01904167.2018.1434541

Sheoran, P., Basak, N., Kumar, A., Yadav, R. K., Singh, R., Sharma, R., Kumar, S., Singh, R. K. & Sharma, P. C. (2021). Ameliorants and salt tolerant varieties improve rice-wheat production in soils undergoing sodification with alkali water irrigation in Indo–Gangetic Plains of India. Agric Water Manag, 243, 106492. https://doi.org/10.1016/j.agwat.2020.106492

Sheoran P, Sharma R, Kumar A, Singh RK, Barman A, Prajapat K, Kumar S, Sharma PC (2022) Climate resilient integrated soil–crop management (CRISCM) for salt affected wheat agri–food production systems. Sci Total Environ 837:155843. https://doi.org/10.1016/j.scitotenv.2022.155843

Singh G (2009) Salinity-related desertification and management strategies: Indian experience. L. Degrad. Dev. 20(4):367–385. https://doi.org/10.1002/ldr.933

Soni PG, Basak N, Rai AK, Sundha P, Narjary B, Kumar P, Yadav G, Kumar S, Yadav RK (2021a) Deficit saline water irrigation under reduced tillage and residue mulch improves soil health in sorghum-wheat cropping system in semi-arid region. Sci Rep. 11(1):1880. https://doi.org/10.1038/s41598-020-80364-4

Soni PG, Rai AK, Basak N, Kumar P, Sundha P (2021b) Productivity and profitability of sorghum-wheat cropping system in saline soils as influenced by conservation agriculture practices. Range Management Agrofor 42(2):277–285

Soni PG, Rai AK, Basak N, Sundha P, Narjary B, Kumar P, Kumar R, Yadav G, Kumar S, Yadav RK, Sharma PC (2023) Effect of deficit saline irrigation, tillage and rice straw mulch in wheat-sorghum cropping system on yield and nutritive value of rainfed forage sorghum in salt-affected soils. Land Degrad Dev:1–12. https://doi.org/10.1002/ldr.4869

Suarez DL (2012) Irrigation water quality assessments. A. In: Wallender WW, Tanji KK (eds) ASCE Manual and Reports on Engineering Practice No. 71 Agricultural Salinity Assessment and Management, 2nd edn. ASCE, Reston, VA, pp 343–370

Sundha P, Basak N, Rai AK, Yadav RK, Sharma DK, Sharma PC (2017) N and P release pattern in saline-sodic soil amended with gypsum and municipal solid waste compost. J Soil Salin Water Qual 9(2):145–155

Sundha P, Basak N, Rai AK, Yadav RK, Sharma PC, Sharma DK (2020) Can conjunctive use of gypsum, city waste composts and marginal quality water rehabilitate saline-sodic soils? Soil Tillage Res 200:104608. https://doi.org/10.1016/j.still.2020.104608

Surapaneni A, Olsson KA (2002) Sodification under conjunctive water use in the Shepparton Irrigation Region of northern Victoria: a review. Aust J Exp Agric 42(3):249–263

Tavakkoli E, Rengasamy P, Smith E, McDonald GK (2015) The effect of cation–anion interactions on soil pH and solubility of organic carbon. Eur J Soil Sci 66(6):1054–1062. https://doi.org/10.1111/ejss.12294

Ulrich B (1991) In: Ulrich B, Sumner ME (eds) An Ecosystem Approach to Soil Acidification BT - Soil Acidity. Springer, Berlin Heidelberg, pp 28–79. https://doi.org/10.1007/978-3-642-74442-6_3

Vaidya PH, Pal DK (2002) Microtopography as a factor in the degradation of Vertisols in central India. L. Degrad. Dev. 13(5):429–445. https://doi.org/10.1002/ldr.525

Wong VNL, Greene RSB, Dalal RC, Murphy BW (2010) Soil carbon dynamics in saline and sodic soils: a review. Soil Use Manag 26(1):2–11. https://doi.org/10.1111/j.1475-2743.2009.00251.x

# Breeding Salt-Tolerant Rice Varieties in Puducherry (U. T.)

12

S. Thirumeni, Krishnasamy Paramasivam, and Mayandi Subramanian

**Abstract**

Salinity is the second major abiotic stress, limiting rice production worldwide and in India. Infusing salt tolerance in rice is economically and socially justifiable option to combat the menace of salinity. Nevertheless, it has to be augmented in conjunction with appropriate management practices. Complexity of the trait is manifested by its polygenic nature (quantitative), interaction with environment (soil-water-temperature-relative humidity), and ontogeny of the rice plant. Past experimental evidences indicate that salt tolerance is a multigenic trait and the components of salt tolerance in turn be determined by many genes. With the use of molecular markers, chromosomal regions associated with component physiological traits of salt tolerance (Quantitative Trait Loci, QTL) could be identified. Marker-assisted selection (MAS) offers immense hope for improvement of salt tolerance in rice, and QTL mapping studies are prerequisite for MAS that can be used to pyramid several QTLs. In Puducherry (Union Territory), salinity is caused due to both primary and secondary salinization; the latter is so critical nowadays in the first season. Considering the gravity of the menace of salt stress, various research projects were carried out to develop improved salt-tolerant rice varieties. Besides, postgraduate scholars were involved for their thesis research focusing on unraveling genetics of salt tolerance and breeding salt-tolerant varieties. We present the research findings made in the institute both by the postgraduate scholars (as part of their thesis research) and by the faculty carrying out the externally funded projects. The various aspects of the germplasm characterization/diversity analysis, genetic studies (classical and molecular), and breeding salt-tolerant varieties are presented and discussed.

S. Thirumeni (✉) · K. Paramasivam · M. Subramanian
Department of Plant Breeding and Genetics, Pandit Jawaharlal Nehru College of Agriculture and Research Institute, Karaikal, Puducherry (U.T.), India





R. K. Singh et al. (eds.), *Genetic Improvement of Rice for Salt Tolerance*,
https://doi.org/10.1007/978-981-99-3830-8_12

**Keywords**

Salinity in Puducherry · Genetics · QTL and MAS

## 12.1 Historical Perspective

The Union Territory of Puducherry comprises of four regions, viz., Puducherry, Karaikal, Mahe, and Yanam, which are not geographically contiguous, and all these regions are coastal. Three regions, viz., Puducherry, Karaikal, and Yanam, are on the east coast while Mahe is on the west coast. The territory of Puducherry was merged with the Indian Union on 1st November 1954 in terms of the *de facto* agreement signed between the Government of India and Government of France on 21st October 1954. Subsequently, the *de jure* transfer of Puducherry took place on 16th August 1962. The Union Territory of Puducherry is administered under the provisions of Government of Union Territories Act 1963. Soil types found in the Union Territory are red, loamy, coastal alluvium, delta alluvium, red laterite, deep black, and red sandy. The mean maximum temperature is 38.2 °C and mean minimum temperature is 24 °C, and northeast monsoon is the major source of rainfall in the Union Territory.

## 12.2 Problem of Salt Stress

In this territory, salinity is caused due to the process of their formation under marine influence and subsequently due to periodical inundation of seawater (primary salinization) during October to December coinciding with northeast monsoon and irrigation with poor quality water (secondary salinization). Out of 30,000 ha of rice area of Puducherry (U.T.), 10,000 ha are affected by coastal salinity. The coastal saline soils of Puducherry and Karaikal districts had pH in the range of slightly acidic to alkaline, i.e., 6.5 to 9.5 and 6.0 to 9.0, respectively. However, these districts were comparatively less saline, the ECe values being 2.0 to 4.0 dS/m and 0.2 to 5 dS/m, respectively (Bandyopadhyay 1999). As most of these areas are either left fallow or converted to industrial purpose, area planted to rice is drastically reduced now. However, in other areas away from coastal line where rice is cultivated using Cauvery water during *kuruvai* season (May to June sowing), it is affected and area is reduced, due to either late receipt or less quantum of canal water received. In order to alleviate the Cauvery water problem, a greater number of bore wells were installed to raise *kuruvai* (May to June) rice. This resulted in salt-water intrusion, owing to nearness to sea, into the alluvial aquifer and hence deep bore wells pump only poor quality water (1.15 and 4.50 dS/m for EC; 7.50 and 8.90 for pH; 8.60 and 25.10 for SAR and 1.0 and 13.37 meq/l for RSC). The quality of most of the bore wells falls under highly saline sodic as per Gupta et al. (1994) classification. Therefore, secondary salinization poses greater threat to rice farmers of this region during first crop (May to June sowing) than primary salinization during second crop

(August to September sowing). To alleviate the problem of salinity, high-yielding, salt-tolerant (vegetative and reproductive) rice varieties, maturing in 105–115 days with semidwarf nature and possessing medium slender white rice, are to be developed. Such varieties can be planted in both the seasons in the salt-prone areas.

## 12.3 Research Initiative Undertaken

Pandit Jawaharlal Nehru College of Agriculture (PAJANCOA) was established in 1987 as an autonomous educational institution by the Government of Puducherry with the prime objectives of imparting agricultural education and meeting the agricultural research and extension needs of the Union Territory. This college was upgraded as Research Institute (PAJANCOA & RI) in 1997. As all the regions of Puducherry (U.T.) are coastal areas, the problem of salinity is very predominant and rice, being the principal food crop, is most affected. Hence, research work on development of salt-tolerant rice varieties was initiated as early as 1994 by participating in the AICRIP as voluntary center and testing salt-tolerant lines in the hotspots. As a result of intensive research work on salt tolerance in rice, seven research projects, funded by various government agencies like ICAR (NATP), Board of Research on Nuclear Sciences (BRNS), IRRI-BMGF (Bill and Melinda Gates Foundation), DBT, Ministry of Agriculture and Cooperation, Government of India, were proposed, approved, and executed (Table 12.1). By utilizing the funds available in these projects, full-fledged molecular marker laboratory and greenhouse facility for screening salt-tolerant line in hydroponics were established.

Besides, postgraduate scholars were involved for their thesis research focusing on unraveling genetics of salt tolerance and breeding salt-tolerant varieties. In any crop-breeding program, variability is the raw material not only to practice selection but also to study genetics through hybridization of contrasting parents (identified through diversity analysis). Genetics, as science empowers plant breeding, therefore, the knowledge of gene action (through classical), QTL analysis (through molecular) associated with salt tolerance, and relevant traits are important to practice selection under salt stress environment (Gregorio et al. 2013). Consequently, various breeding methods like induced mutagenesis, pedigree selection, marker-assisted backcross breeding, marker-assisted pedigree selection, and honeycomb selection design were adopted to develop improved salt-tolerant rice varieties. In this chapter, we present the research findings/accomplishments made in the institute both by the postgraduate scholars (as part of their thesis research) and by the faculty holding externally funded projects. The contents of the chapter are arranged in the order of germplasm characterization/diversity analysis, genetic studies (classical and molecular), and breeding salt-tolerant varieties.

**Table 12.1** Details of externally funded research projects

| No. | Title of project | Outlay (Rs. In lakhs) | Funding agency |
|---|---|---|---|
| 1. | Exploitation of *G. herbaceum* cotton for improving Agricultural output and Economy of the Coastal Agro Ecosystem ***(2000–2003)*** | **8.46** | ICAR (NATP) |
| 2. | Improving salt tolerance in rice (*Oryza sativa*) through induced mutagenesis and recombination breeding ***(2007–2011)*** | **16.40** | BRNS, (DAE), Govt. of India |
| 3. | Stress-tolerant rice for poor farmers in Africa and South Asia (STRASA). **In** 3 ***Phases: 2008–2019*** | **10.42** | BMGF IRRI, Philippines |
| 4. | Phenotyping of Recombinant Inbred Lines (RILs) for salt tolerance ***(2010–2012)*** | **3.50** | (IARI-NRCPB) |
| 5. | From QTL to Variety: Marker Assisted Breeding of Abiotic Stress Tolerant Rice Varieties with Major QTLs for drought, submergence and salt tolerance ***(2010–2016)*** | **83.35** | DBT, Govt. of India |
| 6. | From QTL to Variety: Genomics Assisted Introgression and Field Evaluation of Rice Varieties with genes/QTLs for Yield under drought, flood and salt stress ***(2018–2022)*** | **58.392** | DBT, Govt. of India |
| 7. | DNA Finger Printing of crop varieties and hybrids (Establishment of Lab) ***(2009–2013)*** | **25.00** | Ministry of Ag. & coop. Govt. of India |
| 8. | Identification of traits, genes, physiological mechanisms to develop climate smart varieties for unfavourable environment *(ICAR project no. 2 and sub-project 1)* ***(2017–2022)*** | **1800 USD/Yr** | ICAR-IRRI |
| 9. | Exploring alternate aroma pathways in rice through multi-location evaluation of short grain aromatic rices ***(2022–2023)*** | **2.35** | (ICAR-NIPB) |
| 10. | Phenotyping of 3K rice germplasm panel for salt tolerance ***(2023–2024)*** | **4.00** | ICAR-IIRR |

## 12.4 Genetic Diversity Analysis

To infuse salt tolerance in rice, knowledge of the extent of variability, association of different traits, direct and indirect effects of various traits toward grain yield, and genetic divergence among the genotypes is very much useful. As construction of core collections is an effective tool to extensively characterize germplasm collections, a detailed investigation, at phenotypic and molecular levels, was undertaken by Diwakar (2011), with an objective of constructing rice core collection for salt tolerance involving 236 germplasm accessions using eight morphological traits and 14 SSR markers linked to salt tolerance. A core collection with 32 accessions was constructed, retaining the genetic diversity found in the entire collection of rice taken for study. The 32 accessions in the core collection had a similar distribution for the traits under investigation with all the accessions in the

**Table 12.2** (**a**) Levene's test for homogeneity of variances. (**b**) Evaluating parameters for core collection

**(a)**

| Sl. no | Character | Entire collection variance | Core collection variance | F value | Probability (pr > f) |
|---|---|---|---|---|---|
| 1 | Days to flowering (N) | 240.70 | 247.07 | 0.01 | 0.94 |
| 2 | (S) | 211.75 | 42.70 | 0.71 | 0.40 |
| 3 | Plant height (cm) (N) | 477.06 | 766.60 | 2.18 | 0.14 |
| 4 | (S) | 283.52 | 104.98 | 1.53 | 0.22 |
| 5 | Number of tillers (N) | 16.71 | 13.56 | 3.13 | 0.08 |
| 6 | (S) | 48.96 | 8.83 | 0.22 | 0.64 |
| 7 | Panicle length (cm) (N) | 5.51 | 7.23 | 1.34 | 0.25 |
| 8 | (S) | 4.75 | 6.64 | 0.08 | 0.78 |
| 9 | Spikelet fertility (%) (N) | 9.59 | 8.40 | 0.36 | 0.55 |
| 10 | (S) | 41.55 | 32.61 | 0.02 | 0.89 |
| 11 | Panicle weight (g) (N) | 0.46 | 0.61 | 2.50 | 0.12 |
| 12 | (S) | 0.25 | 0.37 | 0.54 | 0.46 |
| 14 | Single plant yield (g) (N) | 54.87 | 9.74 | 0.36 | 0.55 |
| 14 | (S) | 27.27 | 16.64 | 0.76 | 0.39 |
| 15 | Saline score (S) | – | – | 0.50 | 0.48 |

(N) Normal condition, (S) saline-alkaline condition

**(b)**

| *Sl. no* | *Parameters (%)* | *Value* | *Description* |
|---|---|---|---|
| 1 | Mean difference | 5.19 | Significant |
| 2 | Variable difference | 25.05 | Significant |
| 3 | Coincidence rate | 91.78 | Significant |
| 4 | Variable rate | 111.16 | Significant |

entire collection. The resultant core had 5.19% of mean difference (MD), 25.05 of variance difference (VD), 111.16% of variable rate (VR), and 91.78% of coincidence rate (CR) with the entire collection, which brought about full coverage of all the traits (Tables 12.2a, b). Homogeneity of means was observed between core and entire collection by Newman-Keuls test and all the traits had homogeneous variances between the two collections as revealed by the Levene's test.

Molecular characterization with 14 SSR markers linked to salt tolerance had generated 135 alleles and a maximum of 14 alleles was observed at SSR marker loci RM 7075. Maximum PIC value of 0.928 was observed in RM 17. All 135 alleles identified by the 14 markers in the entire collection were captured in the core, which maximized allelic richness up to 100% and resulted in an average genetic diversity (Nei index) of 0.77, ranging from 0.49 to 0.88 among the markers. This core set of 32 accessions includes highly tolerant to highly susceptible ones through intermediate ones. The core collection, thus constructed, was highly suitable and representative subset of the entire rice collection and the knowledge gained through the core collections allowed the choice of optimal crosses for classical genetic (mating designs) and molecular genetic approaches (QTL studies and association analysis) for better understanding and exploitation of salt-tolerant genes in rice.

## 12.5 Genetics of Salt-Tolerance in Rice

Knowledge on genetics of salt tolerance and its associated traits is of paramount importance to breed tolerant rice varieties. Identification of number of genes governing the traits, their action and interaction, location on the chromosome, and phenotypic variance contributed by each loci/gene/QTL is done in the past through classical methods like combining ability analysis (Fig. 12.1) and now, through molecular methods like QTL analysis, genome-wide association studies (GWAS), etc.

### 12.5.1 Genetics of Salt Tolerance at Seedling Stage

#### 12.5.1.1 Diallel Analysis

Thirunavukkarasu et al. (2015) studied the genetics of seedling-stage salt tolerance using 6 x 6 half diallel mating design. Six parents of high (FL 478, VTL 1), moderate (Kuzhivedichan and ADT 36), and low tolerance (ADT 45 and IR 29) were chosen from screening 118 rice accessions under hydroponics culture system at salt stress of EC 12 $dSm^{-1}$. Data on 19 traits associated with seedling stage salt tolerance recorded on hybrids and six parents were subjected to combining ability analysis following Griffing (1956) and Hayman's (1954) numerical and graphical approaches. Both additive and nonadditive gene actions were observed for all the 19 traits, as evident from significant ***gca*** and ***sca*** effects. However, predominance of non-additive was noticed for most of the traits. The parents VTL 1, FL 478 and Kuzhivedichan were identified as best combiners based on per se performance, ***gca*** effects for majority of the traits. A set of six hybrids each was identified based on per se performance and ***sca*** effects for most of the traits separately. Of these, two hybrids, viz., VTL 1/FL 478 and IR 29/VTL 1, were found superior for both per se and ***sca*** effects.

Further, the study revealed the significance of both additive (D) and dominance ($H_1$ and $H_2$) components for SES score and leaf chlorophyll content. The

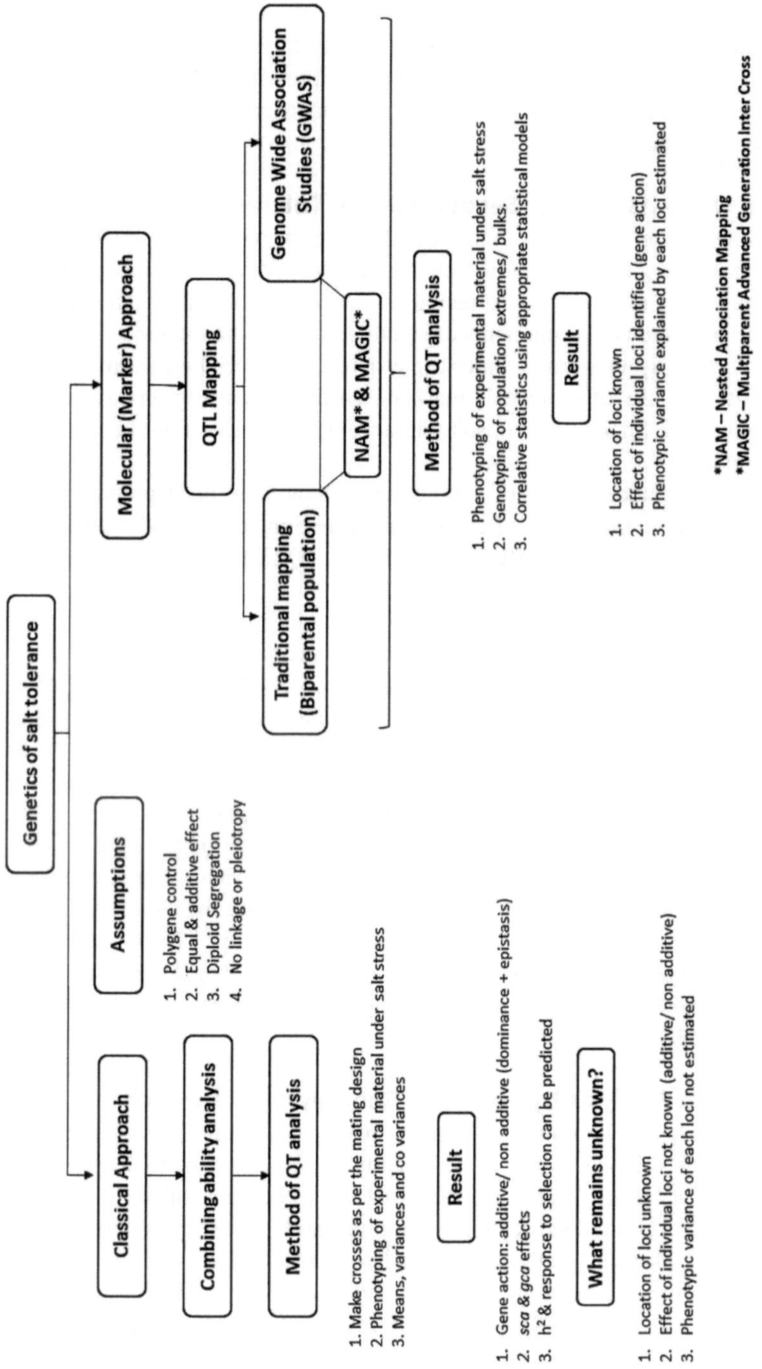


**Fig. 12.1** Comparison of classical and molecular approaches of studying genetics of salt tolerance

non-additive gene action influenced the characters, viz., shoot Na content, shoot Na:K ratio, shoot Na:Ca ratio, root Na content, root Na:K ratio, and root Na:Ca ratio. The overall degree of dominance that was measured by the ratio $(H_1/D)^{1/2}$ was above unity for all the characters that indicated the presence of overdominance. The positive and negative effects of genes were asymmetrically distributed in the parents for all the characters.

#### 12.5.1.2 QTL Analysis for Seedling Stage Tolerance

The complexity of the trait is manifested by its polygenic nature (quantitative), interaction with environment (soil-water-temperature-relative humidity), and ontogeny of the rice plant. Past experimental evidences indicate that salt tolerance is a multigenic trait and the components of salt tolerance (tissue tolerance, Na:K ratio, K accumulators, early vigor, and water use efficiency) in turn be determined by many genes. With the use of molecular markers, chromosomal regions associated with component physiological traits of salt tolerance (Quantitative Trait Loci, QTL) could be identified.

Using these markers, high-density genetic maps have been prepared, which in turn enabled the identification of chromosomal regions associated with quantitative trait, the so-called **"Quantitative Trait Loci" (QTL),** which was not possible, hitherto, through classical genetic approach.

QTL analysis for seedling stage salt tolerance in hydroponics (at EC 12 dS/m) was carried out (Sukanya 2015) using 11 phenotypic data (agro-morphological data-seedling traits) and 40 SSR marker data recorded on 200 $F_2$ plants of ADT 45 x Nona Bokra cross to identify chromosomal regions harboring gene(s) controlling salt tolerance at seedling stage. A total of 17 QTLs were detected, of which five QTLs were for shoot length, seedling height, shoot Na:K ratio, root K content, and root Na:K ratio with LOD scores of 8.4, 6, 5.1, 4.4, and 4.1, explaining 18, 13, 11, 10, and 9% of phenotypic variance, respectively. All these five QTLs were localized on chromosome 1, and among them, QTLs for shoot Na:K ratio, root K concentration, and root Na:K ratio were clustered in the *Saltol* region on the short arm of chromosome 1. As the average marker interval is very high (52.87 cM), owing to less number of markers used, promising QTLs have to be fine mapped with more markers in the QTL region so as to improve the mapping resolution and subsequent use of marker-assisted selection.

#### 12.5.1.3 QTL Analysis by Selective Genotyping Approach

Selective genotyping approach was used to map salt-tolerant QTLs at seedling stage using ADT 45 x Nona Bokra cross (Reddy et al. 2014) in hydroponics culture containing EC 12 $dSm^{-1}$. The screening of 1156 $F_2$ plants resulted in the identification of 131 tolerant and 48 susceptible extremes, and these were selfed to produce $F_3$ progeny. Four physiological traits related to salt tolerance and six yield and its components were recorded in the $F_2$ selected extremes. Thirty-six of the most tolerant and 24 most susceptible individuals were identified based on the phenotyping in hydroponics at salt stress in the $F_2$ and $F_3$ progeny populations. The 60 selected (36 + 24) $F_2$ plants, representing the two extreme tails, were genotyped for 70 polymorphic markers, and a linkage map, spanning 1795 cM, was constructed with an average marker interval of 25.64 cM. Two approaches have been used for QTL mapping in the study (1) comparing marker allele frequencies among groups of individuals with distinct phenotypes, trait-based analysis (TBA) and (2) testing the average phenotypic difference between groups of individuals with distinct marker genotypes, marker-based analysis (MBA).

Trait-based analysis approach led to the identification of four genomic regions with a minimum of two closely placed markers showing significant deviations suggesting the putative location of QTLs for salt tolerance. Of the four QTLs identified, two QTLs were from *ADT 45* on chromosomes 1 and 12 while the other from *Nona Bokra* on chromosomes 3 and 5, respectively. A total of 16 QTLs were detected by employing MBA. A maximum of seven QTLs controlling shoot Na:K ratio, followed by five QTLs for salinity score, two QTLs for leaf chlorophyll content, and one each for shoot $Na^+$ and shoot $K^+$ content, respectively, were detected in the present study. The identified QTLs in the study were compared with the already reported QTLs (Table 12.3).

One of the *Saltol* loci tightly linked to RM3412 for shoot Na:K ratio showed an LOD score of 5.43 with 34% phenotypic variation, and QTLs for shoot Na:K ratio, shoot $Na^+$ content, and leaf chlorophyll content were colocalized in the *Saltol* region.

QTLs identified based on two methods, TBA and MBA, on chromosomes 1, 3, 5, 6, and 12 correspond to the previously located regions for salt tolerance. Hence, these QTLs are potential candidate QTLs for marker-assisted selection (MAS). Marker density has to be increased so as to improve the resolution of the genetic map that can then be used for fine mapping the identified QTLs.

#### 12.5.1.4 Targeted Association Mapping for Salt Tolerance

Targeted association mapping for salt tolerance in rice at seedling stage was attempted by Vinodha (2016) using 124 rice accessions comprising landraces, high-yielding varieties, and breeding lines. These lines were phenotyped in hydroponic system in normal and saline environments (EC 12 dS/m) for seedling stage tolerance, and data on seven physiological traits with SES score (saline only) were recorded. Phenotypic values of all the traits investigated were found to be normally distributed indicating polygenic inheritance. Correlation analysis revealed strong and positive correlation for shoot $Na^+$ content, shoot $Na^+$:$K^+$ ratio, root $Na^+$ content, and root $Na^+$:$K^+$ ratio with SES score while strongly negative correlation with leaf chlorophyll, shoot $K^+$, and root $K^+$ content.

Twenty-two unlinked SSR markers were used for analysis of population structure. Eight SSR markers around *Saltol* region of chromosome 1 were chosen for targeted association mapping. A total of 81 alleles were recorded in 22 SSR markers in the 124 accessions with an average of 3.68 alleles per locus and PIC value of 0.4447. A total of 35 alleles were recorded in eight *Saltol* linked SSR markers in the 124 accessions with an average of 4.38 alleles per locus and PIC value of 0.4800. Population structure of 124 rice accessions, analyzed by Bayesian-based approach, grouped the entire 124 accessions into two subgroups (SG1 and SG2) with 46 and 44 accessions in SG1 and SG2, respectively, and 34 accessions as admixture (AD). The Analysis of Molecular Variance (AMOVA) based on 22 SSR marker loci revealed that 93% of genetic variation was attributed to differences among

**Table 12.3** Comparison of identified QTLs with that of previously reported ones

| Marker position | No. of QTL for salt tolerance | Reports in other studies |
|---|---|---|
| *Chromosome 1* | | |
| RM3412 (11.56) | 5: SNC, SKC, SNKR, RKC, RNKR. | Same position |
| | 3: SKC, SNC, SNKR | Nearly same position |
| | 3: SKC, SDS, RNC | Nearly same position |
| | 3: Low $Na^+$ concentration, high $K^+$ concentration, low $Na^+$:$K^+$ concentration | Nearly same position |
| | 1: Seedling salinity tolerance | Nearly same position |
| | 1: Shoot length in four seasons | – |
| RM11757 (33.46) | Leaf chlorophyll content | Flanking locus (RM11757) |
| | Days from seedling to death, $Na^+$ content in roots | – |
| RM472 (37.88) | 2: STOL | Same position |
| RM12293 (43.23) | – | Not reported |
| *Chromosome 2* | | |
| RM12363 (1.19) | – | Not reported |
| RM318 (29.65) | 1: Relative tiller number | – |
| *Chromosome 3* | | |
| RM7332 (0.39) | – | Nearly same position |
| RM14307 (1.02) | – | Not reported |
| RM3525 (30.19) | Salt injury | Nearly same position |
| *Chromosome 4* | | |
| RM7585 (0.21) | 3: $K^+$ uptake, $Na^+/K^+$ concentration, shoot $Na^+$ concentration | Nearly same position |
| | Shoot $Na^+$ concentration, shoot $Na^+/K^+$ concentration | Nearly same position |
| RM3170 (27.79) | Relative biomass | |
| RM5907 (29.39) | Salt injury | Grossly similar position |
| *Chromosome 6* | | |
| RM20378 (23.94) | Days from seedling to death | Same position |
| | Relative dry weight, relative fresh weight | |
| *Chromosome 12* | | |
| RM27650 (4.15) | – | Not reported |
| RM27767 (6.45) | Concentration | |

individuals while 4% attributed to differences among the populations. The pairwise $F_{ST}$ value of 0.541 was observed between SG1 and SG 2.

Targeted association mapping analysis revealed that 12 SSR markers were found to be correlated with salt tolerance traits. Of these, six markers belonged to *Saltol* region of chromosome 1, while others belonged to chromosomes 2, 4, 6, 7, 10, and 11. Among the eight *Saltol* markers, three markers, viz., RM 3412, *SalT* 1, and *SKC* 1, were associated with six key physiological traits of seedling stage salt tolerance.

Haplotypes analysis using eight *Saltol* markers, in comparison with Pokkali and IR 29 alleles, revealed 44 haplotypes in 124 accessions studied. Novel alleles, in *Saltol* region of chromosome 1, were found in many accessions for various marker loci. Among them, four accessions, viz., Azgo, CSR-1, CSR-2, and Khuzhivedichan (Vettaikkaraniruppu) were possessing novel alleles for *SKC*1, AP 3206, RM 3412, RM 8094, and *SalT* 1 in three marker combinations. Other potential donors for novel alleles were VTL-1, VTL-2, Athurkitchadi, Chityankottai, Emmulisamba, and Jodumani for two marker combinations. These accessions could be exploited in future hybridization programs and novel alleles in marker-assisted selection (MAS) for designing salt-tolerant germplasm.

## 12.5.2 Genetics of Salt Tolerance at Reproductive Stage

### 12.5.2.1 Studies on Gene Action

Line X Tester analysis, involving 4 lines and testers and their 16 hybrids, was attempted to study genetics of salt tolerance in reproductive stage under salt stress field environment (soil: EC 0.90 dS/m, pH 8.65; and irrigation water: EC 1.75 dS/m, pH 8.50, SAR 18.72, RSC 14.14 (meq/l). Observations were recorded on nine traits, viz., days to flowering, plant height, productive tillers, panicle length, grain number, spikelet sterility, 1000 grain weight, single plant yield, and Na:K ratio. The study revealed significant GCA and SCA variances for all the traits studied indicating presence of both additive and nonadditive gene action (Thirumeni et al. 2001). However, predominance of nonadditive gene action was observed for all traits except grain number and panicle length for which additive gene action is predominant.

In generation mean analysis, phenotyping of six generations (P1, P2, F1, F2, BC1, and BC2) of 8 out of 16 crosses under salt stress in field environment, as above, revealed predominance of nonadditive gene action (Thirumeni et al. 2003) with all the three types of epistasis for all the eight characters studied.

### 12.5.2.2 QTL Mapping for Reproductive Stage Salt Tolerance

Mapping salt-tolerant QTLs in rice at reproductive stage was attempted by Mathyvanan (2019), using 186 RIL populations, derived from the cross ADT 45/Nona Bokra, 51 SRR marker data, and phenotypic data on 18 morphophysiological characters. Linkage map covering all the 12 chromosomes of rice was constructed spanning over 1465.54 cM with an average interval of 28.74 cM between marker loci. However, there were huge gaps on most of the chromosomes due to low number of markers used. Thirty QTLs for 7 out of 18 characters related to salt tolerance at reproductive stage were detected on seven chromosomes (2, 3, 4, 5, 7, 10, and 11). Of 30 QTLs, 9 are major QTLs and almost all the 30 QTLs detected in the study were grouped into 5 clusters on chromosome 2 (2 clusters), chromosomes 3, 4, and 7 with one each. The first cluster, flanked by RM 12713 and RM13141 on the short arm of chromosome 2, contained three QTLs for pollen fertility (*qPFR-2-1*), spikelet fertility (*qSPR-2-1*), and salinity score (*qSES-2-1*).

Non-detection of QTL for majority of traits was due to very poor coverage of the genome with very less number of SSR markers (51). In almost all the identified QTLs, the marker intervals are so wide that there are ample chances of detecting QTLs for those 11 undetected traits if more markers are incorporated in the QTL mapping study. Of the 30 QTLs identified in the present investigation, 11 were reported by previous workers. They are four on chromosome 2, two each on chromosome 3 and chromosome 10, one each on chromosome 4, 7, and 11.

## 12.6 Breeding Salt-Tolerant Rice

Conventional plant breeding, though time-consuming and slow especially in stress-prone areas, owing to Genotype X Environment interaction, produced sizable number of salt-tolerant varieties. In this context, hybridization followed by pedigree/bulk/SSD selection methods and induced mutagenesis were the promising approaches employed. Now, with advances in marker technology and the availability of robust QTLs for salt tolerance, identified in many mapping populations, marker-assisted selection (MAS) can be practiced to develop salt-tolerant rice varieties quickly and efficiently. In this context, the attempts made at this center are detailed below.

### 12.6.1 Induced Mutagenesis for Salt Tolerance

A project entitled "Improving salt tolerance in rice through induced mutagenesis and recombination breeding" funded by BRNS, Mumbai, was secured to isolate salt-tolerant mutants of Pokkali and Improved White Ponni.

*Pokkali* is highly salt tolerant but low yielding, poorly tillering, tall, photoperiod-sensitive, susceptible to lodging, and possess red pericarp with poor cooking quality. *Improved White Ponni* is a locally well-adapted, high-yielding variety with good cooking qualities but highly sensitive to salinity. These varieties were subjected to gamma irradiation with objectives of identifying (1) Pokkali mutants retaining salt tolerance with improved traits like semidwarf, photo-insensitivity, and white pericarp and (2) Improved White Ponni mutants with salt tolerance, non-lodging nature, and early maturity but retaining desirable kernel traits and cooking qualities.

Seeds of Pokkali and Improved White Ponni were irradiated with gamma rays at doses of 200 Gy, 300 Gy, and 400 Gy, respectively. Generations were advanced by

**Fig. 12.2** (**a**) Promising mutants of Pokkali for pericarp colour and grain type. (**b**) Promising mutants of Improved White Ponni for maturity and plant

adopting panicle to row method, and among the surviving plants under salt stress, plants with desirable agronomic traits like plant height, early maturity, etc., were selected and forwarded to next generations from $M_2$ to $M_5$. It was observed that mutagenic treatments generated substantial magnitude of genetic variability for all characters and influenced the mean, range, and coefficient of variation independently in both the varieties (Fig. 12.2a, b).

Fourteen promising mutants ($M_6$), nine of Improved White Ponni and five of Pokkali, were found to be superior to the check variety TRY 2. These 14 promising mutants were screened in greenhouse (salinized at EC = 12 dS/m) for physiological parameters like tissue tolerance and Na-K ratio.

Based on these physiological parameters, eight mutants, six of Improved White Ponni and two of Pokkali, were found to be better than TRY2 (Thirumeni et al. 2017). Among the 14 promising mutants, four early maturing salt-tolerant mutants, fulfilling the objectives of the study, were selected and nominated for multilocation trials through AICRIP, and among them, KR 09003 and KR 09009 are likely to be released shortly (Table 12.4).

### 12.6.2 Marker-Assisted Backcross Breeding to Introgress *Saltol* QTL into ADT 45

Rice, though salt-sensitive, is the preferred crop owing to its ability to thrive in standing water that dilutes the harmful salts. A promising approach for improving salt tolerance is the identification of markers closely linked to it and subsequent utilization of it through indirect selection, popularly referred as Marker-Assisted Selection (MAS). A major QTL "*Saltol*" derived from Pokkali was identified, fine mapped, and characterized on chromosome 1 and is stated to be a major contributor (40%) to seedling stage tolerance.

**Fig. 12.2** (continued)

With the availability of large-effect QTL *Saltol* and set of linked markers, salt-tolerant version of mega rice varieties could be synthesized through targeted marker-assisted backcrossing (MABC). Keeping this in view, DBT-funded project on "From QTL to Variety: Marker-Assisted Breeding of Abiotic Stress-Tolerant Rice Varieties with Major QTLs for Drought, Submergence, and Salt Tolerance" was undertaken (Singh et al. 2016) to develop salt-tolerant version of widely grown (in Tamil Nadu and Puducherry) rice variety ADT 45 through MABC.

Hybridizations were made and three backcrosses were generated (Fig. 12.3a) between ADT 45 and FL478, which is a tolerant *indica* recombinant inbred line, derived from "IR29," and Pokkali was used as the donor of *Saltol* (Thirumeni et al. 2014). A set of SSR markers, viz., RM 3412, RM 8094, and AP3206 for foreground selection and flanking markers RM 1287 and RM 493 for recombinant selection, was used in the MABC of *SALTOL* QTL from FL478 into ADT 45. Fifty-five double recombinant homozygous lines (FL478 alleles at *SALTOL* locus flanked by ADT 45 alleles; Fig. 12.3b) were isolated in $BC_3F_2$ generation. These 55 NILS were screened for salt tolerance in hydroponics at EC of 12 dS/m in greenhouse (Karthick et al. 2019) and 20 highly tolerant NILs, with mean SES score less than 5.0 and

**Table 12.4** Salt-tolerant rice cultures nominated for AICRIP

| Sl. no | Culture/IET no. | Project in which it was developed | Pedigree | Duration (days) | Yield (kg/ha) | Grain type | Year nominated |
|---|---|---|---|---|---|---|---|
| 1 | KR 00015 (IET 21947) | PAJANCOA | SSRC 92076/TRY-1 | 105–110 | 3500 | MS white | 2010 |
| 2 | KR 09003 (IET 24538) | BRNS | IWP-200-59-2-1-1-1 (γ-irradiated mutant of improved white Ponni) | 115–120 | 3200 | MS white | 2014 |
| 3 | KR 09009 (IET 24542) | BRNS | IWP-200-263-3-1-1-1 (γ- irradiated mutant of improved white Ponni) | 115–120 | 3500 | MB white | 2014 |
| 4 | KR 09011 (IET 27063) | BRNS | P-200-124-4-1-1-1-1 (γ-irradiated mutant of Pokkali) | 105–110 | 3200 | SS white | 2017 |
| 5 | KR 15006 (IET 28037) | DBT | ADT 45*3/FL478 (NIL) | 105–110 | 2772 | LS white | 2018 |
| 6 | KR 15010 (IET28038) | DBT | ADT 45*3/FL478(NIL) | 105–110 | 2850 | MS white | 2018d |
| 7 | KR 15014 (IET 28039) | DBT | ADT 45*3/FL478(NIL) | 105–110 | 2607 | MS white | 2018 |
| 8 | KR 15016 (IET 28040) | DBT | ADT 45*3/FL478(NIL) | 105–110 | 2946 | MS white | 2018 |
| 9 | KR 15066 (IET 27807) | STRASA/DBT | ADT 45/ FL478 (224-4-3-1-1) | 115–120 | 3450 | LS white | 2018 (approved for release in 2023) |
| 10 | KR 15076 (IET 27817) | STRASA/DBT | ADT 45/ FL478 (542-6-1-1-1) | 110–115 | 3158 | MS white | 2018 |
| 11 | KR 15101 (IET 27822) | STRASA/DBT | ADT 45/ FL478 (869-4-19-1-1) | 110–115 | 3019 | LB white | 2018 |

**Fig. 12.3** (**a**) Generation of backcrosses for introgression of Saltol from FL 478 into ADT45. (**b**) Gel profile of 10 Introgression lines. (**c**) Phenotyping of promising NILs in hydroponics. (**d**) Field view of ADT 45 Saltol introgressions lines

seedling survival percentage above 70, were identified and the gel profile of 10 NILs is shown in Fig. 12.3c.

Background genome recovery of the NILs using 50 K SNP chip revealed that recovery ranged from 97.29 to 98.54%. Out of 20 NILs tested, nine were found highly tolerant that scored less than 3 and more than 80% survival. DUS characterization of these promising tolerant ILs along with parents revealed that parents differed only for 12 out of 27 essential descriptors. All these 12 descriptors of the introgression lines are similar to that of ADT 45 and field view of homogeneous lines are shown in Fig. 12.3d. Among them, four promising introgression lines, viz., KR 15014, KR 15004, KR 15010, and KR 15016, were nominated for AICRIP testing in 2018 under AL and ISTVT-NIL trial and was repeated in 2019 also. The results revealed that none of the entries, KR 15014 (IET 28037), KR 15004(IET 28038), KR 15010(IET 28039), and KR 15016 (IET 28040), performed better than the tolerant checks.

As *Saltol* QTL imparts salt tolerance at seedling stage only, tolerance at reproductive stage is important to reduce yield loss. It has been shown that most of *Saltol* introgressed lines in different genetic backgrounds (mega varieties), tested under AICRIP trials, failed to perform better than the tolerant checks owing to absence of major effect QTL for reproductive stage salt tolerance. Therefore, another *QTL SSISFH 8.1*, identified in CSR 27, is introgressed into ADT 45 *Saltol* lines to improve salt tolerance at reproductive stage (Bhandari et al. 2019) in the phase II of the DBT-funded QTL to variety project and the work is in progress.

### 12.6.3 Combined MAS for Improving Salt Tolerance

As part of STRASA project of IRRI, Marker-assisted pedigree selection was carried out (Satyadevi et al. 2014) using combined MAS approach for improving salt tolerance in rice. In this approach, the phenotyping was combined with MAS wherein two additional QTLs apart from *Saltol* were included to improve the level of salt tolerance. A total of 1212 $F_2$ population of *ADT45 x FL478* cross was

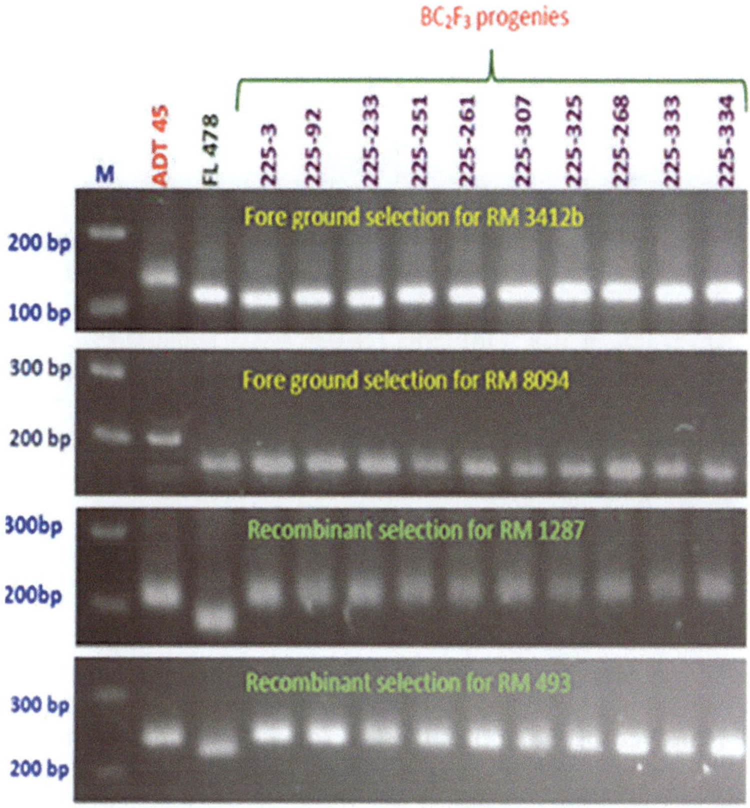


**Fig. 12.3** (continued)

evaluated for seedling stage tolerance in hydroponics culture at EC 12 $dSm^{-1}$ and found 185 tolerant plants. Based on single plant yield, 101 plants (out of 185) were selected and forwarded to $F_3$ as plant to progeny rows. These selected 101 families were further screened in $F_3$ generation in hydroponics system as in $F_2$ and 280 plants with scores 1 and 3 selected. Phenotypically tolerant plants were genotyped using five markers associated with three QTLs: two on chromosome 1 and one on chromosome 3. Another functional marker RID 12 was used to select white seeded rice. In this way, the generations were advanced to $F_5$ and identified 35 homozygous lines for three QTLs and one RID 12 locus for pericarp color. These 35 lines were further tested in replicated row trial along with TRY 2 as local check for two seasons and selected nine promising early maturing lines with medium slender white rice (Thirumeni et al. 2016). Of these, three were nominated to AICRIP for multilocation testing (Table 12.4) and the entry IET 28797 (KR 15066) is approved for release as KKL(R)3 for the states of Haryana, Puducherry and Tamil Nadu.

**Fig. 12.3** (continued)

### 12.6.4 Honeycomb Selection Design for Early Generation Selection

Yoglakshmi et al. (2022) conducted an experiment to know the applicability of honeycomb selection design for early generation selection for salt tolerance in $F_2$ generation of two crosses ADT45 x FL478 and ADT45 x Nona Bokra. A total of 988 plants in each cross consisting of 847 $F_2$ plants and 141 CSR10 were raised in honeycomb selection design at wider spacing/nil competition (100 x 90 cm). The data recorded on yield and its components on the entire population on single plant basis were subjected to HSD analysis (Fig. 12.4). In ADT 45 x Nona Bokra cross, all the characters except panicle weight had higher mean compared to ADT 45 x FL 478 cross. As the number of plants superior to CSR10 were less in ADT45 x FL478 cross, the cross was not forwarded to next generation. In ADT 45 x Nona Bokra cross, out of 110 plants that ranked best, 10 were CSR10 and 100 were $F_2$ plants. From grain character analysis, it was found that 30 plants were short bold, 32 were long bold, 32 were medium slender, 3 were long slender, 2 were basmati type, and 1 was short slender and 77 were red and 23 were white for the pericarp color out of 100 $F_2$ plants of ADT 45 x Nona Bokra cross.

A total of 35 plants were selected based on varietal requirements (early medium maturing, semidwarf with desirable grain qualities) suitable for saline environments in the *Kuruvai* (June to July sowing) season, accounting for 4.13% selection intensity leading to rapid fixation of additive alleles that are indispensable for genetic advance through selection. The 35 selected plants were forwarded to next generation

**Fig. 12.3** (continued)

($F_3$) and they are expected to perform better as a result of enhanced phenotypic expression leading to selection of plants with homozygote advantage in $F_2$ generation. ADT 45 x Nona Bokra cross had higher percentage increase of $F_2$ best plants over CSR10 (108.07%) and total mean of $F_2$ (165.71%) compared to ADT 45 x FL 478 cross. The 35 selected plants in ADT 45 x Nona Bokra cross showed a percentage increase of 213.86% over total mean population. Therefore, early generation selection for yield and other agronomic traits improves breeding efficiency by reducing the number of genotypes to be tested in subsequent expensive yield trials, thereby increasing genetic gain per unit cost. The selection is in advanced stage at $F_5$ and expected to identify promising salt-tolerant, high and stable yielding lines in the years to come.

## 12.7 Linkage with IRRI

Since the initiation of the salinity tolerance breeding in rice, PAJANCOA and RI was a partner of IRRI in the Salinity Breeding Network from 1997 and conducting breeding trials specifically International Rice Soil Stress Tolerance Nursery (IRSSTN) and International Rice Stress (FLOOD) Tolerance Nursery (IRSTN-FP). Subsequently, a new initiative in the name of "Stress Tolerance Rice for Poor Farmers of Africa and South Asia" (STRASA) funded by Bill and Melinda Gates Foundation (BMGF) came into being. This project was carried out in three phases from 2008 to 2019. In this program, Salinity Breeding Network got fresh impetus with participation of many rice breeding centers for breeding stress-tolerant rice varieties under the leadership of IRRI. This provided a very good platform for the participating institute like PAJANCOA and RI to gain knowledge in the frontier

**Fig. 12.4** Honey comb breeding practiced in early generation of rice

areas of rice research besides testing the salt-tolerant lines developed in IRRI and other centers. Advanced salt-tolerant breeding lines of rice were assembled from different research stations and evaluated across saline and alkaline locations in India. The experiments were carried out in different salt stress environments across India with main objective to identify the best tolerant and high-yielding genotypes. The details of the STBN trials conducted are given in the Table 12.5. Over these years, stability of various breeding lines for grain yield under salinity/sodicity and saline-sodic environments was assessed and promising ones were identified for specific environments in phase I (Krishnamurthy et al. 2015), phase II (Krishnamurthy et al. 2017), and phase III (Krishnamurthy et al. 2021).

As part of STRASA project, Participatory Varietal Selection (PVS) was practiced to identify best salt-tolerant genotypes (Rajalakshmi 2017) out of 37 entries. This included 27 breeding lines (15 lines generated in two research projects (BRNS and DBT funded) and 10 check varieties. The experiment was conducted in the institute farm under salt stress condition, managed by researcher. A set of farmers (men and women) and PG scholars inspected separately the trial crop two times (active tillering and flowering stages) and also participated in cooking quality analysis.

Eight selected entries were found to be common between men and women, while six each between men and research scholars and women and research scholars (Fig. 12.5). Seventeen entries were preferred by all three groups, and among them,

**Table 12.5** Details of STBN trials evaluated for salt tolerance

| Phase | Year | Entries tested with checks | No. of entries better than TRY 2 |
|---|---|---|---|
| First | 2008–2009 | 34 + 4 | 6 |
| | 2009–2010 | 25 + 1 | 4 |
| | 2010–2011 | 51 + 8 | 2 |
| Second | 2011–2012 | 21 + 2 | 11 |
| | 2012–2013 | 27 + 4 | 2 |
| | 2013–2014 | 28 + 4 | 11 |
| Third | 2014–2015 | 26 + 4 | 3 |
| | 2015–2016 | 25 + 5 | 4 |
| | 2016–2017 | 60 + 6 | 21 |
| | 2017–2018 | 81 + 7 | 33 |
| | 2018–2019 | 85 + 6 | 17 |

five entries, viz., KR15077, KR09003 (mutant of IWP), KR15075, KR09011 (mutant of Pokkali), and CSR27, were found to be common.

Preferential analysis (PA) revealed characters, viz., plant height, maturity, yield, lodging resistance, no. of productive tillers, pest and disease resistance, salt tolerance, were given weightage by all the three groups. Sensory evaluation resulted in the identification of seven entries, viz., KR15075, KR09003, CSR10, KR09009, TRY2, KR15092, and CST 7-1 and preferred by all three groups. Characters preferred were taste, linear elongation, consistency, cohesiveness, aroma, non-stickiness of grains, and tenderness. Spearman's rank correlation coefficient was used to find out the correlation between the farmers and scientist's selection. Positive and moderate significant correlation of 0.518 was observed between farmers and scientist approaches indicating moderate association between farmers and scientist approach.

## 12.8 Outlook

- Since soil heterogeneity gets aggravated in salt stress environments (G x E interaction), practice honeycomb selection to improve precision in conventional breeding for development of salt-tolerant rice varieties.
- Breeding rice varieties possessing higher salt tolerance, disease resistance, and higher yield potential.
- Mapping of salt tolerance genes for seedling and reproductive stage using novel genetic approaches and introgressing and pyramiding them through molecular marker-assisted breeding.
- Development of dual-tolerant (salt and submergence), early maturing rice varieties through marker-assisted breeding.

**Fig. 12.5** Farmers' participatory varietal selection of salt tolerant rice

## 12.9 Conclusion

As much as 11 salt-tolerant rice strains were developed using conventional and molecular breeding methods with release of first Salt Tolerant variety: KKL(R) 3, few are expected to be released in the coming years. With the available salt-tolerant germplasm and genetic resources, novel QTLs for salt tolerance will be identified employing newer molecular methods. To develop improved salt-tolerant rice varieties, two approaches are being focused in PAJANCOA: One is to develop salt-tolerant rice varieties through honeycomb selection by improving precision in selection in the conventional breeding method. The other approach is to develop MAS-derived line quickly by pyramiding of salt-tolerant QTLs through marker-assisted backcrossing.

## References

Bandyopadhyay AK (1999) Coastal areas and their management. In: 50 years of natural resource management research, Singh GB, Sharma BR (Eds), Division of Natural Resources Manangement, ICAR, New Delhi, pp 635–644

Bhandari A, Jayaswal P, Yadav N, Singh R, Singh Y, Singh B, Singh N, Singh S, Sevanthi A, Rai V, Verulkar S, Rao PVR, Rani MG, Anuradha T, Satyanarayana PV, Krishnamurthy SL, Sharma P, Singh D, Singh PK, Nilanjay Kumar R, Chetia S, Ahmad T, Rai M, Katara J, Marandi B, Swain P, Sarkar RK, Singh DP, Reddy JN, Mandal N, Paramsivam K, Nadarajan S, Thirumeni S, Badri J, Padmavathi G, Ram T, Singh N (2019) Genomics-assisted backcross breeding for infusing climate resilience in high-yielding green revolution varieties of rice. Indian J Genet 79(1):160–170

Diwakar R (2011) Construction of core collection in rice for salt tolerance in rice. Unpublished M. Sc., (Ag.) thesis. Tamil Nadu Agricultural University, Coimbatore

Gregorio GB, Islam MR, Vergara GV, Thirumeni S (2013) Recent advances in rice science to design salinity and other abiotic stress tolerant rice varieties. SABRAO J Breed Genet 45(1): 31–41

Griffing B (1956) Ageneralized treatment of the use of diallel crosses in quantitative inheritance. Heredity 10:31–50

Gupta RK, Singh NT, Sethi M (1994) Ground water quality for irrigation in India. CSSRI Publication Technical Bulletin No.19, pp. 1–13

Hayman BI (1954) The theory and analysis of diallel crosses. Genetics 39:789–809

Karthick J, Thirumeni S, Paramasivam K, Nadarajan S (2019) Phenotyping of *Saltol* introgression lines in rice (cv. ADT 45) for salt tolerance at germination stage. In: *Proceedings of UGC-SAP National Seminar on climate resilient agriculture for abiotic stress*, held at Faculty of Agriculture, Annamalai university, Annamalainagar 608 002, between March 12th and 13th, p 157

Krishnamurthy SL, Sharma SK, Sharma DK, Sharma PC, Singh YP, Mishra VK, Burman D, Maji B, Bandyopadhyay BK, Mandal S, Sarangi SK, Gautam RK, Singh PK, Manohara KK, Marandi BC, Singh DP, Padmavathi G, Vanve PB, Patil KD, Thirumeni S, Verma OP, Khan AH, Tiwari S, Shakila M, Ismail AM, Gregorio GB, Singh RK (2015) Analysis of stability and G × E interaction of Rice genotypes across saline and alkaline environments in India. Cereal Res Commun 44(2):349–360

Krishnamurthy SL, Sharma PC, Sharma DK, Ravikiran KT, Singh YP, Mishra VK, Burman D, Maji B, Mandal S, Sarangi SK, Gautam RK, Singh PK, Manohara KK, Marandi BC, Padmavathi G, Vanve PB, Patil KD, Thirumeni S, Verma OP, Khan AH, Tiwari S, Geetha S, Shakila M, Gill R, Yadav VK, Roy SKB, Prakash M, Bonifacio J, Ismail A, Gregorio GB, Singh RK (2017) Identification of mega-environments and rice genotypes for general and specific adaptation to saline and alkaline stresses in India. Sci Rep 7(1):1–14

Krishnamurthy SL, Sharma PC, Sharma DK, Ravikiran KT, Singh YP, Mishra VK, Burman D, Maji B, Mandal S, Sarangi SK, Gautam RK, Singh PK, Manohara KK, Marandi BC, Chattopadhyay K, Padmavathi G, Vanve PB, Patil KD, Thirumeni S, Verma OP, Khan AH, Tiwari S, Geetha S, Shakila M, Gill R, Yadav VK, Roy SKB, Prakash M, Ananadan A, Bonifacio J, Ismail A, Singh RK (2021) Additive main effects and multiplicative interaction analyses of yield performance in rice genotypes for general and specific adaptation to salt stress in locations in India. Euphytica 217(2):1–15

Mathyvanan A (2019) Mapping reproductive stage salt tolerance in rice. Unpublished M.Sc., (Ag.) thesis. Tamil Nadu Agricultural University, Coimbatore

Rajalakshmi JM (2017) Farmers participatory varietal selection. Unpublished M.Sc., (Ag.) thesis. Tamil Nadu Agricultural University, Coimbatore

Reddy PS, Thirumeni S, Paramasivam K, Narayanan AL (2014) Mapping salt tolerance QTL in rice (*Oryza sativa L.*) by selective genotyping. In: Proceedings of National Seminar on challenges and innovative approaches in crop improvement, organized by Indian society of plant breeders, TNAU, Coimbatore, held at agricultural college & research institute, Madurai, between December 16th and 17th, p 199

Satyadevi G, Thirumeni S, Paramasivam K, Rettinassababady C (2014) Marker assisted pedigree selection in rice (*Oryza sativa* L.) for salt tolerance. In: Proceedings of National Seminar on challenges and innovative approaches in crop improvement, organized by Indian society of plant breeders, TNAU, Coimbatore, held at agricultural college & research institute, Madurai, between December 16th and 17th, p 203

Singh R, Singh Y, Xalaxo S, Verulkar S, Yadav N, Singh S, Singh N, Prasad KSN, Kondayya K, Ramana Rao PV, Girija Rani M, Anuradha T, Suraynarayana Y, Sharma PC, Krishnamurthy SL, Sharma SK, Dwivedi JL, Singh AK, Singh PK, Nilanjay N, Singh K, Kumar R, Chetia SK, Ahmad T, Rai M, Perraju P, Pande A, Singh DN, Mandal NP, Reddy JN, Singh ON, Katara JL, Marandi B, Swain P, Sarkar RK, Singh DP, Mohapatra T, Padmawathi G, Ram T, Kathiresan RM, Paramsivam K, Nadarajan S, Thirumeni S, Nagarajan M, Singh AK, Vikram P, Kumar A, Septiningshih U, Singh S, Ismail AM, Mackill D, Singh NK (2016) From QTL to variety-harnessing the benefits of QTLs for drought, flood and salt tolerance in mega rice varieties of India through a multi-institutional network. Plant Sci 242:278–287

Sukanya KA (2015) QTL mapping for salt tolerance in rice at seedling stage. Unpublished M.Sc., (Ag.) thesis. Tamil Nadu Agricultural University, Coimbatore

Thirunavukkarasu E, Thirumeni S, Paramasivam K, Rettinasababady C, Karthick J (2015) Genetics of salt tolerance in rice. In: Proceedings of "International Rice Symposium on Rice Science for global food and nutritional security", held at IIRR, Hyderabad, between November 17th and 19th

Thirumeni S, Subramanian M, Paramasivam K (2001) Combining ability of rice genotypes under coastal salinity. Oryza 38(1&2):72–75

Thirumeni S, Subramanian M, Paramasivam K (2003) Genetics of salt tolerance in rice (*Oryza sativa* L.). Indian. J Genet 68(1):73–74

Thirumeni S, Paramasivam K, Natarajan S, Karthick J, Singh NK (2014) Marker assisted introgression of *Saltol* QTL into rice mega variety ADT 45. In: Proceedings of 4th international rice congress organized by IRRI, Philippines, held at BITEC, Bangkok, Thailand, between October 27th and 1st November

Thirumeni S, Seetharam K, Paramasivam K, Souframanien J (2017) Induced mutagenesis in rice (*Oryza sativa* L.) for improving salt tolerance. Oryza 53(4):385–390

Thirumeni S, Satyadevi G, Paramasivam K, Karthick J, Platten JD, Singh RK, Singh US (2016) Efficiency of molecular marker tags in improving salt tolerance in rice using forward breeding MAS approach. In: "Proceedings of 20th EUCARPIA Plant breeding Congress" organized by EUCARPIA, Agroscope and ETH held at Zurich, Switzerland between August 28th and September 1st

Vinodha C (2016) Targeted association mapping for salt tolerance in rice. Unpublished M.Sc., (Ag.) thesis. Tamil Nadu Agricultural University, Coimbatore

Yoglakshmi C, Pearl RI, Fasoula VA, Karthick J, Thirumeni S (2022) Efficiency of honeycomb selection design in early segregating generation in ADT45× Nona Bokra cross under salt stress. IJEB 60(07):463–470

# Mechanisms of Saline and Submergence Tolerance in Rice for Coastal Ecology

13

Nitish R. Prakash, B. M. Lokeshkumar, Suman Rathor, A. S. Warriach, Satyendra Yadav, N. M. Vinaykumar, S. L. Krishnamurthy, and Parbodh C. Sharma

**Abstract**

Rice is a prime crop of coastal region and is affected by several stresses such as salinity and submergence during crop growth stages. The molecular mechanism of rice, having tolerance to salinity, includes salt balancing in a cell, insensitive enzymes for better physiological functions, and compartmentalization of toxic molecules. Similarly, for submergence tolerance, physiological mechanisms such as managing oxygen level within a cell, detoxifying the reactive oxygen species produced, and saving energy and metabolites are important. The present chapter is dealing with the molecular, genetic, and physiological mechanism for salt response at various growth stages of rice along with response at submergence condition.

**Keywords**

Salinity tolerance · Submergence tolerance · Coastal rice · Molecular mechanism

N. R. Prakash · P. C. Sharma (✉)
ICAR-Central Soil Salinity Research Institute, RRS, Canning Town, India

B. M. Lokeshkumar · S. Rathor · A. S. Warriach · S. Yadav · S. L. Krishnamurthy (✉)
ICAR-Central Soil Salinity Research Institute, Karnal, India

N. M. Vinaykumar
Kuvempu University, Shankaraghatta, Shimogga, India





R. K. Singh et al. (eds.), *Genetic Improvement of Rice for Salt Tolerance*,
https://doi.org/10.1007/978-981-99-3830-8_13

## 13.1 Introduction

Rice is a major cereal of almost half of the human mouth residing on this planet and thus have very special significance for securing livelihood, nutritional security, and financial needs of millions of farmers from middle- and low-income countries (Fukagawa and Ziska 2019). It is a semiaquatic annual cereal crop belonging to grass family (Family: *Poaceae*), grown from temperate to tropical climate (50° N–35° S) and from an elevation below sea level to 2000 MSL (Gnanamanickam 2009). It assumes special significance in driving Asian economies, as it is staple food for people residing here, which accounts for >90% of the total global production of rice. The importance of rice can be seen from the fact that it was the basis of several ancient civilization (Indian, Chinese, and Japanese) in Asia and Africa (Liu et al. 2022). Rice is being grown in India since Vedic era, as several ancient inscription, literature, and cultural rituals mention its diverse usage as food and offering to gods (Spengler et al. 2021). In India, whole rice grain (dehusked paddy) is being eaten as boiled, puffed (*Murhi* in Hindi), and flattened (*Chiura* or *Poha* in Hindi). The rice flour is being used to prepare *chapati, pittha, idli, dosa, uttapam, appam, kheer,* and sweets in India (Dar et al. 2017). It is a rich source of energy and micronutrients such as vitamins (niacin) and minerals (zinc, iron, and calcium), supplementing the dietary needs of billions (Priya et al. 2019).

Soil salinity is one of the major problems affecting rice cultivation in low-lying coastal regions along with stagnating water during rainy months (Pitman and Läuchli 2002). Almost three-quarters of land on earth is submerged with saline water (ocean), and thus coastal regions (around 50–100 km inside into mainland) are generally saline, resulting from seawater intrusion, groundwater seepage, saline mangroves, heavy rainfall, incessant rain, cyclonic disturbances, etc. (Sabareshwari and Ramya 2018). These impacts of salinity are generally enhanced during dry seasons (rabi, December to May) and thus making these lands largely unsuitable for cultivation of rice (Kumar and Sharma 2020). Coastal regions of the world (most notably tropical regions) are receiving more rainfall and thus face dual challenge of excessive water stresses such as deepwater stagnation, flash flood, submergence, or land degradation along with salinity during rainy season (Pitman and Läuchli 2002). Climate change and anthropogenic interventions have already impacted the cultivation in these areas with increasing sea level, aberrant and erratic rainfall, seawater intrusion, land degradation, and ecosystem disturbances (Prakash et al. 2020). According to an estimate and a study conducted by Rabbani et al. (2013) in Bangladesh, part of Sundarbans delta (a coastal saline region), the loss in rice cultivation because of seawater intrusion and salinity due to *Aila* cyclone (2009) was equivalent to USD 1.9 million in 3 years (after cyclone) for a particular village under study. Therefore, because of concurrence of frequent cyclones, climatic abnormalities, and poor infrastructure, it is very much important to understand and devise technologies for sustained livelihood.

Designing rice crops with sustained cultivation in saline areas along with tolerance to submergence is very much important in coastal regions. Rice is the most important crop being grown in coastal regions during rainy season, whereas other crops are grown endemically in certain pockets of coastal regions of the world (Dasgupta et al. 2014). High salinity is generally impacting rice via augmenting osmotic stress resulting from accumulated salts in cells and intracellular spaces, downregulating the rate of metabolism, toxic effects of ions, and desisting nutrients uptake (Verslues et al. 2006). Germplasm characterization in rice was done to identify several tolerant donor landraces such as *Pokkali, Kasalath, Sadri, Tampha, Nona Bokra, Bengal, Jiucaiqing, Kolajoha,* etc. (Rahman et al. 2016). These landraces were used to understand the genetics of salinity tolerance in rice and unraveling the molecular players (Ganie et al. 2019). Therefore, in this chapter, we have described the current understanding of genetic, molecular, and physiological basis of salinity and submergence response in rice. These current understanding will help in devising better strategy for development of crop cultivars tackling these stresses.

## 13.2 Problems in Rice Cultivation under Coastal Ecosystems

Coastal ecosystem for crop growth and maintenance is generally facing challenge of increased temperature, fluctuating rainfall patterns, seawater inundation, seawater intrusion, salinity, as well as submergence in rainy season and very high salinity in dry months (Gopalakrishnan et al. 2019). Two major problems that are altogether directly impacting crop sustainability and not easily manageable are salinity and submergence.

### 13.2.1 Salinity

Regions with salinity in coastal areas are varying across the world and are depending upon the proximity to sea, rainfall pattern, land use pattern, soil types, rivers, and other drainage systems. According to an estimate in Bangladesh, almost 53% of regions designated as coastal are saline (Rasel et al. 2013). The salinity in these regions affect rice cultivation in kharif (rainy) as well as rabi (winter) season (Xu et al. 2020). Excessive rain during rainy season is diluting the impact of salinity in standing crop. As groundwater at shallow depth is saline in these regions, the capillary upward movement of saline water builds saline upper soil layer during dry month (Schneider and Asch 2020). In dry months, tropical coastal regions have seawater intrusion into groundwater resulting into very high salinity. This in turn impact cultivation of rice during dry month as well as salinity stress during reproductive crop stage in rainy season.

Rice is generally sensitive to salinity but stress at seedling and reproductive stages are detrimental to crop growth and yield (Ganie et al. 2019). High salinity results in reduced uptake of water (due to osmotic stress), higher salt accumulation in cell, deposition of secondary metabolites in cellular and intracellular spaces regulating osmotic adjustment, reduced metabolism resulting from damaged enzymatic activities, desisting nutrient uptake, and ionic toxicities (Verslues et al. 2006). The variations in level of salinity across the location, across different years, in different seasons of the years along with other stresses challenge the predictability of problems associated with rice cultivation. The level of salinity in coastal regions is varying within a small area with proximity to creeks and rivers, water bodies, inherent parental soil deposition (over years), rainfall pattern, and other climatic factors. These affects the possibility of critical evaluation of rice varieties, germplasms, and other cultivation practices in coastal regions (Sarangi and Islam 2019). Salinity impacts crop stand and establishment, development of reproductive organs (panicle and spikelet), number of tillers, photosynthetic efficiency (chlorophyll content), grain size, and floral initiation in rice (Dasgupta et al. 2014).

### 13.2.2 Submergence

Rice is adapted to semiaquatic condition and require continuous flooding (puddled condition) in field condition, but it cannot survive generally in prolonged complete submergence (Septiningsih et al. 2009). Very high rainfall in coastal regions is responsible for creating flooded condition in fields as well as water bodies, along with low infiltration of water into soil during rainy months making this condition more worsened. The early seedling and germinated seeds of rice cannot tolerate anaerobic condition developed in field because of heavy downpour after onset of monsoon and thus affect seedling growth and germination and ultimately to crop loss (Sevanthi et al. 2019). Including coastal areas, in India alone, flooding and submergence affect 6.2 million hectares of rice-cultivated area. The flood-prone areas are generally low-lying areas of river basins or coastal regions (Dar et al. 2017). Rice is affected by water stress in different mode, i.e., submergence during germination, flash flooding, deepwater condition, and stagnant flooding situation. If lands are properly levelled or if there is continuous rain during seed germination in nursery/seed bed, a hypoxia situation develops that hampers seed germination, establishment, growth of radicle and plumules, etc. (Magneschi and Perata 2009). Complete submergence is detrimental to rice and a flash flood for 1–2 weeks can have a devastating impact on seedling survival. The rapid increase in water level creates a hypoxic condition in which plant lose all its energy to survive (bring in more oxygen) and at last die (Bin Rahman and Zhang 2016). In stagnant flooding, water level may go beyond 100 cm and water stays under stagnant condition for several weeks. This happens because of continuous rain and high water saturation in soil (Nagai et al. 2010). Sometimes, in certain low-lying areas, water level may go beyond 1 m and up to 4 m, under such condition floating rice is cultivated. Such condition occurs where continuous rainfall is coinciding with crop growth (Hattori et al. 2011).

Rice is a water-loving cereal crop and requires warm and humid climate for cultivation. It can tolerate waterlogging and/or submergence up to a certain extent because of aerenchyma tissue, which helps in respiration of underground plant parts and plant part below water (Sevanthi et al. 2019). Rice is tolerant to submergence at start of germination, but the growth of radicle and plumules is hampered because of absence of well-developed aerenchyma tissues (Ismail et al. 2009). Submergence may bring several biochemical, morphological, and physiological changes in rice resulting from anaerobic condition faced by several plant parts submerged in water (forcing plant to go for anaerobic respiration), pressure on photosynthetic machinery (as most of the plant parts may be submerged), toxic metabolites produced in cells (as part of anaerobic respiration), imbalanced ionic exchanges by roots and absorption of nutrients from soil, etc. (Nagai et al. 2010).

## 13.3 Germplasm Evaluation and Screening

Morphological traits are mostly controlled by several components each contributing differently in positive or negative manner (Ali et al. 2014). Salinity and submergence affect rice plants through several morphological and physiological phenomenon that have been already briefed in earlier sections.

### 13.3.1 Germplasm Resources

As rice is grown in diverse agroecological conditions, the enormous amount of variability exists for salinity and submergence tolerance in rice (Barrera et al. 2019; Rahman et al. 2016; Ali et al. 2014; Gregorio et al. 2002). In coastal regions, rice has been grown since the millennia, and hence different landraces and wild cultivars can still be found having salinity- and submergence-tolerant mechanisms (Hairmansis et al. 2017; Rasel et al. 2021). In India, landraces such as Pokkali, Nona Bokra, Cheriviruppu, Damodar, Getu, Dasal, etc. have been identified and used in breeding programs to develop salt-tolerant, high-yielding rice varieties (Manohara et al. 2021). Pani et al. (2013) had evaluated several landraces found in Sundarban region of India and reported Talmugur, Odasal, Getu, Marisal, Darsal, Dadsal, and Kalonunia as salt tolerant. Similarly, several salt-tolerant landraces in Bangladesh were identified such as Sona Toly, Nakraji, and Komol Bhog (Tahjib-Ul-Arif et al. 2018) and Ghunsi, Hogla, Holdegotal, Vusieri, and Kanchon (Rasel et al. 2021). In India, the southern and western coasts harbor many salt-tolerant landraces such as Anakodan, Ayyampillypokkali, Cherayipokkali, Cheriya Orpandy, Chootupokkali, Elamkulampokkali, Kadamakudipokkali, Karunagapallipokkali, Khuzhippallypokkali, Kozhippillipokali, Kulapandi, Kuthirunellu, Kuzhippulipokkali, Nedungodupokkali, Odachan, Oorpandy, Orkyma, Orumundakan (black), Pallipurampokkali, Vadanakkudipokkali, Vellapokkali, Vettakkalchettivirippu, etc. in Kerala (Latha et al. 2013); Kuzhiadichan, Poonkar, Kallundai, Sornamugi, etc. in Tamil Nadu (Mohanavel et al. 2021); Korgut, Shidde, Kagga, etc. in Goa (Manohara et al. 2021); Bilithopu vadlu, Kasanella, etc. in Karnataka (Bhambure and Kerkar 2016); and Manjarvel, Harkhel, Ratal, Malkudai, Vailechi, Morchuka, Kilanz, Bhadas, Kalarata, Bhurarata, etc. in Maharashtra (Bhambure and Kerkar 2016) have been identified and are locally adapted. Such germplasm can be very effectively used in identifying suitable novel genomic regions governing seedling and salinity tolerance in rice. Similarly, in other countries such as Indonesia, Pakistan, Iran, Philippines, the USA, China, and Japan, different salinity-tolerant landraces are also identified. Among them, Sadri, Moroberekan (Iran), Fakhr-e-Malakand and Pakhal (Pakistan), Siputih, Lahatan Jambu, and Sirendah (Indonesia) are notable ones (Hairmansis et al. 2017; Sakina et al. 2016; Sabouri et al. 2008). Several salt-tolerant wild relatives of rice have been found in *Oryza nivara*, *O. australiensis*, *O. coarctata,* etc. (Nguyen et al. 2021).

Waterlogging is one of the major abiotic stresses in low-lying, rice-growing areas, having very high rainfall, and in regions having soils with poor drainage (Sarkar et al. 2006). "FR13A", a flash flood-tolerant rice landrace (collected from Odisha, India), is the major source of submergence tolerance breeding (using SUB1 QTLs) in the world (Chakraborty et al. 2021). Several deepwater rice landraces such as Kalanamak, Dal, Chakho, Kahi Jool, Kakua, Buruli, Surabhi, Rashpanjor, etc. have been identified in India that are of immense importance in breeding deepwater (standing water of 2 m depth) rice (Kalita and Handique 2021; Chattopadhyay et al. 2021). Germination and seedling establishments in rice are highly affected by water stagnation and are major problems in developing direct seeded rice (Javed et al. 2021). Several landraces such as Nanhi, Khao Hlan On, ASD1, Kharsu 80A, Ma-Zhan Red, Kalojoma, DSL-78-8, Damsi, Putidepa, etc. have been identified to have tolerance to anaerobic condition during germination (Darko Asante et al. 2021; Tnani et al. 2021).

### 13.3.2 Screening for Seedling Stage Salinity Tolerance

Accurate phenotyping imparting full expression of component traits is a prerequisite of any germplasm evaluation and breeding-related activity (Prakash et al. 2020). Seedling stage salinity tolerance can be evaluated based on visual scoring, leaf photosynthetic efficiency, ionic contents, and physiological growth or leaf, root, and shoot (Prakash et al. 2022). Seedlings can be grown under hydroponics with nutrients supplemented and salinity built using addition of required amount of salt (NaCl). The detailed account of nutrient supplement in a standardized hydroponic evaluation tray of 12 liter capacity is given in IRRI manual (Gregoria et al. 1997). Seeds must be used after breaking dormancy by giving heat treatment, and salinity treatment may be given after 12–14 days after sowing (Prakash et al. 2022). Nutrient solution must be changed every alternate days, and the aeration in water must be established to keep plant screening as near to natural. The standard evaluation scoring (SES) can be done on a scale of 1–9 as per Table 13.1 as given by Gregoria et al. (1997). Generally, traits such as leaf chlorophyll content, root length, shoot length, and root and shoot sodium and potassium concentration are measured on control and treated seedlings, which are very effective in differentiating genotypes for salt tolerance at seedling stage.

**Table 13.1** Scoring of standard evaluation score (SES) for seedling stage salinity tolerance in rice (adapted from Gregoria et al. 1997)

| SES score | Observation | Tolerance |
|---|---|---|
| 1 | Normal growth, no leaf symptoms | Highly tolerant |
| 3 | Nearly normal growth, but leaf tips or few leaf whitish and rolled | Tolerant |
| 5 | Growth severely retarded, most leaves rolled, only a few are elongating | Moderately tolerant |
| 7 | Complete cessation of growth, most leaves dry, some plants drying | Susceptible |
| 9 | Almost all plants dead or dying | Highly susceptible |

### 13.3.3 Screening for Reproductive Stage Salinity Tolerance in Rice

Rice is very sensitive to salinity at reproductive stage, whereas vegetative stage seems to be little affected (Ali et al. 2014). In general, field evaluation for traits is not effective because of the heterogeneous nature of environmental influence. Therefore, saline microplot facilities under rain out shelter can be utilized for evaluation under artificial salinization by maintaining soil salinity of EC 8–10 $dSM^{-1}$ (Prakash et al. 2022). Twenty-five days old seedlings grown under normal condition can be transferred to saline microplot, and salinized irrigation can be provided at reproductive stage (Ahmadizadeh et al. 2016; Pundir et al. 2021). The screening protocols under vegetative and reproductive stage can also be done under hybrid system (soil and hydroponics) as per protocol from IRRI (Gregoria et al. 1997). Several traits such as spikelet fertility, yield, biomass, number of tillers, panicle weight, panicle length, leaf sodium and potassium concentration, etc. can be evaluated (Pundir et al. 2021). Sometimes, it is difficult to have synchronous development of genotypes under evaluation, and therefore, plants may show a complex behavior in phenotyping for reproductive stage salinity tolerance. This can be minimized by grouping genotypes based on their duration, days to 50% flowering and plant height (Ahmadizadeh et al. 2016). Therefore, screening for reproductive stage salinity tolerance can be performed with utmost care by keeping in mind the requirement of individual location.

### 13.3.4 Screening for Submergence Tolerance

Screening of rice genotypes during germination stage can be done by immersing the seeds of genotypes under study in water for a definite period of time. This can be done in a plastic tray or a test tube, where immersed seeds are kept undisturbed for 15 days, and later growth parameters such as germination percentage, seed vigor, mortality, coleoptile elongation, seedling length, length of plant above water, etc. can be measured (Umarani et al. 2017). The water depth can vary from 10 cm in initial screening to 50 cm in final screening (Ghosal et al. 2020). Hsu and Tung (2015) had evaluated rice genotypes for coleoptile length under submerged condition and un-submerged condition (control) in a germination chamber under artificial light and dark condition.

Flash flood generally occurs because of more-than-average rainfall within a short span of time and water recedes from these fields generally within a fortnight (Reddy et al. 2015). Therefore, screening of genotypes can be done in a cemented tank with rice seedling transplanted into it. Seedling submergence up to 1.5 meter can be done up to a fortnight after establishment of transplanted seedlings (generally after a week). The desubmergence can be done after a fortnight, and the regrowth of seedling as well as other phenotypic observation is noted and compared (Pradhan et al. 2019). This kind of flooding tolerance depends on a mechanism of senescence to limit the physiological growth by plant up to a bare minimum and regrowth after desubmergence (Reddy et al. 2015).

Deepwater rice are being cultivated in areas having continuous water stagnation for a maximum period of cropping, i.e., generally 2–3 months. Such condition arises in low-lying areas of rivers, areas having seawater inundation and heavy rainfall, and areas with shallow water table (Biswajit et al. 2017). Evaluation for deepwater condition can be done in a tank with 2.5 m depth so that a very deep standing water can be maintained for prolonged condition. Water level in the tank can be increased gradually, and several photosynthetic as well as yield-related traits can be observed (Anandan et al. 2015). Traits such as sheath length, plant height, leaf blade length, photosynthetic efficiency, grain filling rate, etc. are to be targeted for selection.

Some rice genotypes are able to withstand low-lying condition of coastal region having standing water of up to 50 cm for a prolonged period. Such conditions are designated as semi-deep condition, and rice genotypes having tall stature, erect and strong culm, higher biomass, and longer duration are generally suitable for these areas (Reddy et al. 2009). Evaluation of these genotypes can be done by growing them in a field with stagnating water of 50 cm depth and recording data related to grain yield, photosynthetic ability, and plant height. Lodging tolerance is one of the most important requirements for such condition (Chattopadhyay et al. 2018).

## 13.4 Mechanism for Salinity Tolerance in Rice

Salinity of 30 mmol $L^{-1}$ NaCl equivalent to 3 dS $m^{-1}$ is detrimental to rice crop, and in coastal regions, mild salinity of (4–6 dS $m^{-1}$) in rainy season and very high salinity (>10 dS $m^{-1}$) during dry months of rabi season affect it (Lutts et al. 1995). The reduction in yield is attributed by complex response of plants to salinity, inherent capacity of osmotic balancing, mechanism to tolerate ionic imbalances, managing oxidative stress, and thus ultimate tolerance to salinity (Pareek et al. 2020). Salinity is affecting several growth and developmental and physiological phenomenon in crop plants that have been studied in genetic, molecular, and physiological manner (Chen et al. 2021a). Different growth and developmental stages may show differential response of tolerance to salinity of different level, attributed by individual capacity of crop and crop genotype (Prakash et al. 2022). Plants may show few mechanisms that help it to sustain it into growing soil salinity called tolerance (a case with glycophytes) or with mechanism that helps it to perform better under saline condition called resistance (a case with halophytes) (Munns and Tester 2008). Plants generally have special molecules that sense the oxidative stress or salt accumulation and then create a cascade of responses resulting into changes in gene expression of specific salt-responsive genes and modified physiology to cope up with excessive salts in cell and to maintain growth and development.

### 13.4.1 Molecular and Genetic Mechanism

Plants generally sense the salt in the environment through sensing molecule in cell membrane, which initiates a series of signaling pathways ultimately leading to change in expression of salt-responsive genes and synthesis of salt-responsive molecules and in turn change the cell behavior (Hernández 2019).

#### 13.4.1.1 Ionic Sensing of Salt Stress

Uptake of salts (NaCl) results in their increased concentration in cytosol that in turn enhances the $Ca^{+2}$ level that leads to activation of calcineurin B-like (CBL)/CBL-interacting kinase (CIPK) pathway and Salt Overly Sensitive (SOS) signaling pathway (Qiu et al. 2002; Martínez-Atienza et al. 2007). In rice, several of these genes *OsSOS1*, *OsSOS2/OsCIPK24,* and *OsSOS3/OsCBL4* are well characterized. These genes help in $Na^+$ sequestration in vacuoles, root $K^+$ uptake, and $Na^+/K^+$ homeostasis (Li et al. 2014a, 2014b). Apart from these, the change in calcium concentration in cytosol also regulates calcium-dependent protein kinases (CDPKs), which regulates calcium signaling pathways and downstream genes and thus leading to response to salt stress (Saijo et al. 2000). In rice, *OsCDPK7* has been known to be positive regulator in response to salinity. Similarly, *OsCPK21* promotes expression of salt and ABA-responsive genes such as *OsNAC6, OsLEA3*, *OsNHX1,* and *OsSOS1* (Asano et al. 2011).

#### 13.4.1.2 Regulation of Reactive Oxygen Species (ROS)

Several genes responsible in regulating hydrogen peroxide level in the cell, involved in ROS detoxification, regulating turgor pressure, etc. have been reported (Liu et al. 2021). The production of reactive oxygen species (ROS) in cellular machineries are increased because of reduced photosynthesis under high salinity. This start the ROS signaling pathways that are also integrated with calcium-dependent signaling pathways under salinity response and regulate potassium homeostasis (Fetoni et al. 2019). The deposition of more ROS is toxic to cell that acts in degrading several proteins and lipids involved in enzymatic and structural role in cell organelles. Therefore, plants must degrade, scavenge, or sequestrate these ROS to maintain ionic homeostasis, enzymatic functions, and cell structure. In rice, several of genes belonging to ascorbate peroxidase (APX) gene family have been identified that have a role in $H_2O_2$ scavenging, $Na^+/K^+$ homeostasis, and ABA accumulation (Zhang et al. 2013). Ascorbic acid (ABA)-related signaling pathways are also responsible in regulating glutathione reductase (*OsGR1*, *OsGR2,* and *OsGR3*) in rice, which are involved in ROS scavenging (Wu et al. 2015). Several genes related to mitogen-activated protein (MAP) kinase are also responsible in sensing salt stress and regulating the level of toxic reactive oxygen species (ROS) and ethylene level in cell (Na et al. 2019). Lectin receptor-like kinase (*SIT1*) activating rice *OsMPK3* and *OsMPK6* is also reported to mediate salt stress and affect ethylene accumulation and ROS scavenging (Li et al. 2014a, 2014b).

#### 13.4.1.3 Regulation by Specific Transcription Factors

Several transcription factors have their role in regulation of salt response in rice, belonging to dehydration-responsive element (DRE) binding protein (DREB), ABA-responsive element (ABRE) binding protein/factor (AREB/ABF), and NAC (NAM, ATAF1/2, CUC2) (Chen et al. 2021a). DREB transcription factors have their predicted role in ABA-dependent regulation of salt-responsive genes, regulation of genes belonging to MAP kinase family, and activating ROS scavenging (Wang et al. 2008). NAC transcription factor (*OsNAC106, OsNAC5*, *OsNAC045*, *OsNAC2,* and *OsNAC022*) genes are known to regulate several salt-responsive genes such as *OsDREB2A*, *OsbZIP23*, *OsSAPK1,* and *OsLEA3* in response to salt stress that in turn have a role in ionic balancing and ROS scavenging (Jiang et al. 2019).

#### 13.4.1.4 Regulation by Functional Salt-Responsive Genes

Stomatal conductance is generally regulated by ABA-dependent and ABA-independent manner, which are also reported in rice. A key transcription factor gene called Drought and Salt Tolerance (*DST*) regulates stomatal closure by modulating $H_2O_2$ homeostasis in rice (Huang et al. 2009). The DST-binding sequence is present in peroxidase-related gene (*Prx 24*), *leaf panicle 2* (*LP2*), and *SIMILAR TO RCD ONE* (*OsSRO1c*), which are involved in regulating stomatal closure and $H_2O_2$ homeostasis in rice (Cui et al. 2015). The enhanced expression of salt and drought inducible ring finger protein (*OsSDIR1*) is also responsible to enhance water deficit tolerance in rice via stomatal closure (Gao et al. 2011). Reducing stomatal density can also be an important trait for tolerance to salt in glycophytic crops such as rice (Mohammed et al. 2019). Osmotic adjustment is also contributed by water transport across cell membranes, and thus proteins such as aquaporins have very distinct role to play. Overexpression of plasma membrane intrinsic protein (*OsPIP1;1*) is said to have a role in enhancing salt tolerance (Abdelkader et al. 2012). Enhanced synthesis and accumulation of compatible solutes such as proline, glycine betaine, trehalose, polyols, etc. imparts salinity tolerance in rice, and several genes such as *OsTPS1, OsTPS8, OsTPP1, SAPK9, OsBADH1*, and *OsCMO* are contributing to their synthesis (Chen et al. 2021a). Ionic homeostasis in rice under salt stress is contributed by $Na^+$ sequestration, $Na^+$ loading in xylem, $K^+$ retention, and $Na^+$ efflux. Several genes belonging to root absorption mechanisms (i.e., nonselective cation channels (NSCCs) and high-affinity K+ transporters (HKT)) involving $Na^+$ uptake are responsible for this. Rice *HKT* gene family (*OsHKT1*, *OsHKT2*), plasma membrane $Na^+/H^+$ antiporter gene (*OsSOS1*), and vacuolar $Na^+$ sequestration genes (OsNHX family genes) are mainly involved in pathways related to ionic homeostasis (Liu et al. 2022). A list of various genes associated with functions related to salt tolerance is given in Table 13.2.

## 13.4.2 Physiological Mechanism

In general, rice manage its salinity stress by regulating stomatal opening and closing, ionic homeostasis, tissue tolerance, and osmotic adjustment (Reddy et al. 2017). Physiological drought condition is created by low soil-water potential at root zone and reduced stomatal conductance. Opening and closing of stomata is under direct control of ABA and superoxide ($H_2O_2$) signaling (Van Zelm et al. 2020). Ionic balancing of Na+/K+ ratio through transport, efflux/influx regulation, sequestration, exclusion, and compartmentalization is important for maintaining physiological drought and maintaining stomatal function, leaf expansion, and plant growth (Rajendran et al. 2009; Roy et al. 2014). Increasing tissue tolerance is related to ionic sequestration, deposition of solutes, and enzymatic detoxification (Chakraborty et al. 2020).

### 13.4.2.1 Plant Vigor

Plant vigor is an important mechanism to avoid toxic effects of salts. Rice plants having very high vigor and faster growth are having better tolerance to salinity. It is because of the higher the biomass, the higher will be the tissue volume to have salts in plant (Kumar et al. 2013). This is an avoidance mechanism of plant to sequestrate the excess salts in tissue. In general, rice plant at reproductive stage is more tolerant to salinity.

### 13.4.2.2 Restricted Entry of Salts

Absorption of excessive salt/ions are the first stage in development of salt-induced toxicity in plants. Excessive salt accumulation can be prevented by reduced or restricted entry of salts into plants (Krishnamurthy et al. 2009). This can be performed by ionic exclusion for absorption, reduced ionic exchanges at root hairs, and release of excessive organic acid to stabilize metallic ions (Kumar et al. 2013). The excessive entry can also be prevented by enhanced growth of root cap cells, which helps in $Na^+$ exclusion in rice plants. Tolerant genotypes generally have larger root cap cell (Ferdose et al. 2009). The rapid growth of dividing cell can have great impact in diluting the impact of salinity at root zone (Horie et al. 2012).

### 13.4.2.3 Intracellular Compartmentalization

The active transport and exchange of excessive ions in tissues are important in saving the cellular structure and function in rice plants. Rice generally stores excessive salts in older leaves and leaf sheath to maintain the required growing and working condition for young meristematic tissue as well as young leaf (Reddy et al. 2017). Therefore, the rate of compartmentalization and rate of development of

**Table 13.2** List of genes associated with different functions responsible for salinity tolerance in rice (adapted and modified from Reddy et al. 2017)

| S. No. | Salt response element | Molecular function | Reported genes |
|---|---|---|---|
| 1 | Receptor-like kinase | Work in downstream of *DST* | *OsSIK1, OsSIK2, OsRPK1* |
| 2 | Mitogen-activated protein | Osmotic homeostasis; sodium uptake; $K^+$/$Na^+$ ratio | *OsEDR1, OsMAPK5, OsMAPK44, OsMAPK33, OsMKK1, OsMPK4, OsMKK6* |
| 3 | *SNF1*-related protein kinase | Associated with ABA signaling | *SAPK4, SnRK2* |
| 4 | Transcription factor | Expression of various genes related to salt stress | *OsABF2, OsbZIP23, OsbZIP71, OsABI5, OsMYB3R-2, OsMYB2, OsMYB91, OsNAC6, OsNAC5, OSISAP1, DST, OsTZF, OsTZF1, OsWRKY13, OsWRKY42, OsERF922, OsNAC5, OsNAC10, OsNAP, ONAC, ONAC022, OsRAB1, MYC/MYB, OsNAC/SNAC, SERF1, OsDREB1A, OsDREB1F, OsDREB2A, OrbHLH001, OrbHLH, OsNAC6, SERF1, DST, OsTZF1* |
| 5 | Micro-RNA | Regulation of gene expression | *Osa-miR396, Osa-miR820, Osa-miR169, Osa-miR2006, Osa-miR29, MIR393, MIR1425* |
| 6 | ROS-producing/scavenging enzyme | ROS maintenance and degradation | *NADP-ME, katE encoding catalase, OsIM1, PDH45, SERF1, OsCatC, OsAPx2* |
| 7 | SOS pathway | $Na^+$/$H^+$ exchanger | *OsIM1, OsSOS3, OsSOS2* |
| 8 | Calmodulin (CaM) pathway | Associated with ABA signaling | *OsMSR2, OsCaml–1, OsCML4, OsCML5, OsCML8, OsCML11, OsCaM1, OsCaM2* |
| 9 | Phospholipid and $Ca^{2+}$ ion | Phospholipid based signaling | *OsCIPK24, OsCBL4, OsCIPK15* |
| 10 | Ca-dependent protein kinase | Calcium-dependent stress signaling | *OsCPK4, OsCPK12, OsCDPK7* |
| 11 | Kinase | Signaling of salt tolerance | *SIT1* |
| 12 | Calcium transporter | ROS scavenging | *OsACA6* |
| 13 | Nitrate transporter | ABA accumulation | *OsNRT1;2* |
| 14 | Helicase | Maintaining photosynthesis and antioxidant machinery | *OsSUV3* |
| 15 | $Na^+$/$H^+$ antiporter/exchanger | Ionic balancing | *OsSOS1, OsNHX1* |
| 16 | $Na^+$/$K^+$ symporter | Ionic balancing | *OsHKT1;4, OsHKT1;1, OsHKT2:1, OsHKT1;5, HKT2;3, SKC1* |

(continued)

**Table 13.2** (continued)

| S. No. | Salt response element | Molecular function | Reported genes |
|---|---|---|---|
| 17 | $H^+$-ATPase | Ionic balancing | *OSA3, OsVHA-a1, OSA7* |
| 18 | Channel protein | Metallic transport | *OsCLC1, OsAKT1, OsHAK1, OsHAK5, OsHAK21* |
| 19 | $H^+/ca^+$ antiporter | Ionic balancing | *OsCAX1* |
| 20 | Proline | Osmoprotectants | *P5CS, P5CR, SIDP361* |
| 21 | Glycine betaine | Osmoprotectants | *codA, CMO* |
| 22 | Myoinositol | Osmoprotectants | *PINO1* |
| 23 | Trehalose | Osmoprotectants | *OsTPS1-11, OsTPS-1* |
| 24 | Aquaporin | Osmoprotectants | *OsTIP1;1, HvPIP2;1* |
| 25 | Abscisic acid | Stress hormone, key regulation in ROS scavenging | *OsAP2-39, OsNCEDs1, OsNCEDs3, OsNCEDs4, OsNCEDs5, OsPSY1* |
| 26 | Jasmonate | Growth and metabolism | *OsTIFY1, OsTIFY6, OsTIFY9, OsTIFY10, OsTIFY11* |
| 27 | Brassinosteroid | Growth and metabolism | *OsGSK1(BIN2), OsBRI1, OsDWF4, OsSalT* |

young leaves and tissues must be greater than the rate of uptake of toxic ions from root (Chakraborty et al. 2020). This mechanism is also attributed by increased vacuolar size in older leaves (Kanawapee et al. 2012; Chakraborty et al. 2019). The role of cell ionic homeostasis (maintained by antiporter, symporter, and carrier proteins) and stress signaling molecules (through calcium-mediated signaling, ABA responsiveness, ethylene responsiveness) are important in imparting intracellular compartmentalization of ions in rice (Chen et al. 2021a).

#### 13.4.2.4 Antioxidants

Excessive salts in plant cell promotes production of reactive oxygen species (ROS) such as peroxide ($H_2O_2$), superoxide ($O_2^-$, $O_2^+$), hydroxy ($OH^-$), and singlet oxygen. Such reactive oxygen species cause tissue damage, enzymatic dysfunction, reduced photosynthetic activity, and other toxicity (Kibria et al. 2017). These ROS are responsible for ROD-induced signaling for salt stress and regulation of pathways related to ROS scavenging. Several enzymes such as peroxidase, superoxide dismutase (SOD), catalase, and other antioxidants are overexpressed in response to this in salt-tolerant genotypes (Çelik et al. 2019).

#### 13.4.2.5 Osmoprotectants

Osmoprotectants are compatible solutes that enhance tissue tolerance to salinity by maintaining cell turgor and maintaining ionic exchanges (Garcia et al. 1997). Molecules such as proline, mannitol, glycine betaine, and trehalose are reported to work as osmoprotectants in rice. Several genes reported to work in their biosynthesis are already reported earlier in this chapter. Proline is known to balance pH of cytosol and to work in quenching of singlet oxygen molecule (Omari Alzahrani 2021). Trehalose is responsible for maintaining enzymatic function, enhanced production of proline, and improving sugar content of cell (Li et al. 2011).

## 13.5 Mechanism of Submergence Tolerance in Rice

The semiaquatic nature of rice is making it more suitable for regions with high rainfall but rice cannot tolerate complete submergence of crop plant, and thus erratic rainfall and very high rainfall at sensitive stage are much problematic (Nishiuchi et al. 2012). Rice genotypes differ for tolerance to excessive water stress such as submergence during germination, seedling germination, and adult plant stage (Mishra et al. 1996). Submergence leads to development of anaerobic condition, which results in reduced gaseous exchange at leaves, reduced respiration rate, decreased photosynthetic capacity, imbalanced use of metabolites and products from photosynthesis, enhanced occurrence of pest and diseases, and reduced soil nutrients availability and plant mechanical damage (Nishiuchi et al. 2012; Kuroha and Ashikari 2020). Complete submergence occurring during germination stress is the biggest shortfall in acceptance of direct seeded rice, which is because the coleoptile growth and seedling vigor are hampered during flooded condition (Ghosal et al. 2019). At this stage, the mechanisms such as enhanced anaerobic germination, enhanced coleoptile growth, faster induction of anaerobic germination, and pronounced root and shoot length are important (Mondal et al. 2020). Seedling stage salinity tolerance for flash flood (up to 1–2 weeks) is managed by maintenance of growth control, maintaining internal aeration and attaining senescence for submerged period (Nishiuchi et al. 2012). During this submerged period, growth cessation may be achieved by reduced photosynthesis, photosynthate accumulation, and reduced metabolism (to save energy), which may lead to further utilization of these resources for managing aerobic shock, regrowth, degeneration of toxic metabolites, and enhanced plant growth post flooding. This can be achieved by keeping metabolism to a bare minimum, i.e., quiescent strategy (Panda and Sarkar 2011; Guru and Dwivedi 2018). In conditions where rainwater stagnates for a longer period, rice plant manages submergence stress by enhancing growth to have at least top plant part above water (to maintain respiration, aeration, and minimal photosynthesis), i.e., escape strategy (Guru and Dwivedi 2018; Colmer et al. 2019). Escape strategy results in rice plant succumbing to more growth and in turn losing all its energy and ultimately die out of starvation, while some tolerant variety sustains by this.

### 13.5.1 Internal Aeration: A Common Mechanism

Internal aeration is required by rice plant to keep all the tissues alive and in general achieved by aerenchyma tissues (to distribute oxygen to all plant part), development of gaseous film on leaf (to maintain oxygen balance near to leaf surface), formation of radial oxygen barrier (to reduce loss of oxygen to outside) and development of adventitious roots, etc. (Chakraborty et al. 2021).

#### 13.5.1.1 Formation of Aerenchyma Tissue

Rice plants have well-developed aerenchyma tissue (for gaseous transport) in roots for oxygen and gaseous metabolite (ethanol, ethylene, and methane) transport from shoots. The development of these tissue may occur in all plant parts, which are devoid of aeration, and genotypic differences may occur in extent and speed of development of these tissues resulting into genotypic differences in tolerance to submergence (Steffens et al. 2011). Occurrence of anaerobic respiration and its by-product such as peroxide, superoxide, and ethylene stimulates the expression of gene responsible for development of these tissues (Evans 2004).

Under submergence stress, NADPH oxidase (NOX) activity is enhanced, resulting into generation of reactive oxygen species (ROS). These ROS may give rise to reactive nitrogen species (RNS) under nitric oxide stress (developed after anaerobic degradation of nitrogenous fertilizers). These ROS and RNS may induce programmed cell death of vascular internal tissue resulting into development of aerenchyma (Basu et al. 2020). ROS scavenging enzymes such as peroxidase, catalase, superoxide dismutase, etc., work in degrading reactive oxygen species and thus negatively regulate formation of aerenchyma tissue (Chen et al. 2017). Methane is another by-product of anaerobic condition in rice field and the concentration of methane is also reported to have a role in the development of aerenchyma tissue (Kim et al. 2018). In rice, several genes that regulate development of aerenchyma tissue under oxidative stress through ROS scavenging are reported. These include NADPH oxidase (responsible for ROS formation), METALLOTHIONEINS (encode ROS scavenging enzymes), glutamate decarboxylase (induce formation of GABA), heterotrimeric G protein $\gamma$ subunit DEP1 (work in peroxide signaling), etc. (Yamauchi et al. 2017a; Yamauchi et al. 2017b; Chen et al. 2021b; Mohammed et al. 2019; Zhu et al. 2021). Similarly, ethylene biosynthesis, transportation, trafficking, signaling, and regulation within plant also affect development of aerenchyma tissue in rice.

#### 13.5.1.2 Leaf gas Film

The leaf surface of rice has enough stomata for gaseous exchange and can maintain aeration under normal condition, but during submergence, the dissolved oxygen of water gets depleted apart from reduced stomatal conductance with time, and thus plants require extra effort to absorb oxygen (Guru and Dwivedi 2018; Panda and Barik 2021). This gaseous exchange under submergence condition is facilitated by presence of gaseous films, which helps plant in respiration, promotes underwater photosynthesis, and thus maintains growth and metabolism. In tolerant cultivars, leaf gas film is formed to delay the sensing of submergence stress by internal tissue (Chakraborty et al. 2021). These leaf gas film may also enhance stem elongation through enhanced level of oxygen at cellular growth position and can help in snorkeling effect (Mori et al. 2019). Leaf gas films are also formed in genotypes having SUB1 QTL (Chakraborty et al. 2021).

#### 13.5.1.3 Radial Oxygen Loss Barrier

Transport of oxygen from shoot to root through aerenchyma is impacted by radial loss of oxygen because of diffusion into root zone and into stagnant water (called as radial oxygen loss or ROL) because of less concentration of oxygen in water than that of internal aerenchyma (Colmer et al. 2019). The radial loss is augmented by oxygen usage in path, oxygen concentration before entering into aerenchyma, and intracellular diffusion (Nishiuchi et al. 2012). Rice plant generally reduce this loss by deposition of suberin and lignin in cell wall of aerenchyma tissue, in hypodermis, and in sclerenchyma (Ejiri et al. 2021). Several genes responsible for suberin accumulation under waterlogged condition and ethylene response have increased expression to induce formation of ROL barrier (Kulichikhin et al. 2014). Glycerol-3-phosphate acyltransferase (*GPAT*) and cytochrome P450 family 86 (*CYP86*) are upregulated under waterlogged stress condition for development of aerenchyma and barrier to radial oxygen loss (Nishiuchi et al. 2021).

#### 13.5.1.4 Adventitious Root Formation

The dormant adventitious root (AR) primordia start to develop root under water stress induction (Mhimdi and Pérez-Pérez 2020). Under hypoxic condition, these roots function to transport hormones, gases, nutrients, and metabolites (Nishiuchi et al. 2012). Polar transport of cytokinin, auxin, and strigolactone promotes the development of adventitious root primordia and ethylene promotes its differentiation into adventitious root (Zhi-Guo et al. 2012; Sun et al. 2015; Steffens and Rasmussen 2016; Lin and Sauter 2019). In rice, several genes involved in auxin transport are reported that are regulating development of adventitious root in rice and impart submergence tolerance. These include *OsGNOM1*, *OsPIN* family genes, *narrow leaf1* (*nal1*; affect *OSHB*, *YABBY,* and *PIN-FORMED* class of genes by encoding a trypsin-like serine/cysteine protease), and *OsARL1* (*adventitious rootless1*; have LATERAL ORGAN BOUNDARIES domain) (Liu et al. 2009; Liu et al. 2005; Xu et al. 2005; Cho et al. 2014).

### 13.5.2 Quiescent Strategy: For Flash Flood

Under flash flood-submerged condition, it is a wise strategy to go for senescence by reducing plant growth, minimizing metabolism to save energy and resources for post-flood regrowth, detoxification, and development. In general, rice plant will try to elongate its stem (by internodal elongation) to manage aerobic respiration, keep photosynthetic machinery active and in turn utilize all of its energy and resources, and at last die out of starvation and have limited resources for post-flood regrowth (Kumar et al. 2019). In susceptible genotypes, increased anaerobic respiration during submergence, chlorophyll degradation, reduced photosynthesis, accumulation of toxic metabolites, scarcity of resource and energy, etc., hampers post-flood regrowth

(Nishiuchi et al. 2012; Khasna et al. 2020). Quiescent strategy for lifesaving under submergence in rice is reported to be controlled by submergence QTL *SUB1*, encoding several ethylene response factors (APETALA2 (AP2)/Ethylene Response Factor (ERF) superfamily genes) (Kuanar et al. 2019). Submergence-tolerant genotypes have defunct protein coding genes (*SUB1A*) that reduce the response to ethylene and thus reduce the initiation and development of aerenchyma tissue and thus delay change in mode of respiration from aerobic to anaerobic. This slows down respiration and loss of energy (Singh et al. 2017). Such genotypes having tolerant allele of *SUB1A* also show other mechanisms such as minimized energy consumption, reduced elongation rate of leaf and internode, reduced degeneration of chlorophyll, increased sensitivity to ABA and gibberellic acids, delayed flowering, and increased activity of alcohol dehydrogenase (ADH), catalase, peroxidase, superoxide dismutase, and pyruvate decarboxylase (PDC) (Singh et al. 2017; Oladosu et al. 2020). *SUB1A* (tolerant allele) is hampering ethylene-regulated plant development and signaling and therefore enhances stability and function of gibberellin signaling repressor proteins (*SLR1*, *Slender* Rice*1*; and *SLRL1*, *Slender* Rice *Like1*), which ultimately impact shoot elongation (Fukao and Bailey-Serres 2008).

### 13.5.3 Escape Strategy: Deepwater Rice

Deepwater stagnation for long duration and for entire cropping period may create oxygen deficiency under water, reduced overall photosynthesis, anaerobic respiration, etc. (Kende et al. 1998). Several rice genotypes are adapted to deepwater submergence by escaping the hazard caused by anoxic condition by maintaining internal aeration (with very well-developed and pronounced aerenchyma), at least few plant parts above water, mechanism to have efficient gaseous exchange and photosynthesis, tall plant habit, strong culm, etc. (Hattori et al. 2011). To keep plant part above water, rapid elongation of internode and stem as well as lignification of stem and internode occurs, and such mechanisms are common only in genotypes having adaptation to deepwater ecology (Anandan et al. 2015). To maintain aeration, such genotypes develop a "snorkel-like" structures in internodes for gaseous exchange and rapid development of aerenchyma (in leaves, root, and stem) (Nagai et al. 2010; Hattori et al. 2009). The developmental phenomenon for such adaptation in these genotypes is regulated in response to several phytohormones (gibberellins, abscisic acid, and ethylene) (Nagai et al. 2010; Hattori et al. 2009). Ethylene response factor genes such as *SNORKEL1* (*SK1*) and *SNORKEL2* (*SK2*) have been identified as part of possible ethylene-regulated mechanism for deepwater adaptation in rice genotypes (Hattori et al. 2009). Increased gibberellin biosynthesis and signaling is reported in initiation and maintenance of internode elongation (Ayano et al. 2014). Ethylene is reported to increase the gibberellin biosynthesis and reduce the abscisic acid biosynthesis, which may result in increased GA/ABA ration, which is known to enhance shoot elongation (Shu et al. 2018).

### 13.5.4 Anaerobic Germination

With increase in energy cost and environmental concern, direct seeded rice cultivation is targeted and promoted. Germination and plant establishment are very crucial in direct seeded rice cultivation (Javed et al. 2021). Incessant and erratic rainfall during crop establishment cause anoxic condition to germinated seed. As such, germination is not much affected by anoxic condition, but post-germination growth and coleoptile elongation and development of root and shoot are highly impacted (Magneschi and Perata 2009; Ismail et al. 2009). Genotypes having more reserved starch, faster switching into anaerobic germination at germination stage, faster growth, enhanced ethylene production, better amylase activity, and reduced peroxidase activity can have faster growth to emerge above water (Miro and Ismail 2013; Lee et al. 2014). A major QTL (*qAG-9-2*) encoding *trehalose-6-phosphatase* (*OsTPP7*) has been reported to impart anaerobic germination in rice genotype Khao Hlan On (Kretzschmar et al. 2015). Trehalose-6-phosphate acts as key sensing molecule in sucrose translocation and starch mobilization in developing embryo (Pucciariello 2020). Elongating coleoptile grow faster in tolerant genotype and act as snorkel and come above water to create aerobic channel between germinating seed and water surface and lead to further development of plant (Narsai et al. 2015).

## 13.6 Breeding Rice Varieties for Coastal Ecology

Rice breeding in coastal region primarily focuses on salinity and submergence tolerance along with higher yield and lodging tolerance in rice (Prakash et al. 2020). The prime focus is the accumulation of desirable alleles for the beneficial component trait and their evaluation (Rasel et al. 2021). Breeding rice varieties depends on mining germplasm resources for salinity as well as submergence tolerance, identification of superior genotypes, identifying component traits, trait analysis and understanding genetic basis (QTL mapping and gene cloning), mining of superior alleles and haplotype, selection, evaluation for phenotypic performances, crossing and augmenting superior haplotype, genomic-assisted selection, and release of varieties (Prakash et al. 2020).

Coastal regions are rich in germplasm resources for both salinity and submergence. Several local landraces cultivated in these areas across the world are having one or another mechanism to tolerate such stress (Batayeva et al. 2018). In India, Pokkali system of rice cultivation is practiced since ancient times in regions that are having continuous seawater ingression (Latha et al. 2013). The world-famous salt-tolerant QTL, i.e., *SALTOL,* was identified from Pokkali landraces and has been widely used in transferring seedling stage salt tolerance into high-yielding rice varieties (Krishnamurthy et al. 2020). Similarly, flooding-resistant rice varieties have been developed by transferring submergence tolerance QTL (*Sub1*) from a local landrace in Odisha, India, into several rice varieties (Chakraborty et al. 2020). Germplasm evaluation and characterization to identify the genetic and molecular basis of salinity and submergence tolerance in rice have been done in several landraces (Prakash et al. 2020; Manohara et al. 2021; Hairmansis et al. 2017; Pani et al. 2013).

Classical breeding approaches were used to develop rice varieties suitable for coastal Sundarbans regions using several rice landraces. Varieties such as CSR1, CSR4, Bhutnath, Canning 7, SR26B, etc. were developed either through direct selection from landraces or through hybridization-based pedigree breeding program (Kumar et al. 2019). These varieties are having tolerance to submergence, lodging, and salt tolerance, which makes them suitable for these regions. Marker-assisted backcross breeding (MABB) has been widely used to develop salt (*SALTOL* region) and submergence (*Sub1*) tolerant rice varieties for coastal regions (Nair and Shylaraj 2021). Modern concept of genomic selection and haplotype-based breeding approaches can be effectively utilized in evaluating rice genotypes for stresses in coastal ecologies and to develop several salt-tolerant rice cultivars (Prakash et al. 2022).

## 13.7 Conclusion and Future Perspective

Rice is a very versatile crop, is a prime crop of coastal region, and is being affected by multiple stresses such as salinity and submergence augmented with severe impacts of climate change. Therefore, it is very important to understand the molecular mechanism of salinity and submergence tolerance in rice. The novel crop germplasms, crop wild relatives, and wild species are prime source of variation and hence can be effectively used in breeding and genetics. The molecular mechanism of understanding these traits will also help in targeting their component traits and ultimately improvising selection indices. The current understanding of genomics and phenomics tools along with robust and reliable phenotyping is helping breeders in this regard. Genomic selection and haplotype selection considering targeted genes and QTL can be very effectively used to devise crop varieties having climate-resilient features.

## References

Abdelkader AF, El-khawas S, El-Din El-Sherif NAS, Hassanein RA, Emam MA, Hassan RES (2012) Expression of aquaporin gene (*OsPIP1-3*) in salt-stressed rice (*Oryza sativa* L.) plants pre-treated with the neurotransmitter (dopamine). Plant Omics 5(6):532–541

Ahmadizadeh M, Vispo NA, Calapit-Palao CDO, Pangaan ID, Viña CD, Singh RK (2016) Reproductive stage salinity tolerance in rice: a complex trait to phenotype. IJPP 21(4):528–536

Ali M, Yeasmin L, Gantait S, Goswami R, Chakraborty S (2014) Screening of rice landraces for salinity tolerance at seedling stage through morphological and molecular markers. Physiol Mol Biol Plants 20(4):411–423

Anandan A, Pradhan SK, Das SK, Behera L, Sangeetha G (2015) Differential responses of rice genotypes and physiological mechanism under prolonged Deepwater flooding. Field Crop Res 172:153–163

Asano T, Hakata M, Nakamura H, Aoki N, Komatsu S, Ichikawa H, Hirochika H, Ohsugi R (2011) Functional characterization of *OsCPK21*, a calcium-dependent protein kinase that confers salt tolerance in rice. Plant Mol Biol 75(1):179–191

Ayano M, Kani T, Kojima M, Sakakibara H, Kitaoka T, Kuroha T, Angeles-shim RB, Kitano H, Nagai K, Ashikari M (2014) Gibberellin biosynthesis and signal transduction is essential for internode elongation in deep water rice. Plant Cell Environ 37(10):2313–2324

Barrera WB, Viña CBD, Vispo NA, Singh RK (2019) Genetic diversity using single nucleotide polymorphisms (SNPs) and screening for salinity tolerance in rice germplasm at reproductive stage. Plant Genetic Resour 17(6):522–535

Basu S, Kumar G, Kumari N, Kumari S, Shekhar S, Kumar S, Rajwanshi R (2020) Reactive oxygen species and reactive nitrogen species induce lysigenous aerenchyma formation through programmed cell death in rice roots under submergence. Environ Exp Bot 177:104118

Batayeva D, Labaco B, Ye C, Li X, Usenbekov B, Rysbekova A, Leung H (2018) Genome-wide association study of seedling stage salinity tolerance in temperate japonica rice germplasm. BMC Genet 19(1):1–11

Bhambure AB, Kerkar S (2016) Traditionally cultivated rice varieties in coastal saline soils of India. Vasantrao Dempo EducRes J Arts Sci Humanities 2(1):65–75

Bin Rahman AR, Zhang J (2016) Flood and drought tolerance in rice: opposite but may coexist. Food Energy Secur 5(2):76–88

Biswajit P, Sritama K, Anindya S, Moushree S, Sabyasachi K (2017) Breeding for submergence tolerance in rice (*Oryza sativa* L.) and its management for flash flood in rainfed low land area: a review. Agric Rev 38(3):167–179

Chakraborty K, Chattaopadhyay K, Nayak L, Ray S, Yeasmin L, Jena P, Gupta S, Mohanty SK, Swain P, Sarkar RK (2019) Ionic selectivity and coordinated transport of $Na^+$ and $K^+$ in flag leaves render differential salt tolerance in rice at the reproductive stage. Planta 250(5): 1637–1653

Chakraborty K, Guru A, Jena P, Ray S, Guhey A, Chattopadhyay K, Sarkar RK (2021) Rice with *SUB1* QTL possesses greater initial leaf gas film thickness leading to delayed perception of submergence stress. Ann Bot 127(2):251–265

Chakraborty K, Mondal S, Ray S, Samal P, Pradhan B, Chattopadhyay K, Kar MK, Swain P, Sarkar RK (2020) Tissue tolerance coupled with ionic discrimination can potentially minimize the energy cost of salinity tolerance in rice. Front Plant Sci 11:265

Chattopadhyay K, Chakraborty K, Samal P, Sarkar RK (2021) Identification of QTLs for stagnant flooding tolerance in rice employing genotyping by sequencing of a RIL population derived from Swarna×Rashpanjor. Physiol Mol Biol Plants 27(12):2893–2909

Chattopadhyay K, Reddy JN, Pradhan SK, Patnaik SSC, Marndi BC, Swain P, Korada RR (2018) Genetic improvement of Rice for multiple stress tolerance in unfavorable Rainfed ecology. ICAR-NRRI

Chen T, Shabala S, Niu Y, Chen ZH, Shabala L, Meinke H, Venkataraman G, Pareekh A, Xu J, Zhou M (2021a) Molecular mechanisms of salinity tolerance in rice. Crop J 9(3):506–520

Chen Y, Chen Y, Zhang Y, Zhang D, Li G, Wei J, Hua X, Lv B, Liu L (2021b) Heterotrimeric G protein γ subunit *DEP1* is involved in hydrogen peroxide signalling and promotes aerenchyma formation in rice roots. Plant Signal Behav 16(5):1889251

Chen Y, Gao C, Ren B, Hu X, Shen Q, Guo S (2017) Physiological mechanism of nitrogen forms affect aerenchyma formation of rice root under water stress. J Nanjing Agric Univ 40(2): 273–280

Cho SH, Yoo SC, Zhang H, Lim JH, Paek NC (2014) Rice *NARROW LEAF1* regulates LEAF and adventitious root development. PMBR 32(1):270–281

Colmer TD, Kotula L, Malik AI, Takahashi H, Konnerup D, Nakazono M, Pedersen O (2019) Rice acclimation to soil flooding: low concentrations of organic acids can trigger a barrier to radial oxygen loss in roots. Plant Cell Environ 42(7):2183–2197

Cui LG, Shan JX, Shi M, Gao JP, Lin HX (2015) *DCA1* acts as a transcriptional co-activator of *DST* and contributes to drought and salt tolerance in rice. PLoS Genet 11(10):e1005617

Dar MH, Chakravorty R, Waza SA, Sharma M, Zaidi NW, Singh AN, Singh US, Ismail AM (2017) Transforming rice cultivation in flood prone coastal Odisha to ensure food and economic security. Food Security 9(4):711–722

Darko Asante M, Ipinyomi SO, Abe A, Adjah KL, Aculey P, Kwame Bam R, Manneh B (2021) Genetic variability for and tolerance to anaerobic germination in rice (*Oryza sativa* L.). J Crop Improv 35(6):832–847

Dasgupta S, Hossain M, Huq M, Wheeler D (2014) Climate change, soil salinity, and the economics of high-yield rice production in coastal Bangladesh (December 1, 2014). World Bank Policy Research Working Paper No. 7140, Available at SSRN: https://ssrn.com/abstract=2540770

Ejiri M, Fukao T, Miyashita T, Shiono K (2021) A barrier to radial oxygen loss helps the root system cope with waterlogging-induced hypoxia. Breed Sci 71(1):40–50

Evans DE (2004) Aerenchyma formation. New Phytol 161(1):35–49

Ferdose J, Kawasaki M, Taniguchi M, Miyake H (2009) Differential sensitivity of rice cultivars to salinity and its relation to ion accumulation and root tip structure. Plant Prod Sci 12(4):453–461

Fetoni AR, Paciello F, Rolesi R, Paludetti G, Troiani D (2019) Targeting dysregulation of redox homeostasis in noise-induced hearing loss: oxidative stress and ROS signaling. Free Radic Biol Med 135:46–59

Fukagawa NK, Ziska LH (2019) Rice: Importance for global nutrition. J Nutr Sci Vitaminol 65 (Supplement):S2–S3

Fukao T, Bailey-Serres J (2008) Submergence tolerance conferred by *Sub1A* is mediated by *SLR1* and *SLRL1* restriction of gibberellin responses in rice. Proc Natl Acad Sci 105(43): 16814–16819

Ganie SA, Molla KA, Henry RJ, Bhat KV, Mondal TK (2019) Advances in understanding salt tolerance in rice. Theor Appl Genet 132(4):851–870

Gao T, Wu Y, Zhang Y, Liu L, Ning Y, Wang D, Chen S, Chengcai Chu C, Xie Q (2011) *OsSDIR1* overexpression greatly improves drought tolerance in transgenic rice. Plant Mol Biol 76(1): 145–156

Garcia AB, Engler JDA, Iyer S, Gerats T, Van Montagu M, Caplan AB (1997) Effects of osmoprotectants upon NaCl stress in rice. Plant Physiol 115(1):159–169

Ghosal S, Casal C, Quilloy FA, Septiningsih EM, Mendioro MS, Dixit S (2019) Deciphering genetics underlying stable anaerobic germination in rice: phenotyping, QTL identification, and interaction analysis. Rice 12(1):1–15

Ghosal S, Quilloy FA, Casal C, Septiningsih EM, Mendioro MS, Dixit S (2020) Trait-based mapping to identify the genetic factors underlying anaerobic germination of rice: phenotyping, GXE, and QTL mapping. BMC Genet 21(1):1–13

Gnanamanickam SS (2009) Rice and its importance to human life. In: *Biological control of* Rice *diseases*. Springer, Dordrecht, pp 1–11

Gopalakrishnan T, Hasan MK, Haque ATM, Jayasinghe SL, Kumar L (2019) Sustainability of coastal agriculture under climate change. Sustainability 11(24):7200

Gregoria GB, Senadhira D, Mendoza RD (1997) Screening rice for salinity tolerance. IRRI Discussion Paper Series 22:30

Gregorio GB, Senadhira D, Mendoza RD, Manigbas NL, Roxas JP, Guerta CQ (2002) Progress in breeding for salinity tolerance and associated abiotic stresses in rice. Field Crop Res 76(2–3): 91–101

Guru A, Dwivedi P (2018) Physiological, biochemical and molecular mechanism of submergence tolerance in rice (*Oryza sativa* L.). J Pharmacogn Phytochem 7(6):1116–1121

Hairmansis A, Nafisah N, Jamil A (2017) Towards developing salinity tolerant rice adaptable for coastal regions in Indonesia. KnE Life Sci 2:72–79

Hattori Y, Nagai K, Ashikari M (2011) Rice growth adapting to deep water. Curr Opin Plant Biol 14(1):100–105

Hattori Y, Nagai K, Furukawa S, Song XJ, Kawano R, Sakakibara H, Wu J, Matsumoto T, Yoshimara A, Kitano H, Matsuoka M, Mori H, Ashikari M (2009) The ethylene response factors *SNORKEL1* and *SNORKEL2* allow rice to adapt to deep water. Nature 460(7258): 1026–1030

Hernández JA (2019) Salinity tolerance in plants: trends and perspectives. Int J Mol Sci 20(10): 2408

Horie T, Karahara I, Katsuhara M (2012) Salinity tolerance mechanisms in glycophytes: an overview with the central focus on rice plants. Rice 5(1):1–18

Hsu SK, Tung CW (2015) Genetic mapping of anaerobic germination-associated QTLs controlling coleoptile elongation in rice. Rice 8(1):1–12

Huang XY, Chao DY, Gao JP, Zhu MZ, Shi M, Lin HX (2009) A previously unknown zinc finger protein, DST, regulates drought and salt tolerance in rice via stomatal aperture control. Genes Dev 23(15):1805–1817

Ismail AM, Ella ES, Vergara GV, Mackill DJ (2009) Mechanisms associated with tolerance to flooding during germination and early seedling growth in rice (*Oryza sativa*). Ann Bot 103(2): 197–209

Javed T, Afzal I, Mauro RP (2021) Seed coating in direct seeded rice: an innovative and sustainable approach to enhance grain yield and weed management under submerged conditions. Sustainability 13(4):2190

Jiang D, Zhou L, Chen W, Ye N, Xia J, Zhuang C (2019) Overexpression of a microRNA-targeted NAC transcription factor improves drought and salt tolerance in Rice via ABA-mediated pathways. Rice 12(1):1–11

Kalita K, Handique GK (2021) Evaluation of indigenous land races of deep-water paddy of north East India for nutraceutical value. Ann Plant Soil Res 23(1):125–127

Kanawapee N, Sanitchon J, Lontom W, Threerakulpisut P (2012) Evaluation of salt tolerance at the seedling stage in rice genotypes by growth performance, ion accumulation, proline and chlorophyll content. Plant Soil 358(1):235–249

Kende H, Van Der Knaap E, Cho HT (1998) Deepwater rice: a model plant to study stem elongation. Plant Physiol 118(4):1105–1110

Khasna EN, Ardana IKKG, Zakiyah AS, Fikriani CN, Anggraini NB, Listyorini D (2020, September) *Sub1A* gene screening for submergence stress in Indonesian local rice varieties. In: AIP Conference Proceedings, vol 2260(1). AIP Publishing LLC, p 060012

Kibria MG, Hossain M, Murata Y, Hoque MA (2017) Antioxidant defense mechanisms of salinity tolerance in rice genotypes. Rice Sci 24(3):155–162

Kim WJ, Bui LT, Chun JB, McClung AM, Barnaby JY (2018) Correlation between methane ($CH_4$) emissions and root aerenchyma of rice varieties. Plant Breed Biotechnol 6(4):381–390

Kretzschmar T, Pelayo MAF, Trijatmiko KR, Gabunada LFM, Alam R, Jimenez R, Mendioro MS, Slamet-Loedin IH, Sreenivasulu N, Bailey-Serres J, Ismail AM, Mackill DJ, Septiningsih EM (2015) A trehalose-6-phosphate phosphatase enhances anaerobic germination tolerance in rice. Nat Plants 1(9):1–5

Krishnamurthy P, Ranathunge K, Franke R, Prakash HS, Schreiber L, Mathew MK (2009) The role of root apoplastic transport barriers in salt tolerance of rice (*Oryza sativa* L.). Planta 230(1): 119–134

Krishnamurthy SL, Pundir P, Warraich AS, Rathor S, Lokeshkumar BM, Singh NK, Sharma PC (2020) Introgressed saltol QTL lines improves the salinity tolerance in rice at seedling stage. Front Plant Sci 11:833

Kuanar SR, Molla KA, Chattopadhyay K, Sarkar RK, Mohapatra PK (2019) Introgression of *Sub1* (*SUB1*) QTL in mega rice cultivars increases ethylene production to the detriment of grain-filling under stagnant flooding. Sci Rep 9(1):1–12

Kulichikhin K, Yamauchi T, Watanabe K, Nakazono M (2014) Biochemical and molecular characterization of rice (*Oryza sativa* L.) roots forming a barrier to radial oxygen loss. Plant Cell Environ 37(10):2406–2420

Kumar A, Sarangi SK, Sharma PC (2019) Genetic divergence study of rice at seedling stage augments a successful breeding strategy for salinity tolerance. Indian Soc Soil Salinity Water Quality 11(1):45

Kumar K, Kumar M, Kim SR, Ryu H, Cho YG (2013) Insights into genomics of salt stress response in rice. Rice 6(1):1–15

Kumar P, Sharma PK (2020) Soil salinity and food security in India. Front Sustain Food Syst 4:174

Kuroha T, Ashikari M (2020) Molecular mechanisms and future improvement of submergence tolerance in rice. Mol Breed 40(4):1–14

Latha M, Abraham Z, Nair RA, Mani S, Dutta M (2013) Rice landraces of Kerala state of India: a documentation. Int J Biodivers Conserv 5(4):250–263

Lee KW, Chen PW, Yu SM (2014) Metabolic adaptation to sugar/$O_2$ deficiency for anaerobic germination and seedling growth in rice. Plant Cell Environ 37(10):2234–2244

Li CH, Wang G, Zhao JL, Zhang LQ, Ai LF, Han YF, Sun DY, Zhang SW, Sun Y (2014a) The receptor-like kinase *SIT1* mediates salt sensitivity by activating *MAPK3/6* and regulating ethylene homeostasis in rice. Plant Cell 26(6):2538–2553

Li HW, Zang BS, Deng XW, Wang XP (2011) Overexpression of the trehalose-6-phosphate synthase gene *OsTPS1* enhances abiotic stress tolerance in rice. Planta 234(5):1007–1018

Li J, Long Y, Qi GN, Li J, Xu ZJ, Wu WH, Wang Y (2014b) The *OsAKT1* channel is critical for $K^+$ uptake in rice roots and is modulated by the rice *CBL1-CIPK23* complex. Plant Cell 26(8): 3387–3402

Lin C, Sauter M (2019) Polar auxin transport determines adventitious root emergence and growth in rice. Front Plant Sci 10:444

Liu C, Mao B, Yuan D, Chu C, Duan M (2022) Salt tolerance in rice: physiological responses and molecular mechanisms. Crop J 10(1):13–25

Liu H, Wang S, Yu X, Yu J, He X, Zhang S, Shou H, Wu P (2005) *ARL1*, a LOB-domain protein required for adventitious root formation in rice. Plant J 43(1):47–56

Liu R, Pollard M, Schulting R, Rawson J, Liu C (2021) Synthesis of stable isotopic data for human bone collagen: a study of the broad dietary patterns across ancient China. The Holocene 31(2): 302–312

Liu S, Wang J, Wang L, Wang X, Xue Y, Wu P, Shou H (2009) Adventitious root formation in rice requires *OsGNOM1* and is mediated by the *OsPINs* family. Cell Res 19(9):1110–1119

Lutts S, Kinet JM, Bouharmont J (1995) Changes in plant response to NaCl during development of rice (*Oryza sativa* L.) varieties differing in salinity resistance. J Exp Bot 46(12):1843–1852

Magneschi L, Perata P (2009) Rice germination and seedling growth in the absence of oxygen. Ann Bot 103(2):181–196

Manohara KK, Morajkar S, Shanbhag Y, Phadte P, Singh NK (2021) Haplotype analysis of Saltol QTL region in diverse landraces, wild rice and introgression lines of rice (*Oryza sativa* L.). plant genetic. Resources 19(4):289–298

Martínez-Atienza J, Jiang X, Garciadeblas B, Mendoza I, Zhu JK, Pardo JM, Quintero FJ (2007) Conservation of the salt overly sensitive pathway in rice. Plant Physiol 143(2):1001–1012

Mhimdi M, Pérez-Pérez JM (2020) Understanding of adventitious root formation: what can we learn from comparative genetics? Front Plant Sci 11:582020

Miro B, Ismail AM (2013) Tolerance of anaerobic conditions caused by flooding during germination and early growth in rice (*Oryza sativa* L.). Frontiers. Plant Sci 4:269

Mishra SB, Senadhira D, Manigbas NL (1996) Genetics of submergence tolerance in rice (*Oryza sativa* L.). Field Crop Res 46(1–3):177–181

Mohammed U, Caine RS, Atkinson JA, Harrison EL, Wells D, Chater CC, Gray JE, Swarup R, Murchie EH (2019) Rice plants overexpressing *OsEPF1* show reduced stomatal density and increased root cortical aerenchyma formation. Sci Rep 9(1):1–13

Mohanavel V, Selvam Yesudhas A, Sharma A, Ramasamy A, Samy PMA, Subramanian M, Muthusamy R (2021) Haplotype and diversity analysis of indigenous rice for salinity tolerance in early-stage seedling using simple sequence repeat markers. Biotechnol Rep 31:e00666

Mondal S, Khan MIR, Entila F, Dixit S, Cruz PCS, Ali MP, Pittendrigh B, Septiningsih EM, Ismail AM (2020) Responses of *AG1* and *AG2* QTL introgression lines and seed pre-treatment on growth and physiological processes during anaerobic germination of rice under flooding. Sci Rep 10(1):1–15

Mori Y, Kurokawa Y, Koike M, Malik AI, Colmer TD, Ashikari M, Pedersen O, Nagai K (2019) Diel $O_2$ dynamics in partially and completely submerged deep water rice: leaf gas films enhance internodal $O_2$ status, influence gene expression and accelerate stem elongation for 'snorkelling' during submergence. Plant Cell Physiol 60(5):973–985

Munns R, Tester M (2008) Mechanisms of salinity tolerance. Annu Rev Plant Biol 59:651–681

Na YJ, Choi HK, Park MY, Choi SW, Xuan Vo KT, Jeon JS, Kim SY (2019) *OsMAPKKK63* is involved in salt stress response and seed dormancy control. Plant Signal Behav 14(3):e1578633

Nagai K, Hattori Y, Ashikari M (2010) Stunt or elongate? Two opposite strategies by which rice adapts to floods. J Plant Res 123(3):303–309

Nair MM, Shylaraj KS (2021) Introgression of dual abiotic stress tolerance QTLs (*Saltol* QTL and *Sub1* gene) into Rice (*Oryza sativa* L.) variety Aiswarya through marker assisted backcross breeding. Physiol Mol Biol Plants 27(3):497–514

Narsai R, Edwards JM, Roberts TH, Whelan J, Joss GH, Atwell BJ (2015) Mechanisms of growth and patterns of gene expression in oxygen-deprived rice coleoptiles. Plant J 82(1):25–40

Nguyen HTT, Das Bhowmik S, Long H, Cheng Y, Mundree S, Hoang LTM (2021) Rapid accumulation of proline enhances salinity tolerance in Australian wild rice *Oryza australiensis* Domin. Plan Theory 10(10):2044

Nishiuchi S, Watanabe K, Sato S, Takahashi H, Nakazono M (2021) Expression analysis of genes for cytochrome *P450 CYP86* and glycerol-3-phosphate acyltransferase related to suberin biosynthesis in rice roots under stagnant deoxygenated conditions. Plant Root 15:19–35

Nishiuchi S, Yamauchi T, Takahashi H, Kotula L, Nakazono M (2012) Mechanisms for coping with submergence and waterlogging in rice. Rice 5(1):1–14

Oladosu Y, Rafii MY, Arolu F, Chukwu SC, Muhammad I, Kareem I, Salisu MA, Arolu IW (2020) Submergence tolerance in rice: Review of mechanism, breeding and, future prospects. Sustainability 12(4):1632

Omari Alzahrani F (2021) Metabolic engineering of osmoprotectants to elucidate the mechanism (s) of salt stress tolerance in crop plants. Planta 253(1):1–17

Panda D, Barik J (2021) Flooding tolerance in rice: focus on mechanisms and approaches. Rice Sci 28(1):43–57

Panda D, Sarkar RK (2011) Non-structural carbohydrate metabolism associated with submergence tolerance in rice. Genetics Plant Physiol 1(3,4):155–162

Pani DR, Sarangi SK, Subudhi HN, Misra RC, Bhandari DC (2013) Exploration, evaluation and conservation of salt tolerant rice genetic resources from Sundarbans region of West Bengal. J Indian Soc Coastal Agric Res 30:45–53

Pareek A, Dhankher OP, Foyer CH (2020) Mitigating the impact of climate change on plant productivity and ecosystem sustainability. J Exp Bot 71(2):451–456

Pitman MG, Läuchli A (2002) Global impact of salinity and agricultural ecosystems. In: Salinity: environment-plants-molecules. Springer, Dordrecht, pp 3–20

Pradhan SK, Pandit E, Pawar S, Baksh SY, Mukherjee AK, Mohanty SP (2019) Development of flash-flood tolerant and durable bacterial blight resistant versions of mega rice variety 'Swarna' through marker-assisted backcross breeding. Sci Rep 9(1):1–15

Prakash NR, Lokeshkumar BM, Rathor S, Warraich AS, Yadav S, Vinaykumar NM, Dushynthkumar BM, Sharma PC (2022) Meta-analysis and validation of genomic loci governing seedling and reproductive stage salinity tolerance in rice. Physiol Plant e13629

Prakash, N.R., Sheoran, S., Saini, M., Punia, M., Rathod, N.K.K, Bhinda, M.S., Vinesh, B., Choudhary, M.K. and Sarkar, B. (2020). Offsetting climate change impact through genetic enhancement, In: Ch. Srinivasarao et al., (Eds). Climate change and Indian agriculture: challenges and adaptation strategies, ICAR-National Academy of Agricultural Research Management Hyderabad, Telanganapp-71–104

Priya TR, Nelson ARLE, Ravichandran K, Antony U (2019) Nutritional and functional properties of coloured rice varieties of South India: a review. J Ethn Foods 6(1):1–11

Pucciariello C (2020) Molecular mechanisms supporting rice germination and coleoptile elongation under low oxygen. Plan Theory 9(8):1037

Pundir P, Devi A, Krishnamurthy SL, Sharma PC, Vinaykumar NM (2021) QTLs in salt rice variety CSR10 reveals salinity tolerance at reproductive stage. Acta Physiol Plant 43(2):1–15

Qiu QS, Guo Y, Dietrich MA, Schumaker KS, Zhu JK (2002) Regulation of *SOS1*, a plasma membrane $Na^+/H^+$ exchanger in *Arabidopsis thaliana*, by *SOS2* and *SOS3*. Proc Natl Acad Sci 99(12):8436–8441

Rabbani G, Rahman A, Mainuddin K (2013) Salinity-induced loss and damage to farming households in coastal Bangladesh. Int J Global Warming 5(4):400–415

Rahman MA, Thomson MJ, Shah-E-Alam M, de Ocampo M, Egdane J, Ismail AM (2016) Exploring novel genetic sources of salinity tolerance in rice through molecular and physiological characterization. Ann Bot 117(6):1083–1097

Rajendran K, Tester M, Roy SJ (2009) Quantifying the three main components of salinity tolerance in cereals. Plant Cell Environ 32(3):237–249

Rasel HM, Hasan MR, Ahmed B, Miah MSU (2013) Investigation of soil and water salinity, its effect on crop production and adaptation strategy. Int J Water Resour Environ Eng 5(8):475–481

Rasel M, Tahjib-Ul-Arif M, Hossain MA, Hassan L, Farzana S, Brestic M (2021) Screening of salt-tolerant rice landraces by seedling stage phenotyping and dissecting biochemical determinants of tolerance mechanism. J Plant Growth Regul 40(5):1853–1868

Reddy INBL, Kim BK, Yoon IS, Kim KH, Kwon TR (2017) Salt tolerance in rice: focus on mechanisms and approaches. Rice Sci 24(3):123–144

Reddy JN, Sarkar RK, Patnaik SSC, Singh DP, Singh US, Ismail AM, Mackill DJ (2009) Improvement of rice germplasm for rainfed lowlands of eastern India. Sabrao J Breed Genet 41:1–5

Reddy VA, Girija Rani M, Satyanarayana PV, Suryanarayana Y, Chamundeswari N, Ravikumar BNVSR et al (2015) Physiological and molecular response of rice genotypes for different types of flooding. Curr Biotica 8(4):345–350

Roy SJ, Negrao S, Tester M (2014) Salt resistant crop plants. Curr Opin Biotechnol 26:115–124

Sabareshwari V, Ramya A (2018) Coastal saline soils of India: a review. Agric Rev 39(1):86–88

Sabouri H, Rezai AM, Moumeni A (2008) Evaluation of salt tolerance in Iranian landrace and improved rice cultivars. JWSS-Isfahan UnivTechnol 12(45):47–63

Saijo Y, Hata S, Kyozuka J, Shimamoto K, Izui K (2000) Over-expression of a single $Ca^{2+}$-dependent protein kinase confers both cold and salt/drought tolerance on rice plants. Plant J 23(3):319–327

Sakina A, Ahmed I, Shahzad A, Iqbal M, Asif M (2016) Genetic variation for salinity tolerance in Pakistani rice (*Oryza sativa* L.) germplasm. J Agron Crop Sci 202(1):25–36

Sarangi SK, Islam MR (2019) Advances in agronomic and related management options for Sundarbans. In: The Sundarbans: a disaster-prone eco-region. Springer, Cham, pp 225–260

Sarkar RK, Reddy JN, Sharma SG, Ismail AM (2006) Physiological basis of submergence tolerance in rice and implications for crop improvement. Curr Sci:899–906

Schneider P, Asch F (2020) Rice production and food security in Asian mega deltas-a review on characteristics, vulnerabilities and agricultural adaptation options to cope with climate change. J Agron Crop Sci 206(4):491–503

Septiningsih EM, Pamplona AM, Sanchez DL, Neeraja CN, Vergara GV, Heuer S, Ismail AM, Mackill DJ (2009) Development of submergence-tolerant rice cultivars: the *Sub1* locus and beyond. Ann Bot 103(2):151–160

Sevanthi AM, Prakash C, Shanmugavadivel PS (2019) Recent progress in rice varietal development for abiotic stress tolerance. Adv Rice Res Abiotic Stress Tolerance:47–68

Shu K, Zhou W, Chen F, Luo X, Yang W (2018) Abscisic acid and gibberellins antagonistically mediate plant development and abiotic stress responses. Front Plant Sci 9:416

Singh A, Septiningsih EM, Balyan HS, Singh NK, Rai V (2017) Genetics, physiological mechanisms and breeding of flood-tolerant rice (*Oryza sativa* L.). Plant Cell Physiol 58(2): 185–197

Spengler RN, Stark S, Zhou X, Fuks D, Tang L, Mir-Makhamad B, Bjørn R, Jiang H, Olivieri LM, Begmatov A, Boivin N (2021) A journey to the west: the ancient dispersal of Rice out of East Asia. Rice 14(1):1–18

Steffens B, Rasmussen A (2016) The physiology of adventitious roots. Plant Physiol 170(2): 603–617

Steffens B, Geske T, Sauter M (2011) Aerenchyma formation in the rice stem and its promotion by $H_2O_2$. New Phytol 190(2):369–378

Sun H, Tao J, Hou M, Huang S, Chen S, Liang Z, Xie T, Wei Y, Xie X, Yoneyama K, Xu G, Zhang Y (2015) A strigolactone signal is required for adventitious root formation in rice. Ann Bot 115(7):1155–1162

Tahjib-Ul-Arif M, Sayed MA, Islam MM, Siddiqui MN, Begum SN, Hossain MA (2018) Screening of rice landraces (*Oryza sativa* L.) for seedling stage salinity tolerance using morpho-physiological and molecular markers. Acta Physiol Plant 40(4):1–12

Tnani H, Chebotarov D, Thapa R, Ignacio JCI, Israel WK, Quilloy FA, Dixit S, Septiningsih EM, Kretzschmar T (2021) Enriched-GWAS and transcriptome analysis to refine and characterize a major QTL for anaerobic germination tolerance in Rice. Int J Mol Sci 22(9):4445

Umarani E, Hemalatha V, Bhadana VP (2017) Screening of rice (*Oryza sativa* L.) genotypes for anaerobic germination. IJPAB 6(5):1318–1325

Van Zelm E, Zhang Y, Testerink C (2020) Salt tolerance mechanisms of plants. Annu Rev Plant Biol 71:403–433

Verslues PE, Agarwal M, Katiyar-Agarwal S, Zhu J, Zhu JK (2006) Methods and concepts in quantifying resistance to drought, salt and freezing, abiotic stresses that affect plant water status. Plant J 45(4):523–539

Wang Q, Guan Y, Wu Y, Chen H, Chen F, Chu C (2008) Overexpression of a rice *OsDREB1F* gene increases salt, drought, and low temperature tolerance in both Arabidopsis and rice. Plant Mol Biol 67(6):589–602

Wu TM, Lin WR, Kao CH, Hong CY (2015) Gene knockout of *glutathione reductase3* results in increased sensitivity to salt stress in rice. Plant Mol Biol 87(6):555–564

Xu M, Zhu L, Shou H, Wu P (2005) A *PIN1* family gene, *OsPIN1*, involved in auxin-dependent adventitious root emergence and tillering in rice. Plant Cell Physiol 46(10):1674–1681

Xu Z, Shao T, Lv Z, Yue Y, Liu A, Long X, Zhou Z, Gao X, Rengel Z (2020) The mechanisms of improving coastal saline soils by planting rice. Sci Total Environ 703:135529

Yamauchi T, Fukazawa A, Nakazono M (2017a) *METALLOTHIONEIN* genes encoding ROS scavenging enzymes are down-regulated in the root cortex during inducible aerenchyma formation in rice. Plant Signal Behav 12(11):e1388976

Yamauchi T, Yoshioka M, Fukazawa A, Mori H, Nishizawa NK, Tsutsumi N, Yoshioka H, Nakazono M (2017b) An *NADPH oxidase* RBOH functions in rice roots during lysigenous aerenchyma formation under oxygen-deficient conditions. Plant Cell 29(4):775–790

Zhang Z, Zhang Q, Wu J, Zheng X, Zheng S, Sun X, Qui Q, Lu T (2013) Gene knockout study reveals that cytosolic ascorbate peroxidase 2 (*OsAPX2*) plays a critical role in growth and reproduction in rice under drought, salt and cold stresses. PLoS One 8(2):e57472

Zhi-Guo E, Ge L, Wang L (2012) Molecular mechanism of adventitious root formation in rice. Plant Growth Regul 68(3):325–331

Zhu C, Qi Q, Niu H, Wu J, Yang N, Gan L (2021) *γ-Aminobutyric acid* suppresses iron transportation from roots to shoots in rice seedlings by inducing aerenchyma formation. IJMS 22(1):220

# Genetic Improvement of Rice for Salt Tolerance at RRS, Vyttila

14

Veena Vighneswaran, A. K. Sreelatha, Deepa Thomas, and S. Amrithakumary

**Abstract**

Rice Research Station, Vyttila, established 1958, was brought under the Kerala Agricultural University in 1974, with the mandates of crop improvement of saline-tolerant rice varieties and management of coastal saline rice tracts (Pokkali). Till date, 11 saline-tolerant varieties were released, which includes two pure line selection, three mutants, and one marker-assisted introgression.

**Keywords**

Crop improvement · Pokkali rice · Saline tolerance

## 14.1 Introduction

The coastal areas of Kerala have a unique system of rice cultivation on saline soils, locally known as Pokkali cultivation or salt water farming. The term Pokkali refers to salt-tolerant rice cultivars mainly grown in Ernakulam district and is synonymous with Kaipad cultivation in Kannur district and Kariland cultivation in Alappuzha districts of the state. These rice fields are characterized by the fact that rice can only be grown during the low-salinity, monsoon season from June to October. Salinity increases with the cessation of monsoon during November and it becomes too high that rice cannot be cultivated. Fields that become fallow during this time are used for traditional shrimp and fish farming called shrimp filtration. The extent of this unique type of agriculture turns out to be high in the coastal areas of the Ernakulam district

V. Vighneswaran (✉) · A. K. Sreelatha · D. Thomas · S. Amrithakumary
Kerala Agricultural University, Rice Research Station, Ernakulam, India
e-mail: veena.v@kau.in





R. K. Singh et al. (eds.), *Genetic Improvement of Rice for Salt Tolerance*,
https://doi.org/10.1007/978-981-99-3830-8_14

and Vytilla, representing the Pokkali area with high salinity fluctuations, ideal for studying the problems of salinity farming.

The Rice Research Station, Vyttila, occupies a prominent location in the center of the Pokkali tract. The station was established in 1958 on leased land at Kunnala and moved to its current location of Vyttila in 1963. The station was taken over by Kerala Agricultural University (KAU) in 1974, after its inception in 1972. Considering the importance of fish-shrimp farming during the saline phase, the Fisheries Research Unit was started in 1976. The station is located on the shoreline of Kaniyampuzha River. Kaniyampuzha River and its subsidiaries fill and drain saline water into the paddy fields of the station and finally drains out to Vembanadu Lake and then to the Arabian Sea. Thus, the research station is situated in a representative site in the midst of the Pokkali tract. It lies between $9^{0}58$" North latitude and $76^{0}20''$ East longitude at an elevation of 1.7 m above MSL. Currently, the station has a total area of 8.91 hectares out of which 4.25 hectares are wetlands. Rice Research Station, Vyttila, is the only research station dedicated for salinity research on Pokkali, as well as research on salinity of soil and water under Kerala Agricultural University (KAU).

The station has also started marker-assisted selection and backcrossing for the development of improved rice varieties. The enhancement of analytical laboratories with modern equipment for estimating macronutrients, secondary nutrients, and micronutrients in plants, soils, and water is another important development of the station. This station is a volunteer center for the All India Coordination Rice Improvement Project and the All India Coordination Project on Saline Soil Management and Use of Saline Water in Agriculture.

## 14.2 Mandates of the Station

1. To develop high-yielding rice varieties having multiple tolerances to abiotic stresses like salinity, acidity, and submergence suited to the coastal saline agroecosystem.
2. To adopt appropriate management practices for yield enhancement of *Pokkali* rice.
3. Development of an appropriate integrated agricultural system model for increasing income per unit area suitable for Pokkali tract.

## 14.3 Wonder Rice Genotype "Pokkali"

"Pokkali" is the name used to describe a unique ecosystem rich in biodiversity, with lowland marshes and marshes near lowland estuaries and rivers. It also refers to the farming system practiced in the coastal saline ecosystem. The Pokkali farming method is the only natural organic farming method popularized in the state that does not use chemical fertilizers or pesticides.

Pokkali also refers to a unique and exceptional rice variety grown 100% organically in the flooded coastal areas of Alappuzha, Thrissur, and Ernakulam districts with unique characteristics of tolerance to abiotic stresses such as acidity, salinity, and submergence. Moreover, a unique style of rice farming culture being customized for this swathe alludes the same name "Pokkali," as a technique for multi-stress management. Pokkali's land is designated under the state's special agroecological unit based on soil, landforms, and climate characteristics. Agroecological unit AEU 5, short for Pokkali tract, represents the lowlands, often below sea level, in the coastal areas of Ernakulam district, extending to parts of Trissur and Alappuzha districts, including 34 Panchayaths. The region's climate resembles a tropical humid monsoon, with an average annual temperature of 27.60 °C and rainfall of 3049 mm.

The characteristics of this wonder rice are:

- Pokkali is the world's most salt-tolerant rice cultivar and a provider of Saltol QTL to the International Salt-Tolerant Rice Breeding Program.
- A "three in one" wonder variety—a variety blessed with tolerance to three abiotic stresses (salinity, submergence, and acidity) in one variety.
- Superior medicinal benefits—Bran is rich in oils and antioxidants such as oryzanols, tocopherols, and tocotrienols, more than the popular medicinal rice Njavara.
- Suitable for diabetics due to its low glycemic index.
- Rich in iron, potassium, zinc, and manganese.

Pokkali's cultivation system is unique in the world and is vanishing today due to many practical issues and socioeconomic constraints. In order to maintain this specific genotype and unique cultivation system, Vyttila Rice Research Station took the initiative in 2007 to acquire the Geographical Indication Registration Certificate and the logo of Pokkali rice as a branded rice to fetch premium price for this pure organic product. Further, this center had also taken initiative to nominate the Registered *Pokkali* Farmers Group (Varappuzha-Kadamakkudy Jaiva Pokkali ICS) for the Plant Genome Savior Community Award (2011–2012) by the Protection of Plant Variety and Farmers' Right Authority, Government of India, with a cash award of Rs. 10.00 lakhs and a citation. This amount was intended to protect this valuable genotype through cultivation persistently.

A special agroecological unit requires a unique management strategy based on the unparallel and unique ecosystem. A distinctive feature of this unique rice cultivation practice is that rice can only be grown during the low salinity/monsoon season, which begins from June to July to mid-October. From November onward, salinity builds up and the high salinity of seawater makes the paddy cultivation impossible. The fields that are fallow during this time are utilized for a traditional type of prawn and fish culture known as prawn filtration or "chemeen kettu." The traditional Pokkali rice cultivation and shrimp farming practiced here is one of the most eco-friendly farming practices in the world.

## 14.4 Paddy Cultivation

Rice cultivation begins with the strengthening of the outer walls by mid-April after the setting up of sluices to control water levels. The fields are drained at low tide and the sluices are closed and let the fields dry. Once the soil is dry, pile it up to create mounds of 1 m base and 0.5 m high, and allow the mound to dry and weather during summer months. With the onset of monsoon, from May to June, the salt from the top of mounds are washed off and the salt-dissolved water is drained from the fields through channels. The tops of mounds become free from salts very soon and sprouted seeds are sown. These mounds act as elevated in situ nursery and provide protection to seedlings from flash floods.

Pokkali is the most common land race in this cultivation system. Other landraces from this region include Choottu Pokkali, Chettivirippu, Cheruvirippu, Kuruka, Anakodan, Eravapandy, Bali, and Orpandy. Apart from these, 11 saline-tolerant, high-yielding varieties of rice have been released from Rice Research Station, Vyttila, viz., VTL-1 to VTL-11 (Table 14.1). Out of these, VTL-1 to VTL-5 are tall varieties and they cause considerable yield loss due to lodging. The last six varieties, viz., VTL-6 to VTL-11, are semi-tall and nonlodging, and the average yield is 3500 to 4000 kg/ha organically. Further, these varieties are amenable for mechanical harvesting too.

**Table 14.1** Details of high-yielding varieties evolved at Rice Research Station, Vyttila

| Variety | Parent | Method of breeding | Height (cm) | Duration (days) | Yield (t/ha) | Kernel color |
|---|---|---|---|---|---|---|
| VTL-1 | Choottupokkali | Pure line selection | 145–160 | 115 | 1.5–2.0 | Red |
| VTL-2 | Cheruvirippu | Pure line selection | 150–160 | 125 | 1.8–2.0 | Red |
| VTL-3 | VTL1 x TN1 | Hybridization and selection | 160–170 | 115 | 2.5–3.0 | Red |
| VTL-4 | Chettivinppu x IR 4630–22-217 | Hybridization and selection | 160–170 | 120 | 3.0–3.5 | Red |
| VTL-5 | Mahsuri | Mutation breeding | 140–150 | 120 | 3.0–3.5 | White |
| VTL-6 | Cheruvirippu/ IR5//Jaya | Hybridization and selection | 115–120 | 105 | 3.5–4.0 | Red |
| VTL-7 | IR8/Patnai23 | Hybridization and selection | 120–125 | 115 | 3.5–4.0 | White |
| VTL-8 | IR47310–94-4-I/CSR10 | Hybridization and selection | 130–135 | 110 | 4.0–4.5 | Red |
| VTL-9 | Chettivirippu | Mutation breeding | 118 | 115–120 | 4.0 | Red |
| VTL-10 | VTL–3 | Mutation breeding | 130 | 110–115 | 4.2 | Red |
| VTL-11 (Jyotsna) | Jyothi/FL 478// Jyothi | Backcross breeding | 105 | 105–110 | 5.0 | Red |

## 14.5 Salient Research Work Done at the Station

### 14.5.1 Rice Breeding

Salinity, acidity, and submergence are complex abiotic stress problems faced by the Pokkali ecosystem. In Pokkali, biotic stresses such as pest and disease *infestations* are below thresholds.

The breeding for varieties in Pokkali ecosystem is based on the following requirements:

- Salinity, acidity, and submergence tolerance.
- Semi-tall stature (120–130 cm) and nonlodging.
- Plant duration not more than 125 days.
- Should possess a dormancy period from 2 weeks to one month.

#### 14.5.1.1 Germplasm Collection and Utilization

Germplasm collection had been started since the inception of the station, away back in 1958 itself. Prominent landraces from different Pokkali ecosystems like Chellanam, Kumbalangi, Vypin, Kadamakkudy, N. Parur, Cherai, Thuravoor, Pattanakkad, Cherthalai, etc. were procured initially and their performance was evaluated. The landraces like *Pokkali*, Cheruvirippu, Chettivirippu, Kuruka, Karuthakuruka, Ponkuruka, Anakodan, Mundakan, etc. are some among them. Accessions from other states and countries with saline tolerance were identified, introduced, evaluated, and utilized for breeding experiments. Further, they were maintained in the germplasm. The current germplasm includes national accessions and international accessions, breeding lines, cultures, etc. A saline-tolerant germplasm of about 55 accessions is now maintained in the germplasm bank.

#### 14.5.1.2 Selection

Mass selection and pure line selection have been attempted from the beginning with prominent local landraces. Pure line selection was started in the two widely cultivated land races, viz., Choottupokkali and Cheruvirippu. The first high-yielding, salt-tolerant rice cultivar VTL-1 was developed from this Choottupokkali by pure line selection and approved for cultivation by the Ministry of Agriculture. This variety has yields 10–15% higher than local landrace. The average yield of this variety is 1500 kg/ha. However, this variety has the *Saltol* QTL on chromosome 1, the most salt-tolerant QTL gene in the world, and is used in salt-tolerant breeding programs in the International Rice Research Institute, Manila, Philippines, and other countries around the world.

VTL-2 is the second high-yielding, salt-tolerant rice variety developed by pure line selection from Cheruvirippu and is approved for cultivation by the Department of Agriculture. Also, this variety gives 10–15% higher yields than its native parent variety. The average yield for this variety is 1750 kg/ha.

#### 14.5.1.3 Hybridization and Selection

During 1975–1976, a rice hybridization program was started at the Rice Research Institute in Vytilla. A series of breeding experiments were conducted to combine the high yield of the released high-yielding cultivar with the abiotic stress tolerance of the Pokkali landrace. As a result, the first high-yielding variety developed by combination breeding was VTL-3 by crossing VTL-I and Taichung native-1. The average yield of this variety is 2500 kg/ha.

The next series of experiments was initiated at the station between 1980 and 1986, combining the high-yield traits of IRRI rice cultivars with the abiotic stress tolerance of Pokkali landraces. The second high-performance variety developed by combination breeding was VTL-4 by crossing the landrace Chettivirippu with IR 4630-22-2-17. This variety was released in 1994 and has been the predominant variety in the region for 20 years. Even now, this variety has good acceptance and is dominant as a tall variety in the mid-deepwater areas.

#### 14.5.1.4 Mutation Breeding

The station had also initiated a mutation breeding program from 1984 to 1986 to develop white kernelled resistant rice cultivars from Mahsuri. High-yielding mutant VTL 5 rice with white kernel was developed and approved for cultivation. The average yield of this variety is 3500 kg/ha. Later, Chettivirippu, the landrace of Pokkali, was also subjected to mutation breeding to create a high-yielding semidwarf line. In 2015, a semi-tall mutant, VTL-9, with an average yield of 4000 kg/ha was developed from this popular landrace, and in 2018, another semi-tall mutant, VTL-10, with an average yield of 4200 kg/ha was released from VTL-3.

#### 14.5.1.5 Development of Semi-Tall Nonlodging, High-Yielding Variety

With the introduction of mechanization in paddy cultivation during early 90 s, the station also started breeding program to develop a semi-tall, nonlodging, high-yielding variety amenable for mechanical harvesting. Thus, the first semi-tall, nonlodging, high-yielding variety, VTL-6, was developed from the cross Cheruvirippu/IR 5//Jaya. The average yield of this variety is 3500–4000 kg/ha and is accepted widely for cultivation in the P*okkali* ecosystem.

The variety VTL-7 was developed for export purpose and is a semi-tall, nonlodging, high-yielding white kernelled rice variety. The parentage is IR 8 x Patnai 23. The average yield of this variety is 4250 kg/ha. A slightly tall, nonlodging, high-yielding variety for the semi-deepwater-logged areas was evolved from the cross IR 47310–94–4-1 X CSR-10 and named it as VTL-8. The average yield of this variety is 4500 kg/ha.

#### 14.5.1.6 Marker-Assisted Backcross Breeding

Marker-assisted backcross breeding where selection is aided with various markers associated with a particular trait was also attempted at RRS Vyttila from 2005 onward. The popular variety Jyothi was crossed with FL 478 (internationally accepted donor of saline tolerance) and the F1s with the QTL *Saltol* were again backcrossed with Jyothi and the variety VTL-11 (Jyotsna) was released. The average yield of this red bold and long rice is about 5000 kg/ha (Deepa et al. 2017; Rohini and Shylaraj 2017).

### 14.5.2 Nutritional Properties

#### 14.5.2.1 *Pokkali* Rice, an Organic Nutraceutical

The medicinal properties of Pokkali rice have always been adored. However, the actual ingredients that contribute to the medicinal efficiency were unknown. Therefore, a scientific study was conducted at the Vytilla Rice Research Institute to clarify the nutritional supplement properties of Pokkali rice. Pokkali bran has been found to be nutritionally rich in bran oil and antioxidants such as oryzanols, tocopherols, and tocotrienols compared to the most popular varieties such as Njavara, Jyothi, and Uma. Oryzanol is famous for its antioxidant, anticancer, and antilipidemic properties (Table 14.2). Tocopherols and tocotrienols are vitamin E analogues with antioxidant and anticancer properties. These ingredients were also compared with medicinal rice Njavara. These levels were observed to be higher in Pokkali rice bran. Pokkali rice is also rich in micronutrients such as iron, zinc, manganese, and vitamin B (Table 14.3). Therefore, in order to make use of the medicinal properties of rice bran, it is recommended to use mild polishing for Pokkali rice.

**Table 14.2** Antioxidant content of *Pokkali* rice bran

| Sl. no. | Varieties | Oryzanol (mg/100 g bran) | Tocopherol (mg/100 g bran) | Tocotrienol (mg/100 g bran) |
|---|---|---|---|---|
| 1 | Jyothi | 112.1 | 0.44 | 1.22 |
| 2 | Uma | 138.7 | 0.48 | 2.23 |
| 3 | Pokkali | 202.9 | 2.26 | 5.13 |
| 4 | Black Njavara | 149.4 | 1.85 | 4.44 |
| 5 | Golden Njavara | 160.0 | 2.01 | 3.90 |

(Shylaraj et al. 2018)

**Table 14.3** Bran oil and micronutrient content of *Pokkali* rice

| Sl. no. | Varieties | Fat content (%) | Fe (mg/kg) | Zn (mg/kg) | Cu (mg/kg) | Mn (mg/kg) | B (mg/kg) |
|---|---|---|---|---|---|---|---|
| 1 | Jyothi | 6.96 | 16.8 | 13.0 | 0.3 | 7.2 | 25.3 |
| 2 | Uma | 6.93 | 11.2 | 12.2 | 0.8 | 13.0 | 35.0 |
| 3 | Pokkali | 9.83 | 25.0 | 22.6 | 1.4 | 20.4 | 76.7 |
| 4 | Black Njavara | 7.88 | 14.2 | 12.6 | 1.4 | 13.2 | 38.3 |
| 5 | Golden Njavara | 7.43 | 28.8 | 14.9 | 2.6 | 19.4 | 56.7 |

(Shylaraj et al. 2018)

### 14.5.3 Crop Management

#### 14.5.3.1 Soil Characteristics of *Pokkali* Tract

The soils of Pokkali are highly acidic, with 87% of the soils showing acidity, of which 81% of soils with pH < 5.5 exhibit extreme acidity to strong acidity response. Under highly acidic conditions below pH 5.5, Al becomes highly soluble, leading to a toxic state that thus imparts consequent Mg deficiency in the soil. The Pokkali tract, which consists of low-lying, water-filled areas along the saline coastal belt of Kerala, is tidal affected and the salinity of these soils is EC > 4 dS $m^{-1}$, and 34% of the samples and 94% of the soil samples have moderate to high organic carbon content.

In Pokkali lands, 60% of the soils are in high available P status and 32% are in medium P status. Similarly, 68% soils are high in K status and 13% are in medium K status. Therefore, P and K fertilizer applications are not recommended. The secondary nutrients and micronutrients are adequate in majority of the soils in this tract. Soil and plant nutrient status of 30 padasekarams in the Pokkali area were monitored, indicating that the soils were highly fertile and suitable for organic rice production. Due to the high soil fertility and the large number of natural soil microorganisms, there is little response to the addition of biofertilizers. Native flora of azospirillum and P solubilizers were present in Pokkali soil. As the chemical fertilizers had little impact on rice yield in the Pokkali area, Pokkali field is ideal for organic farming. Due to the application of chemical fertilizers to the soil of Pokkali, the yield has decreased. Therefore, the fertilization recommendations of 20:40:0 kg/ha N:$P_2O_5$:$K_2O$ for Pokkali rice have been removed from the Kerala Agricultural University best practice recommendation package based on response studies conducted at Kerala Agricultural University. This declared the entire Pokkali region to be a completely organic zone.

#### 14.5.3.2 AICRP on Management of Salt-Affected Soils and use of Saline Water in Agriculture

Since 2014, the station has been functioning as a voluntary center for AICRP on management of salt-affected soils and use of saline water. Under the project, collection and analysis of groundwater samples from 14 districts of Kerala revealed that 84.33%, 1.14%, 1.14%, 2.28%, 1.42%, and 0.85% fall under good, marginally saline, saline, marginally alkaline, and high alkali category of groundwater quality. Georeferenced maps showing spatial variability of EC, SAR, RSC, and groundwater quality are available for all the 14 districts of Kerala.

Analysis of soil samples also reveals that the soils are acidic with EC values in the good category in most of the districts of Kerala. Only in the coastal belt soils showed high EC values. The percentage of organic carbon in the sample is medium to high. The available phosphorus content is high in almost all samples. Among the secondary nutrients, the content of available magnesium is often deficient, while the deficiencies of calcium are prominent in southern districts. Extensive zinc, copper, and boron deficiencies have been noted.

### 14.5.4 Completed Projects

| Sl. no. | Title of project | Name of PI | Outlay in lakhs | Duration | Funding agency | Year of start |
|---|---|---|---|---|---|---|
| 1 | Ad hoc project on rice-fish/prawn components for an integrated farming systems | Dr. N. K. Sasidharan, Professor (Agron) | 4.2 | 3 | ICAR | 1998–99 |
| 2 | Rice varietal improvement under abiotic stress | Dr. KS Shylaraj, Professor (PBGN) | 3.63 | 3 | KSCSTE, Govt. of Kerala | 2003–04 |
| 3 | In vitro and in vivo mutagenesis in rice and screening for tolerance to abiotic stresses | Dr. KS Shylaraj, Professor (PBGN) | 21.60 | 4 | BRNS, Govt. of India | 2005–06 |
| 4 | Developing protocol for organic rice production in Pokkali tract phase I | Dr. K Anilakumar, Professor (SS&AC) | 25.00 | 3 | RKVY, Govt. of India | 2007–08 |
| 5 | Boosting organic rice production of Kerala through marker-assisted selection and high-quality seed production program | Dr. KS Shylaraj, Professor (PBGN) | 60.15 | 3 | RKVY, Govt. of India | 2008–09 |
| 6 | Developing protocol for organic rice production in Pokkali tract phase II | Dr. Annie Koruth, Assoc. Prof. (SS&AC) | 25.00 | 2 | RKVY, Govt. of India | 2010–11 |
| 7 | Soil-based plant nutrient management plan for AEZ in Kerala | Dr. V Sreekumaran, professor (Agron) and Dr. Annie Koruth, Assoc. Prof. (SS&AC) | 10.92 | 2 | State planning board | 2010–11 |
| 8 | Nutraceutical properties of unique Pokkali rice | Dr. KS Shylaraj, professor (PBGN) | 10.54 | 1 | RKVY, Paddy Mission | 2010–11 |

(continued)

| Sl. no. | Title of project | Name of PI | Outlay in lakhs | Duration | Funding agency | Year of start |
|---|---|---|---|---|---|---|
| 9 | Strengthening rice seed production at RRS, Vyttila | Dr. V Sreekumaran, Professor (Agron) | 1.86 | 1 | RKVY, Paddy Mission | 2010–11 |
| 10 | Development of Saltol sub 1 mega rice varieties for flood-prone, salt-affected areas and standardization of production technology | Dr. KS Shylaraj, Professor (PBGN) | 156.42 | 4 | Kuttanad Devpt XIIIth finance commission GoK | 2012–13 |
| 11 | Elucidation of salt tolerance mechanism and development of saline-tolerant rice variety through marker-assisted backcrossing | Dr. KS Shylaraj, Professor (PBGN) | 50.00 | 1 | ICAR, Govt. of India | 2013–14 |
| 12 | Soil-based plant nutrient management plan for agroecosystems of Kerala, second phase | Dr. Sreelatha AK, Assoc Prof. (SS&AC) | 13.60 | 2 | Planning board | 2014–16 |

## 14.6 Conclusion

This research station was founded in 1958 with the sole purpose of conducting salinity research on *Pokkali* rice and could accomplish research with respect to the crop improvement aspects through development of high-yielding rice varieties, management aspects of saline farming and development of suitable organic management packages for *Pokkali* tract, and development of suitable integrated farming models for *Pokkali* ecosystem. Further, the center has started advanced molecular breeding research like marker-assisted breeding in rice and introgressed the *Saltol* gene to the mega rice variety, Jyothi, and a new variety VTL-11(Jyotsna) was released in 2017. This work is first of its kind in Kerala, and the introgression of abiotic stress-tolerant genes/QTLs like *Saltol* and *Sub1* genes into three popular rice varieties, viz., Uma, Jyothi, and Jaya, was done. The abiotic stress-tolerant QTL introgressed lines are intended to be cultivated in all saline-prone areas of Kerala including salt-intruded areas of Kuttanad and Kole lands. This study of the underlying mechanisms in the different valuable Pokkali landraces and pyramiding of these different mechanisms into a single rice variety is another major ongoing program of the station. The search of new saline-tolerant genes/QTLs of their existing landraces in the germplasm is another area of research.

## References

Deepa J, Rohini PC, Shylaraj KS (2017) Genetic analysis of the salt tolerant rice varieties of Kerala through molecular characterization using SSR markers. Green Farming 8(5):1012–1017

Rohini PC, Shylaraj KS (2017) Introgression of Saltol QTL into an elite rice variety Jyothi through marker assisted selection. Green Farming 8(3):512–517

Shylaraj KS, Nadh SG, Chandran SK (2018) Comparative analysis of grain quality and nutraceutical properties of selected rice varieties from Kerala. Curr Sci 114:1088–1094

MIX
Papier aus verantwortungsvollen Quellen
Paper from responsible sources
FSC® C105338

If you have any concerns about our products,
you can contact us on
**ProductSafety@springernature.com**

In case Publisher is established outside the EU,
the EU authorized representative is:
**Springer Nature Customer Service Center GmbH**
**Europaplatz 3, 69115 Heidelberg, Germany**

Printed by Libri Plureos GmbH
in Hamburg, Germany